Modern General Relativity

Modern General Relativity

Editor: August Hall

NY RESEARCH PRESS
New York

Published by NY Research Press
118-35 Queens Blvd., Suite 400,
Forest Hills, NY 11375, USA
www.nyresearchpress.com

Modern General Relativity
Edited by August Hall

International Standard Book Number: 978-1-64725-465-0 (Hardback)

Cataloging-in-Publication Data

Modern general relativity / edited by August Hall.
p. cm.
Includes bibliographical references and index.
ISBN 978-1-64725-465-0
1. General relativity (Physics). 2. Relativity (Physics). 3. Physics. I. Hall, August.
QC173.6 .M63 2023
530.11--dc23

Contents

Preface

The general theory of relativity, also called the Einstein's theory of gravity and the general relativity theory, is the geometric theory of gravitation. It provides the description of gravity as a geometric property of space and time. There are several astrophysical applications of the general theory of relativity including gravitational lensing, cosmology, gravitational-wave astronomy, and black holes. The theory of general relativity forms the basis for disciplines of cosmology and relativistic astrophysics. Cosmology is a branch of astronomy that studies the origin and evolution of the universe. The contemporary researches in cosmology are based on the field equations of general relativity. Under cosmology, the universe is considered on extremely large scales that are evolving over extremely long times that are comparable to the age of the universe. Some researchers have also proposed cosmological models based on space-times with dimensions, $D > 4$. Some of the important applications of general relativity in cosmology and astrophysics include gamma-ray bursts, neutron stars, black holes and gravitational waves. This book includes some of the vital pieces of works being conducted across the world, on various topics related to modern general relativity. Researchers and students studying this theory and its applications will be assisted by this book.

This book unites the global concepts and researches in an organized manner for a comprehensive understanding of the subject. It is a ripe text for all researchers, students, scientists or anyone else who is interested in acquiring a better knowledge of this dynamic field.

I extend my sincere thanks to the contributors for such eloquent research chapters. Finally, I thank my family for being a source of support and help.

Editor

1

General Relativistic Wormhole Connections from Planck-Scales and the ER = EPR Conjecture

Fabrizio Tamburini [1,*,†] and Ignazio Licata [2,3,4,†]

1 ZKM—Zentrum für Kunst und Medientechnologie, Lorentzstr. 19, D-76135 Karlsruhe, Germany
2 Institute for Scientific Methodology (ISEM), Via Ugo La Malfa 153, I-90146 Palermo, Italy; ignazio.licata3@gmail.com
3 School of Advanced International Studies on Theoretical and Nonlinear Methodologies of Physics, I-70124 Bari, Italy
4 International Institute for Applicable Mathematics and Information Sciences (IIAMIS), B.M. Birla Science Centre, Adarsh Nagar, Hyderabad 500 463, India
* Correspondence: fabrizio.tamburini@gmail.com
† These authors contributed equally to this work.

Abstract: Einstein's equations of general relativity (GR) can describe the connection between events within a given hypervolume of size L larger than the Planck length L_P in terms of wormhole connections where metric fluctuations give rise to an indetermination relationship that involves the Riemann curvature tensor. At low energies (when $L \gg L_P$), these connections behave like an exchange of a virtual graviton with wavelength $\lambda_G = L$ as if gravitation were an emergent physical property. Down to Planck scales, wormholes avoid the gravitational collapse and any superposition of events or space–times become indistinguishable. These properties of Einstein's equations can find connections with the novel picture of quantum gravity (QG) known as the "Einstein–Rosen (ER) = Einstein–Podolski–Rosen (EPR)" (ER = EPR) conjecture proposed by Susskind and Maldacena in Anti-de-Sitter (AdS) space–times in their equivalence with conformal field theories (CFTs). In this scenario, non-traversable wormhole connections of two or more distant events in space–time through Einstein–Rosen (ER) wormholes that are solutions of the equations of GR, are supposed to be equivalent to events connected with non-local Einstein–Podolski–Rosen (EPR) entangled states that instead belong to the language of quantum mechanics. Our findings suggest that if the ER = EPR conjecture is valid, it can be extended to other different types of space–times and that gravity and space–time could be emergent physical quantities if the exchange of a virtual graviton between events can be considered connected by ER wormholes equivalent to entanglement connections.

Keywords: wormholes; entanglement; ER = EPR; relativistic quantum information; Planck scales

---

1. Introduction

The formulation of an effective theory of quantum gravity can be considered the holy grail of modern physics. Gravitation was the first force to be mathematically described by Newton and it is the last force of nature that has yet to be quantized. As pointed out by DeWitt in his early pioneering works [1–5], since the introduction of quantum field theory around 1930 by Heisenberg, Dirac, Pauli, Fock, Jordan and others, many attempts were made to find a robust and logically closed method of quantizing the gravitational field, without success, even if Einstein's equations are known to remain valid down to the Planck scales. Rosenfeld [6,7] realized the difficulty of finding general methods to quantize gravity and that the quanta of the field, if they do exist, cannot give observational effects until reaching a very high energy $E_p = \sqrt{\hbar c^5/G} \simeq 1.22 \times 10^{19}$ GeV that corresponds to the so-called Planck length, $L_p = \sqrt{\hbar G/c^2}$. Thus, Planck scales were somehow "artificially" introduced in the framework of general relativity (GR), a classical theory, in the attempt of building a quantum theory of gravitation based on the finite quantum of action h linked with the gravitational constant G and the speed of light c.

In principle, following the works by Pauli, De Witt and other pioneers in this field, the fundamental building block for a quantum theory of gravity is the graviton, a spin-2 massless particle with the well-known limitations in the building of a QG theory due to the coupling constant of the gravitational field that depends on the inverse square of the mass. This coupling constant makes the Einstein–Hilbert Lagrangian of quantum gravity divergent at the loop level. It is a non-renormalizable theory, unless introducing additional concepts such as supersymmetry like in string theory and supergravity. Up to now no supersymmetric partners of the known quanta have been found from the Large Hadron Collider and other experiments presented by the Particle Data Group [8]. At the present moment one can consider different approaches, including string theory scenarios that do not require supersymmetric partners at the explored energies, or to consider a model of the Universe without strings, or adopt the approach of loop quantum gravity [9], where gravitons do not represent the building blocks of the theory. The interactions between events that can be ascribed to graviton exchanges can be recovered in a weak field limit approximation. The exchange of a virtual graviton between two particles does not have the support of an actual theory of quantum gravity. As an example, in string theory and in most quantum field theory (QFT) scenarios, in the building of the theory, one must introduce the quanta of the associated field. The quanta are introduced in terms of quantized excitations on a classically fixed background. The main conceptual problem for the formulation of a consistent theory of QG is that this theory must unify and contain as special cases both GR and quantum mechanics (QM). Unfortunately, GR has concepts and mathematical structures that are incompatible with those of QM and vice versa, with the result that the two theories do not communicate between each other. GR is a local deterministic theory based on point-to-point connections of events and observers that define a four-dimensional manifold. Einstein in 1947, his latest memoirs, stated that space–time is made with connections between events and, more precisely, with coincidences of events. On the other hand, QM presents non-locality and the well-known probabilistic behavior from the deterministic equations that rule the quanta.

This contrast between GR and QG can find a fusion in the simple heuristic approach formulated by Susskind and Maldacena who, starting from the quantum mechanical language, set an hypothetical equivalence between non-traversable wormhole connections of two (or more) particles or events in space–time through Einstein–Rosen (ER) bridges and entangled states (the idea that wormholes and flux tubes can play a role in quantum mechanics and quantum field theory is not new, in particular for systems with electric and/or magnetic charges and their renormalization has earlier work in [10,11]), and the quantum properties of the "spooky action at distance" of Einstein–Podolski–Rosen (EPR) states [12–15]. The ER=EPR equivalence was first defined in Anti-de-Sitter (AdS) space–times in their equivalence with CFTs [16–19]. In other words, EPR entangled particles are supposed to be equivalent to connections obtained through ER wormholes involving the concept of entanglement entropy to describe these many-body quantum state/wormhole connections, even if the ER = EPR equivalence is more evident with monogamous entangled pairs [20]. Spacetime is supposed to emerge from quantum entanglement, as discussed in [21] where, from some examples where gauge theory/gravity duality is valid, one finds that the emergence of space–time is related to the quantum entanglement of the degrees of freedom present in these quantum systems. Superpositions of quantum states corresponding to disconnected space–times can give rise to states that are interpreted in terms of classically connected space–times. In this vision, gravity can be also interpreted as an entropic force, a thermodynamic property of physical systems defined in an holographic scenario: gravity and space–time connections are emergent phenomena from the degrees of freedom of a physical system encoded in an holographic boundary or to emerge from a background-free approach by using quantum entanglement [22,23].

At all effects, one can conclude that space–time is built with the quantum information shared between EPR states that are equivalently connected with an ER wormhole.

While ER wormholes are classical solutions of GR, a local deterministic theory, quantum entanglement, is instead one of the most intriguing quantum physical aspects of nature characterized by non-locality and the stochastic properties of quantum mechanics. Entanglement occurs when a pair of particles (or a group of quanta) is generated in a way that the quantum state of each particle of the pair cannot be described independently of the state of the others even if they are separated by large distances. For a deeper insight see [24–29]. In ER = EPR, causality is not violated. ER bridges do not violate causality because of the topological censorship, which forbids ordinary traversable wormholes; EPR states, instead, prevent causality violations because of the properties of entangled states described by Bell's inequalities—no information is transferred between the two entangled states during the wavefunction collapse of the entangled pair as each quantum state in an EPR pair cannot be described independently of the other states [24,26]. The ER = EPR equivalence is valid if there are no traversable wormhole solutions that do not require the violation of the strong and/or weak energy conditions [30–33], they may instead behave as quantum communication channels between the quantum fields there defined [34].

The ER = EPR conjecture was initially formulated in the gravity/gauge theory equivalence between Anti-de Sitter (AdS) space–times and conformal field theories (CFTs) by Maldacena (gauge/gravity duality) within a relationship between the entanglement entropy of a set of black holes and the cross-section area of ER bridges connecting them. AdS space–times represent an elegant solution of Einstein's equations with negative curvature where the outer boundary is a surface and, in the CFT, correspondence quanta can interact and generate the holographic universe there contained. The AdS/CFT correspondence provides a complete non-perturbative definition of gravity with quantum field theory, extending this correspondence also to space–time scenarios of quantum gravity where the asymptotic behavior of the space–time is that of AdS space–time. The AdS/CFT correspondence plays a key role in the calculations of strong coupled quantum field theories. When the boundary theory is strongly coupled, the bulk theory is weakly coupled, and vice versa. A strongly coupled field theory can have an AdS dual gravity description weakly coupled and therefore calculable and vice versa. As an example, if the curvature of the AdS space increases, the gravitational coupling becomes stronger and the boundary coupling is weaker.

To extend ER = EPR conjecture to space–times different from the Anti-de Sitter solution, one has to investigate how much ER = EPR depends strictly on AdS/CFT correspondence and from the properties of wormholes also in de Sitter (dS) space–times. First of all we must consider that AdS/CFT is a structural correspondence between bulk and screen, but does not contain in itself any specific indication of the possible dynamics of wormhole formation. Wormholes require a cosmological scenario. For example, it is plausible that the wormholes were formed in the initial chaotic phases of the universe with a rate similar to that of the formation of mini black holes (BHs), with very specific traces as regards the event horizon, as discussed in [35–37]. This aspect is decisive because all the problems related to the quantum aspect of the wormholes imply a cosmological background capable of providing a plausible scenario for their existence described by the Ryu–Takanayagi's entropy that relates the entanglement entropy in CFT and the geometry of AdS space–times. The Ryu–Takanayagi formula is a generalization of the BH entropy formula by Bekenstein–Hawking [38,39] to a whole class of holographic theories [40,41] where gravitational models with dimension D are dual to a gauge theory in dimension $D-1$.

In these recent years, the interest in the maximum symmetry properties of de Sitter's space gave this structure a new centrality with respect to AdS. One of the most relevant problems was to project a hologram of a quantum particle that lives in the infinite future of AdS, which makes it difficult to describe real-time space in holographic terms. In particular, the main classes of essential results must be mentioned here: the CPT Universe [42] and the numerous results on the non-locality in dS space–times [43–47]. Of relevant importance is the so-called "uplifting" technique by Dong et al. [48] where two

Anti-de Sitter space–times are transformed into a de Sitter space–time. The uplifting changes the curvature of two "saddle-shaped" AdS space–times that, once warped, are glued together along their rims and turned into a "bowl-shaped" dS space–time via entanglement or more general two-throated Randall–Sundrum systems [49,50] and the CFTs relative to both hemispheres become coupled with each other. In this way one forms a single quantum system that is holographically dual to the entire spherical de Sitter space, defined on its boundary located at a finite distance away. The technique of uplifting two AdS into a dS permits us to modify the curvature in a more general way than that offered by the set of local transformations obtained through Wick rotations that can only act locally—the curvature changes everywhere by introducing extra fields whose energy density acts as an extra source of curvature to landscape the AdS space–time into a dS one. The cosmological constant in the bulk space is then transformed from negative to positive and the holographic projection of the space–time into its boundaries is changed. Some examples are reported by Silverstein and Polchinski [51] or in Vasiliev's higher-spin gravity—in AdS, the boundary theory is an O(N)-vector field theory, while in dS space it becomes an Sp(N) scalar field theory, where N is the number of the vector (scalar) fields of the boundary theories [52,53].

In this work, we analyze Einstein's equations in a finite volume of space–time down to the Planck scale, finding wormhole connections that avoid the singularity problem and an indetermination relationship that involves the Riemann curvature. This finds application to the ER = EPR conjecture—in this framework, geometry behaves as a geodesic tensor network that defines the quantum state properties of a fundamental quantum state of a given metric [54] and a virtual graviton exchange becomes equivalent to entanglement to which one can apply the concept of Penrose's decoherence of a quantum state [55]. In this crossing between locality of GR and the emergence of non-locality of QM as in [56,57], where de Sitter space–time is taken as the geometric structure of vacuum, the analysis of Einstein's equations can provide an additional support to the ER = EPR conjecture extended from AdS to dS [58] and to locally Euclidean space–times. This can be interpreted as the route to ER = EPR from general relativity.

2. Wormhole Connections down to Planck Scales from Einstein's Equations

In the ER = EPR scenario, wormhole connections are fundamental in the building of space–time. Consider an entangled quantum system. The emergence of space–time in terms of ER connections, in the gravity picture, is intimately related to the quantum entanglement of degrees of freedom in the corresponding conventional quantum system, building up space–time with quantum entanglement. The ER = EPR equivalence suggests that space–time and gravity may emerge from the degrees of freedom of the field theory. On the other hand, space–time becomes the optimal way to build entanglement starting from wormhole connections. At Planck scales, the Planck area is defined as the area by which the surface of a Schwarzschild black hole increases when in the black hole is injected one bit of information. In a Riemannian manifold (M, g) the scalar curvature in an $(n-1)$ hyperplane relates GR with the entanglement of quantum states in an arbitrary Hilbert space without reference to AdS = CFT or any other holographic boundary construction.

2.1. *Einstein's Equations in the Neighborhood of an Event*

Einstein's equations are the core of GR—they describe gravity in terms of the curvature of space–time. Spacetime geometry and the metric tensor g_{ik} are determined from Einstein's equations, given the distribution of energy, mass and momentum in space–time encoded in the stress–energy tensor T_{ik}.

In our approach, by assuming that Einstein's equations remain valid down to the Planck scale, we find that the connection between events are achieved through wormhole connections, avoiding the gravitational collapse and the presence of singularities at Planck scales. To this aim, we adopt the approach by Schwinger in the analysis of classical fields [59]—to determine the properties of a field, one cannot measure the field in a point, otherwise, because of the equivalence principle, one finds only a local Minkowskian space–time tangent to the manifold (M, g) in the given point event.

Following Schwinger, from his studies on electromagnetic theory [59,60], the analysis of a classical field must be made in a neighborhood of the event. This approach is clearly valid in electromagnetism and antenna theory: when sensing the electromagnetic field with an antenna, one cannot measure in a point the fluctuations of the electric field. The antenna must have a finite length in space and the field must be measured in a finite time interval to be revealed. As happens for any antenna, as well as for the gravitational field, one has to consider a finite length or a finite hypervolume in which to determine the properties of the field. For the same reasons, one cannot deduce the properties of the gravitational field in a single point and at a given time because of the equivalence principle, as discussed in the free-falling particle paradox [61].

Let us find the main properties of the gravitational field when a finite length L of measure is fixed down to the Planck scale. Given a Riemannian manifold (M, g), where M is the manifold and g the metric tensor, $g \in \otimes^2 \dot{T}$ of tensorial order 2, (written as $g_{(2)}$), Einstein's equations are

$$R_{ik} - \frac{1}{2} R g_{ik} + \Lambda g_{ik} = 8\pi T_{ik} \tag{1}$$

where Λ is the cosmological constant and $R_{ik} = g^{lm} R_{lmik}$ and $R = g^{ik} g^{lm} R_{lmik}$ represent the tensorial and scalar space–time curvature terms obtained from the Riemann tensor R_{iklm} that, in the tensorial index notation, takes the usual well-known form [62]

$$R_{iklm} = \frac{1}{2} \left(\frac{\partial^2 g_{lm}}{\partial x^k \partial x^l} + \frac{\partial^2 g_{kl}}{\partial x^i \partial x^m} - \frac{\partial^2 g_{il}}{\partial x^k \partial x^m} - \frac{\partial^2 g_{km}}{\partial x^i \partial x^l} \right) + g_{np} \left(\Gamma^n_{kl} \Gamma^p_{im} - \Gamma^n_{km} \Gamma^p_{il} \right) \tag{2}$$

the tensor is an element of the rank-four tensors $R_{iklm} \in \otimes^4 \dot{T}$ in the cotangent bundle $\dot{T}$ of the manifold (M, g) that we will indicate with the symbol $R_{(4)}$, where 4 is the tensorial index.

The gravitational field has the fundamental property that, any body, independently from their mass, moves in the same way. This is described by the strong equivalence principle, which suggests that gravity is a geometrical quantity and one cannot measure the gravitational field in an event, as the field becomes locally Galilean and diagonalizable, and the energy of the field cannot be uniquely defined.

In a neighborhood of a given event, space–time is built by chains of events and observers. The building of space–time is obtained by causally transferring information encoded locally in one event of the field to form events and coincidences of events described by punctual (point-to-point) correlations in a four-dimensional manifold (M, g).

Because of the equivalence principle, these observables must be generated within a volume of finite spatial extent from a given spatial length L and propagated—through a chain of events—in the form of four-volumetric densities (or in geometrical sub-varieties) in the four-dimensional manifold (M, g) to a remotely located finite region of space–time of likewise finite spatial/temporal extent—the observation (hyper-)volume $V \subseteq M$ over which they are volume integrated into observables, allowing the information carried by them to be extracted and decoded. More specifically, we will use the local split of $3 + 1$ in space and time where we will consider the integration of the field properties over a three-dimensional volume $V = L^3$. The volumetric density of every gravitational observable carried by the gravitational field is a linear combination of quantities that are second order (quadratic/bilinear) in the metric of the field, and/or of the derivatives of the field. To obtain these quantities one must integrate the density of a conserved quantity e.g., over a given 3D space-like hypersurface σ or over a four-dimensional interval, characterized by a given finite length L; there, the field equations are integrated and averaged to obtain the field observables we need to analyze the properties of space–time down to the Planck scales.

Let us consider a Lorentzian manifold as example, with cosmological constant Λ and then introduce a characteristic length L. This quantity is the proper length associated to the generic coordinate variation

Δx written in terms of the metric tensor g, viz., $L \sim g^{1/2}\Delta x$. The Riemann tensor is then written in terms of the metric variations Δg, the covariant metric tensor g and the contrarvariant one, g^{-1}

$$R_{(4)}(g, L) \sim \frac{g^2}{L^2}\left(\Delta\left(\Delta g(g)^{-1}\right) + \left(\Delta g(g)^{-1}\right)^2\right) \tag{3}$$

where the term $\left(\Delta g\,(g)^{-1}\right)^2_{\{kl,im-km,il\}} = \Gamma^n_{kl}\Gamma^p_{im} - \Gamma^n_{km}\Gamma^p_{il}$ represents the affine connection and $\Delta\left(g^{-1}\Delta g\right)$ is the second derivatives of the metric tensor with respect to the coordinates, being $\partial^2_{kl}\, g = \frac{g}{L^2}\Delta_k\Delta_l\, g = \frac{g^2}{L^2}\Delta_k\left(\Delta_l g(g)^{-1}\right)$. The Ricci tensor and scalar are $R_{(2)} \sim g^{-1}R_{(4)}$ and $R \sim g^{-2}R_{(4)}$, the Einstein tensor is $G = R_{(4)}(g)^{-1}$ and Einstein's equations are $G + g\Lambda = T$, where T is the energy–momentum tensor.

By introducing a characteristic length L, if Einstein's equations hold down to the Planck scales, from the basic formulation of the Riemann tensor and Einstein equations we find that the field equations, integrated and averaged over a 3D space-like hypersurface σ with unit normal vector $n \sim g^{-1/2}$, obey an indetermination relationship that recalls Heisenberg's. Instead of focusing on the more general energy–tensor quantity (or the momentum vector), we consider for the sake of simplicity the scalar proper energy E, averaged over a proper volume L^3, which is given by the integral of the energy momentum tensor over a given proper volume element of a space-like 3D–hypersurface. This leads to the following formulation of the proper energy averaged over the given proper volume

$$\langle E - g\Lambda\rangle = \bar{E} \sim \frac{g^2}{L}R_{(4)} = L\left(\Delta\left(\Delta g(g)^{-1}\right) + \left(\Delta g(g)^{-1}\right)^2\right) \tag{4}$$

If we rescale this relationship down to the Planck scale L_p, by defining the light crossing time as $\tau = L$ and the Planck Time τ_p, the Einstein equations retain their validity down to the Planck scale, even if metric fluctuations over a scale larger than L_p can occur. We find that these fluctuations can give rise to a relationship

$$\left(\frac{\tau_p}{\tau}\right)^2\left(\frac{\bar{E}\times\tau}{\hbar}\right) = \left(\frac{L_p}{L}\right)^2\left(\frac{\bar{E}\times\tau}{\hbar}\right) = \frac{L^2}{g^2}R_{(4)}(g, L) \tag{5}$$

that holds down to the Planck scales. Fixing a characteristic spatial scale (or time), the relationship in Equation (5) corresponds to the introduction of fluctuations of the averaged quantity over L^3 of the proper energy $\bar{E}$. If we set $\bar{E} = \Delta E^*$ and $\tau = \Delta t$, we can write Equation (5) in a more familiar Heisenberg relationship that involves the Riemann tensor and the contribution from the dark energy

$$\Delta E^* \times \Delta t = \hbar\left(\frac{\tau}{\tau_p}\right)^2\frac{L^2}{g^2}R_{(4)}(g, L) = \frac{\hbar}{g^2}\left(\frac{L^2}{L_p}\right)^2 R_{(4)}(g, L) \tag{6}$$

that at Planck scales becomes

$$\Delta E^* \times \Delta t = \hbar\frac{L_p^2}{g^2}R_{(4)}(g, L) = \hbar\left(\Delta\left(\Delta g(g)^{-1}\right) + \left(\Delta g(g)^{-1}\right)^2\right) \tag{7}$$

where $\Delta E^* = \Delta E + \Delta g\;\Lambda + g\Delta\Lambda$ averaged on the volume L^3 of the 3D space-like hypersurface σ.

2.2. The Energy of the Gravitational Field

Dark energy and other different vacua are parameterized by the cosmological constant Λ. When $\Lambda > 1$, the equations describe an AdS space–time. The gravitational fluctuations are mainly expressed by the affine connection term $\left(\Delta g(g)^{-1}\right)^2$ for any space–time. To describe the energy of the gravitational field, which is not defined as a global and conserved quantity, one has to introduce pseudotensorial quantities describing the energy trapped non-locally in the geometry. One example is the non-symmetric Einstein pseudotensor, which is constructed exclusively from the metric tensor and its first derivatives but is not suitable for our purposes. Instead, the Landau–Lifshitz pseudotensor t_{ik} [62] permits us to write for the integrated non-local gravitational energy E_g that includes the contribution of the cosmological constant in terms of the curvature tensor. This quantity is quadratic in the connection and, for a general covariant component of the pseudotensor, averaged on the volume $V = L^3$ one obtains

$$\left\langle g^{-1}\left(t+\Lambda g\right)\right\rangle_{V=L^3} \sim \frac{E_g + E_\Lambda}{L^3} \tag{8}$$

where $E_\Lambda = g^{-1}\Lambda g L^3$ is the energy associated to the value of the cosmological constant and to dark energy. Considering that

$$|g|\, g^{-2}\left(t+\Lambda g\right) \sim \left(\frac{\Delta g}{\Delta x}\right)^2, \tag{9}$$

this relationship leads to a background curvature with fluctuations having wavelength $\lambda = L$ that can be interpreted as connections between events due to an exchange of virtual gravitons with wavelength λ and energy $\hbar/\lambda$ or, in the ER = EPR scenario, to the connection through an ER wormhole,

$$\left\langle \left(\Delta g (g)^{-1}\right)^2 \right\rangle_{V=L^3} \sim \frac{E_g + E_\Lambda}{L} = \left(\frac{\tau_p}{\tau}\right) \frac{E_g + E_\Lambda}{E_p} \tag{10}$$

where E_p is Planck's energy. In the ER = EPR hypothesis, these energy fluctuations would be considered as equivalent to the connection between two entangled events separated by the distance L, giving a paradoxical meaning to the exchange of a virtual graviton in terms of entanglement connections between events like in an emergent gravity scenario.

Following the already cited works by De Witt and the classical QG interpretation found in the literature [1–5], this term would describe a virtual graviton exchange between two events within a space–time connection. On the other hand, this term—that can be also interpreted in terms of a wormhole connection between the two events—with the exchange of at least 1 qbit of information (in the ER = EPR conjecture) would correspond to the entanglement of two particles. If ER wormholes are equivalent to a monogamous connection between the two events [20], as realized through a virtual graviton exchange, one could state that entanglement of EPR states can derive from the exchanges of virtual gravitons between two events. From another perspective, entangled states should be provided by a mixed state between the two entangled pairs with that of the virtual graviton. The question is what is the correct perspective?

From Einstein's equations, the observable averaged total energy of a metric fluctuation over the volume V on a scale L, becomes

$$E \sim \left(\frac{\tau}{\tau_p}\right) \left\langle \Delta\left(\Delta g (g)^{-1}\right)\right\rangle_V E_p + E_g + E_\Lambda \tag{11}$$

and is made with the energy of geometry and vacuum and energy of interaction expressed in terms of gradients of the geometry fluctuations, second order derivatives of the metric tensor, as in the Riemann tensor that make the connection between observers.

2.3. *Planck-Scale Wormhole Connections*

In a local neighborhood of a given event $\{x^i\}_0$, one performs a discrete infinite denumerable $3+1$ local slicing of the space–time with time steps a Planck time unit. To the initial event $\{x^i\}_0$ corresponds the slice $N = 0$. The N-th slice corresponds to the time $\tau_N = (N+1)\tau_p$, building up a symbolic dynamics of space–time events. The energy of the gravitational perturbation in the N-th slice is

$$\left(E_g + E_\Lambda\right)_N \sim \frac{E_p}{N+1} \tag{12}$$

the total energy is instead

$$E_N \sim (N+1) \left\langle \Delta \left(\Delta g(g)^{-1} \right) \right\rangle_{V,N} E_p + \frac{E_p}{N+1} \tag{13}$$

for $N \to 0$ the energy fluctuation becomes $E_N \to E_p$ for which, by definition, $E_N \tau_N \sim \hbar$ and the metric tidal fluctuations tend to zero—space–time at Planck lengths is homogeneous and isotropic and the local geometry depends only on the energy of fluctuations in space–time and from the energy of the cosmological constant. This because $\Delta g/g \to 1$ are both on the order of the Planck scale. This is the reason why the field does not diverge and no singularities are present.

For $N \to \infty$ the dominant energy is that of tidal fluctuations at scales larger than L_p accompanied with that of the cosmological constant when integrated over the metric and remains as a constant function over the volume of integration; E_g becomes instead negligible. This means that for a process connecting two events lasting a time τ, the amount of energy does not entirely contribute to the vacuum energy but it is partially spent in geometry in this process, involving the dark energy contribution expressed by the cosmological constant Λ. Recalling Heisenberg principle from Equation (5), the larger is the energy fluctuation, the smaller results the space/time interval fluctuation.

In the neighborhood of Planck scales, when $N > 0$, the curvature of space–time remains finite and the Riemann tensor can be written as

$$R_{(4)}(g, L) \sim \frac{E_p}{\hbar} \left(\frac{\tau_p}{\tau} \right)^2 \frac{g^2}{L^2} \tag{14}$$

and the Ricci scalar is

$$R(g, L) \sim \frac{1}{L^2} \frac{E_p}{\hbar} \left(\frac{\tau_p}{\tau} \right)^2 \tag{15}$$

that for $L \to L_p$ we have $R(g, L) \to 1/L_p^2$ with the result that at Planck scales there is no singularity in the curvature and the gravitational radius becomes

$$R_g = 2 \frac{E\, \tau\, L_p^2}{\hbar\, L} \tag{16}$$

that is written as $R_g = 2L_p$, which corresponds to elementary wormhole connections at the Planck scale and finding a trivial equivalence with the corresponding Penrose diagrams. Directly from Einstein's equations we find that, at Planck scales, the singularities expected from quantum gravity can be interpreted in terms of wormhole connections between the events, as required in the ER = EPR conjecture and obtain an indetermination relationship shown in Equation (6) involving the Riemann tensor and geometry fluctuations. In this view, wormhole and equivalent EPR connections can also be formally equivalent to an exchange of a virtual graviton at scales larger than Planck scale, whilst any group of superimposed states below Planck scales, instead, will be indistinguishable and therefore entangled.

2.4. *Tests for the ER = EPR Conjecture*

If we suppose the validity of the ER = EPR conjecture, the geometry fluctuations present at Planck scales may be revealed with quantum entanglement. By applying the indetermination relationship that involves the Riemann tensor expressed in Equations (6) and (7), we argue that one can obtain information about the fluctuations of space–time and determine whether a characteristic scale like the Planck's one is present, as expected in QG.

If space–time is discrete, its discreteness is expected to be characterized by a typical scale of space and/or time: There exist a minimum time interval t_p and a minimum length L_p where wormholes connections—equivalent to entangled states between two or more regions of space–time—connect different events or space–times. If events/space–times are connected with intervals smaller than L_p and t_p, they would be entangled and actually be the same event or the same space–time. Their quantum superposition can exist and collapse after a finite time interval and the properties of wormhole connections are reflected in the properties of the corresponding EPR states also when they connect events at scales larger than the Planck scale. This scenario is different from Penrose's assumptions [55], where space–time is thought to be continuous and the quantum superposition of space–times result unfeasible leading to the gravitational collapse.

Noe we propose to test the ER = EPR scenario by using the Heisenberg uncertainty principle applied to pairs (or groups) of entangled particles and including the additional indetermination introduced by quantum gravity effects. The generalized Heisenberg's uncertainty principle for the momentum p and the position x that includes the existence of a characteristic length L_p such as the Planck scale, or any other scale typical that can be found in certain quantum gravity models, is given by [63,64]

$$\Delta x = \Delta x_{QM} + \Delta x_{GR} \geq \frac{\hbar}{2\Delta p} + k\,\Delta p \tag{17}$$

the existence of a minimum interval in space–time is revealed by a deviation from the classical term due to quantum mechanics only, Δx_{QM}. The quantum gravity term, Δ_{GR}, due to the existence of a characteristic length L_p and to the properties of the gravitational field, can be characterized instead by a parameter k, a constant characteristic of the quantum theory of gravitation here considered. To give an example, in a string theory scenario, $k = \alpha Y$, where α is the string tension and Y a constant that depends on the theory. In our case, following [65–67], one can find that $k = 2L_p^2/\hbar$. From our calculations that involve the Riemann tensor, we find that $\Delta x_{QG} = 2E_p/\Delta E^*$ and thus $\Delta p = \hbar E_p/\Delta E^* L_p^2$, a term that includes the effects of dark energy in the term ΔE^* too.

We write now the Heisenberg relationship for sets of $N-$particle entangled states. Following [68–72], consider first a couple of entangled particles with positions x_1 and x_2 and momenta p_1 and p_2, respectively. For $N = 2$, the classical indetermination principle is

$$\Delta(x_1, x_2)^2_{QM} = \left[\Delta(x_1)^2 + \Delta(x_2)^2\right] \times \left[\Delta(p_1)^2 + \Delta(p_2)^2\right] \geq \frac{\hbar^2}{4}. \tag{18}$$

In the simplest case, where $\Delta(x_1) = \Delta(x_2) = \Delta x_e$ and $\Delta(p_1) = \Delta(p_2) = \Delta p_e$, the uncertainty relationship becomes $(\Delta x_e)^2(\Delta p_e)^2 \geq \hbar^2$. For N identical entangled states, the extended indetermination principle becomes

$$(\Delta x_e)^2(\Delta p_e)^2 \geq \frac{N^2\hbar^2}{4} \tag{19}$$

and when we include the effects of the gravitational field one obtains

$$\Delta x = \Delta(x_1, x_2)_{QM} + \Delta(x_1, x_2)_{GR} \geq \frac{N\hbar}{2\Delta p} + \frac{2NL_p^2\Delta p}{\hbar}. \tag{20}$$

By assuming that Einstein's equations retain their validity down to the Planck scales and that wormhole connections represent the building blocks of the physics of the gravitational field at and below Planck scales (t_p and L_p), ER = EPR links connecting any space–time (or event) with a difference smaller than t_p and L_p mean that different space–times and events are physically identical, and then in principle undetectable and entangled. Instead, in a region with radius R, the spatial difference of two space–times/events is $\Delta L = 2L_p^2\Delta E^*/\hbar c$, and the difference of their corresponding space–times is the difference of the proper spatial sizes of the regions occupied by them and the time of the wavefunction collapse is on the order of $\tau_c \sim 2\hbar E_p/(\Delta E^*)^2$.

If the properties of ER = EPR links remain valid from the Planck up to the macroscopic scales, where entanglement can be observed in the lab, the term Δx_{QG} in the Heisenberg relationship expressed in Equation (20) is expected to reveal the properties of the wormhole structure of space–time from a deep analysis of the wavefunction collapse of an entangled pair. In other words, the deviation from the quantity Δx_{QM} of the classical Heisenberg principle would reveal the fuzziness space–time or, better, of the point-by-point identification of the spatial section of the two events/space–times, better evident with a set of a large number of N entangled quanta like a Schrödinger cat.

From the point of view of relativistic quantum information discipline, entanglement and wormholes are expected to create space–time and entanglement events (and space–times) through quantum information—information that emerges from the connection of quantum bits. In fact, from a quantum-computational interpretation of space–time entanglement, in a foliation of space–time, the quantum fluctuations of the metric present on the slice n can be interpreted as wormhole connections between one Planckian pixel in the slice n with that one present in the $n-1$ slice. Following [73], the holographic principle suggests that such a geometrical connection is space–time entanglement. If not entangled, following Penrose's argumentation only the quantum superposition of two space–times with a difference larger than the minimum sizes can not exist, and should collapse instantaneously. If they are connected by an ER wormhole they should obey the indetermination relationship expressed in Equation (20).

To verify possible additional anomalies in the indetermination principle introduced by the ER = EPR conjecture one may instead want to consider to measure the time/energy entangled states and study the time of collapse as a function of their energy differences. This may explain why the wavefunction collapse of an EPR pair is not always instantaneous, as it may depend on the geometry fluctuations. Moreover, one has to also consider the effects introduced by the presence of the cosmological constant, of the information encoded and shared between the entangled quantum states and their relationship with the gravitational information entropy that go beyond the purpose of the present work.

Of course an experimenter has to consider that EPR states depend on the choice of reference frames and that Bell's inequalities are preserved in certain reference frames only, and should also consider the effects of simultaneity and include in the experiment the additional macroscopic effects induced by the gravitational field at large scales in the presence of massive bodies. As an example, simultaneity is responsible for the uncertainty of the ordering of non-local wavefunction collapse when the relativistic effects cannot be neglected. In any case, if a time measurement performed with an entangled pair of photons is seen as simultaneous in one shared reference frame, then the result of this measure can be considered simultaneous to all measuring observers who do not share a reference frame. The inversion of the temporal order due to simultaneity is impossible to determine, the attempt to measure this effect will unavoidably introduce an uncertainty in the result. There is no need to have any preferred reference frame for the wavefunction collapse of entangled states. If an experimenter tries to determine the exact reference frame where the wavefunction collapsed, the measurement process will unavoidably introduce an uncertainty that would make impossible the identification of the "exact" reference frame. Obviously, if one can determine the order of the measurement in a shared reference frame it can result like that in certain reference frames and, instead, indeterminate in the other reference frames [74,75].

3. Discussion and Conclusions

It is one of the great merits of Albert Einstein to have investigated the possibility of a multiple-connected space–time and in theoretical physics there is a long tradition of studying quantum behavior in spaces of this type [76]. These lines of research have progressively merged into the quantum study of wormholes, assuming a decisive relevance not only for the study of the structure of the GR and its cosmological implications, but has given the question a decisive configuration as regards the relations of "coexistence" peaceful between QM and GR. In this work we proposed a formal technique for the study of the quantum effects of a wormhole within the conjecture ER = EPR. We then considered different scenarios from the original Susskind and Maldacena one, in particular those related to the dS space, which seems to be a much more promising ground for the study of the emergence of classical information starting from a quantum background where time is not defined [56,77–79].

These reflections suggest that an effective generalization of the physical meaning of ER = EPR requires a different and more complex philosophy on the emergence of physical space–time as a holographic "settlement" of temperature/energy scales, and the use of well-known techniques in QFT [80,81].

In other words, these scenarios suggest that the idea of transition of the metric suggested by Sacharov may be the most "natural" way to characterize non-locality in a metric formalism. The assumption of ER = EPR would be only one of the aspects of a more general phenomenon of Raum–Zeit–Materie production starting from a non-local Euclidean background through quantum computation procedures. The observable part of space time would therefore, in a rather literal sense, result in a thin layer of ice emerging from an ocean of non-locality and the extension of ER = EPR conjecture to Euclidean non-locality may extend its domain from the original AdS/CFT scenario.

Finally, we suggest readers consider the conjecture ER = EPR within the scenario of de Sitter's projective cosmology, described by Hartle-Hawking boundary conditions as Nucleation by Sitter Vacuum [57]. In this cosmological approach one can define the localization conditions in time of the particles starting from an Euclidean pre-space that models a non-local phase. Using the Bekenstein relation, it is possible to identify the area of the micro-horizon $A = (c\theta_0)^2 \simeq 10^{-26}$ cm^2, where theta is the chronon, chosen as time scale of the baryonic location. In this case the construction of wormholes applies to a scale much larger than the Planck length. In this case the wormholes are defined by a transition of the metric similar to that hypothesized in the classical work by Sacharov in 1984 [82].

Anyway, the wormhole structure of space–time could in principle be characterized by the extended Heisenberg principle through a deep study of the wavefunction collapse of entangled particles and reveal possible scenarios of QG and cosmology or emergent gravity theories where the exchange of a virtual graviton could also be interpreted in terms of entanglement. At Planck scales wormhole connections would avoid the gravitational collapse and singularities. Moreover, the exchange of a virtual graviton would become equivalent to a wormhole connection and/or entanglement between two or more events. From this we can argue that the ER = EPR conjecture alone, as it is, cannot fully explain without experimental results whether Planck-scale phenomenology can be revealed through entanglement or that gravity and space–time are emergent physical quantities.

Author Contributions: Writing—original draft, F.T. and I.L. All authors have read and agreed to the published version of the manuscript.

Acknowledgments: We dedicate this work to Dino and Gep. F.T. acknowledges ZKM and Peter Weibel for the financial support.

References

1. DeWitt, B.S. Quantum Theory of Gravity. I. The Canonical Theory. *Phys. Rev.* **1967**, *160*, 1113. [CrossRef]
2. DeWitt, B.S. Quantum Theory of Gravity. II. The Manifestly Covariant Theory. *Phys. Rev.* **1967**, *162*, 1195. [CrossRef]
3. DeWitt, B.S. Quantum Theory of Gravity. III. Applications of the Covariant Theory. *Phys. Rev.* **1967**, *162*, 1239. [CrossRef]
4. DeWitt, B.S.; DeWitt, C.M. The Quantum Theory of Interacting Gravitational and Spinor Fields. *Phys. Rev.* **1952**, *87*, 116. [CrossRef]
5. DeWitt, B.S. Approximate Effective Action for Quantum Gravity. *Phys. Rev. Lett.* **1981**, *47*, 1647. [CrossRef]
6. Rosenfeld, L. Zur quantelung der wellenfelder. *Ann. Phys.* **1930**, *5*, 113. [CrossRef]
7. Rosenfeld, L.Z. Über die Gravitationswirkungen des Lichtes. *Physik* **1930**, *65*, 589. [CrossRef]
8. Tanabashi, M.; Hagiwara, K.; Hikasa, K.; Nakamura, K.; Sumino, Y.; Takahashi, F.; Tanaka, J.; Agashe, K.; Aielli, G.; Amsler, C.; et al. Review of Particle Physics. *Phys. Rev. D* **2018**, *98*, 030001. [CrossRef]
9. Rovelli, C. Loop Quantum Gravity. *Living Rev. Relativ.* **2008**, *11*, 5. [CrossRef] [PubMed]
10. Dzhunushaliev, V.D. Multidimensional Ggeometrical Model of the Renormalized Electrical Charge with Splitting off the Extra Coordinates. *Mod. Phys. Lett. A* **1998**, *13*, 2179. [CrossRef]

11. Dzhunushaliev, V.D.; Singleton, D. Wormholes and Flux Tubes in 5D Kaluza-Klein Theory. *Phys. Rev. D* **1999**, *59*, 064018. [CrossRef]
12. Maldacena, J.; Susskind, L. Cool horizons for entangled black holes. *Fortschr. Phys.* **2013**, *61*, 781–811. [CrossRef]
13. Einstein, A.; Rosen, N. The Particle Problem in the General Theory of Relativity. *Phys. Rev.* **1935**, *48*, 73–77. [CrossRef]
14. Einstein, A.; Podolsky, B.; Rosen, N. Can Quantum-Mechanical Description of Physical Reality Be Considered Complete? *Phys. Rev.* **1935**, *47*, 777–780. [CrossRef]
15. Susskind, L. Copenhagen vs Everett, Teleportation, and ER = EPR. *Fortschr. Phys.* **2016**, *64*, 551–564. [CrossRef]
16. Maldacena, J. The large N limit of superconformal field theories and supergravity. *Adv. Theor. Math. Phys.* **1998**, *2*, 231. [CrossRef]
17. Aharony, O.; Gubser, S.S.; Maldacena, J.; Ooguri, H.; Oz, Y. Large N Field Theories, String Theory and Gravity. *Phys. Rep.* **2000**, *323*, 183–386. [CrossRef]
18. Biquard, O. *AdS/CFT Correspondence: Einstein Metrics and Their Conformal Boundaries. EMS IRMA Lectures in Mathematics and Theoretical Physics*; European Mathematical Society: Strasbourg, France, 2005; ISBN 978-3-03719-013-5.
19. Witten, E. Anti de Sitter space and holography. *Adv. Theor. Math. Phys.* **1998**, *2*, 253–291. [CrossRef]
20. Gharibyan, H.; Penna, R.F. Are entangled particles connected by wormholes? Support for the ER = EPR conjecture from entropy inequalities. *Phys. Rev. D* **2014**, *89*, 066001. [CrossRef]
21. van Raamsdonk, M. Building up space–time with quantum entanglement. *Gen. Relativ. Gravit.* **2010**, *42*, 2323–2329. [CrossRef]
22. Verlinde, E.P. On the Origin of Gravity and the Laws of Newton. *JHEP* **2011**, 29. [CrossRef]
23. Cao, C.; Carroll, S.M. Bulk entanglement gravity without a boundary: Towards finding Einstein's equation in Hilbert space. *Phys. Rev. D* **2018**, *97*, 086003. [CrossRef]
24. Bell, J.S. On the Einstein-Poldolsky-Rosen paradox. *Physics* **1964**, *1*, 195–200. [CrossRef]
25. Gottfried, K. *John S. Bell on the Foundations of Quantum Mechanics*; Bell, M., Gottfried, K., Veltman, M., Eds.; World Scientific Publishing Company: Singapore, 2001.
26. Horodecki, R.; Horodecki, P.; Horodecki, M.; Horodecki, K. Quantum entanglement. *Rev. Mod. Phys.* **2009**, *81*, 865. [CrossRef]
27. Bouwmeester, D.; Ekert, A.; Zeilinger, A. (Eds.) *The Physics of Quantum Information*; Springer: Berlin/Heidelberg, Germany, 2000; ISBN 978-3-642-08607-6.
28. Bertlmann, R.; Zeilinger, A. (Eds.) *Quantum [Un]Speakables, from Bell to Quantum Information*; Springer: Berlin/Heidelberg, Germany, 2002; ISBN 978-3-662-05032-3.
29. Bertlmann, R.; Zeilinger, A. (Eds.) *Quantum [Un]Speakables II*; Springer Nature AG: Basel, Switzerland, 2017; ISBN 978-3-319-38987-5.
30. Visser, M. *Lorentzian Wormholes: From Einstein to Hawking*; AIP: Woodbury, NY, USA, 1995; 412p.
31. Maldacena, J.; Milekhin, A.; Popov, F. Traversable wormholes in four dimensions. *arXiv* **2018**, arXiv:1807.04726.
32. Maldacena, J.; Qi, X.-L. Eternal traversable wormhole. *arXiv* **2018**, arXiv:1804.00491.
33. Horowitz, G.T.; Marolf, D.; Santos, J.E.; Wang, D. Creating a Traversable Wormhole. *arXiv* **2019**, arXiv:1904.02187.
34. Bao, N.; Chatwin-Davies, A.; Pollack, J.; Remmen, G.N. Traversable wormholes as quantum channels: Exploring CFT entanglement structure and channel capacity in holography. *JHEP* **2018**, 71. [CrossRef]
35. Bueno, P.; Cano, P.A.; Goelen, F.; Hertog, T.; Vercnocke, B. Echoes of Kerr-like wormholes. *Phys. Rev. D* **2018**, *97*, 024040. [CrossRef]
36. Cardoso, V.; Franzin, E.; Pani, P. Is the gravitational-wave ringdown a probe of the event horizon? *Phys. Rev. Lett.* **2016**, *116*, 171101; Erratum *Phys. Rev. Lett.* **2016**, *117*, 089902. [CrossRef]
37. Konoplya, R. How to tell the shape of a wormhole by its quasinormal modes. *Phys. Lett. B* **2018**, *784*, 43–49. [CrossRef]
38. Bekenstein, J.D. Black holes and entropy. *Phys. Rev. D* **1973**, *7*, 2333–2346. [CrossRef]
39. Hawking, S.W. Particle creation by black holes. *Commun. Math. Phys.* **1975**, *43*, 199–220. [CrossRef]

40. Hubeny, V.E.; Rangamani, M.; Takayanagi, T. A Covariant Holographic Entanglement Entropy Proposal. *JHEP* **2007**. [CrossRef]
41. Sethi, S.S. Notes at URL. Available online: http://theory.uchicago.edu/~sethi/Teaching/P483-W2018/Intro%20to%20the%20RT.pdf (accessed on 18 December 2019).
42. Boyle, L.; Finn, K.; Turok, N. *CPT*-Symmetric Universe. *Phys. Rev. Lett.* **2019**, *121*, 251301. [CrossRef]
43. Narain, G.; Zhang, H.Q. Non-locality effect on the entanglement entropy in deSitter. *JCAP* **2019**, *06*, 012. [CrossRef]
44. Chena, P.; Hua, Y.-C.; Yeoma, D.-H. Fuzzy Euclidean wormholes in de Sitter space. *JCAP* **2017**, *1707*, 001. [CrossRef]
45. Maldacena, J.; Pimentel G.L. Entanglement entropy in de Sitter space. *JHEP* **2013**, *1302*, 038. [CrossRef]
46. Narayan, K. de Sitter entropy as entanglement. *Int. J. Mod. Phys. D* **2019**. [CrossRef]
47. Arias, C.; Diaz, F.; Sundell, P. De Sitter Space and Entanglement. *arXiv* **2019**, arXiv:1901.04554.
48. Dong, X.; Silverstein, E.; Torroba, G. De Sitter holography and entanglement entropy. *JHEP* **2018**, 50. [CrossRef]
49. Dimopoulos, S.; Kachru, S.; Kaloper, N.; Lawrence, A.; Silverstein, E. Generating small numbers by tunneling in multithroat compactifications. *Int. J. Mod. Phys. A* **2004**, *19*, 2657–2704. [CrossRef]
50. Dimopoulos, S.; Kachru, S.; Kaloper, N.; Lawrence, A.E.; Silverstein, E. Small numbers from tunneling between brane throats. *Phys. Rev. D* **2001**, *64*, 121702. [CrossRef]
51. Polchinski, J.; Silverstein, E. Dual Purpose Landscaping Tools: Small Extra Dimensions in AdS/CFT. In *Strings, Gauge Fields, and the Geometry Behind. The Legacy of Maximilian Kreuzer*; Rebhan, A., Katzarkov, L., Knapp, J., Rashkov, R., Scheidegger, E., Eds.; World Scientific: Singapore, 2011; ISBN-13: 978-9814412544.
52. Vasiliev, M. Higher-Spin Gauge Theories in Four, Three and Two Dimensions. *Int. J. Mod. Phys. D* **1996**, *5*, 763–797. [CrossRef]
53. Anninos, D.; Hartman, T.; Strominger, A. Higher Spin Realization of the dS/CFT Correspondence. *CQG* **2016**, *34*, 015009. [CrossRef]
54. Dong, X.; Zhou, L. Spacetime as the optimal generative network of quantum states: A roadmap to QM=GR? *arXiv* **2018**, arXiv:1804.07908.
55. Penrose, R. On gravity's role in quantum state reduction. *Gen. Relativ. Gravit.* **1996**, *28*, 581–600. [CrossRef]
56. Licata, I.; Chiatti, L. Event-Based Quantum Mechanics: A Context for the Emergence of Classical Information. *Symmetry* **2019**, *11*, 181. [CrossRef]
57. Feleppa, F.; Licata, I.; Corda, C. Hartle-Hawking boundary conditions as Nucleation by de Sitter Vacuum. *Phys. Dark Universe* **2019**, *26*, 100381. [CrossRef]
58. Strominger, A. The dS/CFT Correspondence. *arXiv* **2001**, arXiv:hep-th/0106113v2.
59. Schwinger, J.; Deraad, L.L., Jr.; Milton, K.A.; Tsaiyang, W. *Classical Electrodynamics*; Perseus Books: Reading, MA, USA, 1965; ISBN 0-7382-0056-5.
60. Thidé, B. *Electromagnetic Field Theory*, 2nd ed.; Dover Publications, Inc.: Mineola, NY, USA, 2011.
61. Tamburini, F.; De Laurentis, M.; Licata, I. Radiation from charged particles due to explicit symmetry breaking in a gravitational field. *Int. J. Geom. Meth. Mod. Phys.* **2018**, *15*, 1850122. [CrossRef]
62. Landau, L.D.; Lifshitz, E.M. *The Classical Theory of Fields*, 4th ed.; Butterworth-Heinemann: Oxford, UK, 1975; Volume 2, ISBN 978-0-7506-2768-9.
63. Kempf, A.; Mangano, G.; Mann, R.B. Hilbert space representation of the minimal length uncertainty relation. *Phys. Rev. D* **1995**, *52*, 1108. [CrossRef] [PubMed]
64. Scardigli, F. Generalized Uncertainty Principle in Quantum Gravity from Micro-Black Hole Gedanken Experiment. *Phys. Lett. B* **1999**, *452*, 39–44. [CrossRef]
65. Adler, R.J.; Santiago, D.I. On Gravity and the Uncertainty Principle. *Mod. Phys. Lett. A* **1999**, *14*, 1371–1381. [CrossRef]
66. Garay, L.J. Quantum gravity and minimum length. *Int. J. Mod. Phys. A* **1995**, *10*, 145. [CrossRef]
67. Shan, G. A Model of Wavefunction Collapse in Discrete Space-Time. *IJTP* **2006**, *45*, 10. [CrossRef]
68. Blado, G.; Herrera, F.; Erwin, J. Quantum Entanglement and the Generalized Uncertainty Principle. *arXiv* **2017**, arXiv:1706.10013.
69. Zeng, J.; Lei, Y.; Pei, S.Y.; Zeng, X.C. CSCO Criterion for Entanglement and Heisenberg Uncertainty Principle. *arXiv* **2013**, arXiv:1306.3325.

70. Rigolin, G. Uncertainty relations for entangled states. *Found. Phys. Lett.* **2002**, *15*, 293.:1021039822206. [CrossRef]
71. Rigolin, G. Entanglement, Identical Particles and the Uncertainty Principle. *Commun. Theor. Phys.* **2016**, *66*, 201. [CrossRef]
72. Prevedel, R.; Hamel, D.R.; Colbeck, R.; Fisher, K.; Resch, K.J. Experimental investigation of the uncertainty principle in the presence of quantum memory and its application to witnessing entanglement. *Nat. Phys.* **2011**, *7*, 757–761. [CrossRef]
73. Zizzi, P. Entangled Space-Time. *Mod. Phys. Lett. A* **2018**, *33*, 1850168. [CrossRef]
74. Resconi, G.; Licata, I.; Fiscaletti, D. Unification of Quantum and Gravity by Non Classical Information Entropy Space. *Entropy* **2013**, *15*, 3602–3619. [CrossRef]
75. Olson, S.J.; Dowling, J.P. Probability, unitarity, and realism in generally covariant quantum information. *arXiv* **2007**, arXiv:0708.3535.
76. Ho, V.B.; Morgan, M.J. Mathematical and General Quantum mechanics in multiply–connected spaces. *J. Phys. A* **1996** *29*, 7.
77. Vistarini, T. Holographic space and time: Emergent in what sense? *Stud. Hist. Philos. Sci. Part B Stud. Hist. Philos. Mod. Phys.* **2017**, *59*, 126–135. [CrossRef]
78. Rovelli, C. Quantum mechanics without time: A model. *Phys. Rev. D* **1990** *42*, 2638. [CrossRef]
79. Qi, X.-L. Exact holographic mapping and emergent space–time geometry. *arXiv* **2013**, arXiv:1309.6282v1.
80. Barvinsky, A.O. Aspects of nonlocality in quantum field theory, quantum gravity and cosmology. *Mod. Phys. Lett. A* **2015**, *30*, 1540003 . [CrossRef]
81. Garrett Lisi, A. Quantum mechanics from a universal action reservoir. *arXiv* **2006**, arXiv:physics/0605068.
82. Sacharov, A. Cosmological Transitions with a Change in Metric Signature. *Usp. Fiz. Nauk* **1991**, *161*, 94–104. [CrossRef]

2

Yang–Mills Theory of Gravity

Malik Al Matwi

Department of Mathematical Science, Ritsumeikan University, 4-2-28, Kusatsu-Shi, Shiga 525-0034, Japan; gr0395ki@ed.ritsumei.ac.jp

Abstract: The canonical formulation of general relativity (GR) is based on decomposition space–time manifold M into $R \times \Sigma$, where R represents the time, and Σ is the three-dimensional space-like surface. This decomposition has to preserve the invariance of GR, invariance under general coordinates, and local Lorentz transformations. These symmetries are associated with conserved currents that are coupled to gravity. These symmetries are studied on a three dimensional space-like hypersurface Σ embedded in a four-dimensional space–time manifold. This implies continuous symmetries and conserved currents by Noether's theorem on that surface. We construct a three-form $E_i \wedge DA^i$ (D represents covariant exterior derivative) in the phase space (E_i^a, A_a^i) on the surface Σ, and derive an equation of continuity on that surface, and search for canonical relations and a Lagrangian that correspond to the same equation of continuity according to the canonical field theory. We find that Σ_i^{0a} is a conjugate momentum of A_a^i and $\Sigma_i^{ab} F_{ab}^i$ is its energy density. We show that there is conserved spin current that couples to A^i, and show that we have to include the term $F_{\mu\nu i} F^{\mu\nu i}$ in GR. Lagrangian, where $F^i = DA^i$, and A^i is complex $SO(3)$ connection. The term $F_{\mu\nu i} F^{\mu\nu i}$ includes one variable, A^i, similar to Yang–Mills gauge theory. Finally we couple the connection A^i to a left-handed spinor field ψ, and find the corresponding beta function.

Keywords: gravitational Lagrangian; complex connection A^i; left-handed spinor field

---

1. Introduction

Gravity can be formulated based on gauge theory by gauging the Lorentz group $SO(3,1)$ [1]. For this purpose, we need to fix some base space and consider that the Lorentz group $SO(3,1)$ acts locally on Lorentz frames which are regarded as a frame bundle over a fixed base space. We can consider this base space as an arbitrary space–time manifold M with coordinates x^μ, and consider the local Lorentz frame as an element in the tangent frame bundle over M. By that we have two symmetries; invariance under continuous transformations of local Lorentz frame, $SO(3,1)$ group, and invariance under diffeomorphism of the space–time M, which is originally considered as a base space [2].

Since the group $SO(3)$ is a subgroup of $SO(3,1)$, the Lagrangian of gravity has an internal gauge symmetry group $SO(3)$. Thus the elements of $SO(3)$ act locally on some spacial Lorentz orthonormal frames (e^1, e^2, e^3), we consider these frames as a basis of the tangent vector bundle on the three-dimensional space-like hypersurface Σ. Using some local coordinate system σ^a, $a = 1,2,3$ on Σ, we expand these basis vectors into $e^i = e_a^i d\sigma^a$, so defining the gravitational field e_a^i with metric $g_{ab} = \delta_{ij} e_a^i e_b^j$ on Σ. Using the isomorphism between Lie algebra of $SU(2)$ and that of $SO(3)$, one can regard e_a^i also as a local $su(2)$-valued one-form. So we have an $SU(2)$ vector bundle with real spin connection ω^{ij} [3]. These facts can be generated into self-dual and anti-self-dual formalism of general relativity (GR) with complex connection A^i and complex conjugate one-form field E_i.

This paper proceeds as follows: We start with the four-dimensional (4D) Palatini Lagrangian and perform a $3+1$ decomposition based on the decomposition $M \to R \times \Sigma$, where R represents the time, and Σ is the three-dimensional (3D) space-like surface, thus we specify the Lagrangian part $L_1(g_{ab})$ on Σ. Then we try to show that $L_1(g_{ab})$ is independent of time on this surface, we try to prove this fact using the fact that there are no dynamics in the space-like region of M. After that we use $\frac{d}{dt}L_1(g_{ab}) = 0$ to get an equation of continuity on Σ, and search for canonical relations and Lagrangian that correspond to the same equation of continuity according to the canonical field theory. We find that Σ_i^{0a} is a conjugate momentum of A_a^i and $\Sigma_i^{ab}F_{ab}^i$ is its energy density. We obtain a Lagrangian for the connection A^i in 4-manifold M, then we couple it to a left-handed fermion field and find the beta function.

2. Decomposition Space–Time Manifold M into $R \times \Sigma$

The formulation of GR based on decomposition space–time manifold M into $R \times \Sigma$ is needed for expressing the metric of space–time as a solution of an equation for time evolution, such as in the Hamiltonian formulation. Thus the time evolution is the changing of the geometry of this surface. This decomposition preserves the continuous symmetries (gauge invariance and diffeomorphism invariance) of GR and its canonical quantization, so we can use it for the gauge theory of GR [4–7].

We define gravitational field as a one-form $e^I = e^I_\mu(x)dx^\mu$ that is related with metric $g_{\mu\nu}(x)$ on an arbitrary space–time manifold M by $g_{\mu\nu} = \eta_{IJ}e^I_\mu e^J_\nu$, with spin connection $\omega^{IJ}(x) \in \Omega^1(M, so(3,1))$, where $so(3,1)$ is Lie algebra of Lorentz group $SO(3,1)$. The spin connection defines covariant derivative D_μ that acts on all fields which have Lorentz indices $(I, J, ...)$:

$$D_\mu v^I = \partial_\mu v^I + \omega^I_{\mu J} v^J.$$

We start with the GR Lagrangian of the form

$$L(e,\omega) = (16\pi G)^{-1} e^\mu_I e^\nu_J (R_{\mu\nu})^{IJ} e, \tag{1}$$

where

$$R^{IJ} = d\omega^{IJ} + \omega^I{}_K \wedge \omega^{KJ}$$

is the Riemannian curvature tensor and e^μ_I satisfies $e^I_\mu e^\mu_J = \delta^I_J$.

By the decomposition $M \to R \times \Sigma$, we decompose this Lagrangian into

$$L(e,\omega) = (16\pi G)^{-1} e^a_i e^b_j (R_{ab})^{ij} e + (16\pi G)^{-1} e^a_0 e^b_J (R_{ab})^{0J} e + (16\pi G)^{-1} e^0_I e^a_J (R_{0a})^{IJ} e, \tag{2}$$

where i and j are Lorentz indices for $I = i = 1, 2, 3$, and $a = 1, 2, 3$. The part

$$L_1 = (16\pi G)^{-1} e^a_i e^b_j (R_{ab})^{ij} e$$

has the gauge symmetry of the group $SO(3)$, which is a subgroup of $SO(3,1)$, it also relates to the geometry of the surface $\Sigma(\sigma^a)$ under the variation in the direction of $\Sigma(\sigma^a)$ subject to $\delta e^a_0 = 0$, $\delta\omega^{ij}_0 = 0$, since it depends only on the metric $g_{ab} = \delta_{ij} e^i_a e^j_b$ which is defined on $\Sigma(\sigma^a)$, which is intrinsic geometry.

Since $(R_{ab})^{ij}$ is an anti-symmetric tensor, we can introduce a new one-form field E^k_c, the Hodge dual of $e^i \wedge e^j$ in the internal spin space on the surface $\Sigma(\sigma^a)$; $E^i = E^i_a d\sigma^a$, it is called the gravitational electric field [8,9]

$$\frac{1}{2}\left(e^a_i e^b_j - e^a_j e^b_i\right) = \varepsilon^{abc} \varepsilon_{ijk} E^k_c.$$

In self-dual formalism of GR, E^a_i is complex given by [10,11]

$$E^{ia} = \frac{1}{2}\varepsilon^{abc}P^i_{IJ}e^I_b e^J_c, \text{ with } E^i_{ab} = \Sigma^i_{ab} = P^i_{IJ}e^I_a e^J_b \tag{3}$$

where P^i_{IJ} is a self-dual projector given by

$$P^i_{IJ} = \frac{1}{2}\varepsilon^i_{jk}, \text{ for } I = i, J = j, \text{ and } P^i_{0j} = -P^i_{j0} = \frac{i}{2}\delta^i_j, \text{ for } I = 0, J = j \neq 0. \tag{4}$$

For example $E^{1a} = \frac{1}{2}\varepsilon^{abc}(e^2_b e^3_c + ie^0_b e^1_c)$.

For the Lagrangian part L_1 on $\Sigma(\sigma^a)$, we use the first one:

$$L_1 = (16\pi G)^{-1}E^k_c\varepsilon^{abc}\varepsilon_{ijk}(R_{ab})^{ij}e. \tag{5}$$

The remaining part

$$L_2 = (16\pi G)^{-1}e^a_0 e^b_J(R_{ab})^{0J}e + (16\pi G)^{-1}e^0_I e^a_J(R_{0a})^{IJ}e,$$

associates with the time evolution under the variation in the normal direction of $\Sigma(\sigma^a)$, it is subject to $\delta e^a_i = 0$, $\delta\omega^{ij}_a = 0$, and changes the geometry of the surface $\Sigma(\sigma^a)$ during the time. Dynamics, such as propagation $g_{ab}(x) \to g_{ab}(x')$ in the time-like region $(\Delta x)^2 = g_{\mu\nu}(x)\Delta x^\mu\Delta x^\nu < 0$, and determines how the surface $\Sigma(\sigma^a)$ is embedded into the 4D manifold M, which is extrinsic geometry. But the surface $\Sigma(\sigma^a)$ is embedded in a space-like region in M; $V_1, V_2 \in T_p\Sigma(\sigma^a), g(V_1, V_2) > 0$, so there are no dynamics on $T_p\Sigma(\sigma^a)$. We can see this fact by noting that $\nabla_t g_{ab} = 0$ (the covariant derivative of the metric is zero), so

$$\frac{dL_1(g_{ab})}{dt}dt = \frac{\partial L_1(g_{ab})}{\partial g_{ab}}\nabla_t g_{ab}dt = 0.$$

We will rewrite $\nabla_t g_{ab} = 0$ as $t^a t^b\nabla_t g_{ab} = 0$, for $t^a \in T_p\Sigma_t(\sigma^a)$. The formula $t^a t^b\nabla_t g_{ab} = 0$ is more general than $\nabla_t g_{ab} = 0$ since there is $\nabla_t g_{ab} \notin \Gamma\left(T^*\Sigma_t(\sigma^a) \times T^*\Sigma_t(\sigma^a)\right)$, and so its projection onto $T_p\Sigma(\sigma^a)$ is zero. This case appears in the diffeomorphism maps of $\Sigma(\sigma^a)$ into another space-like surface, as we will see.

We can study the embedding of $\Sigma(\sigma^a)$ by letting the time derivative of a position vector on its tangent space be in the direction of the normal to this tangent space $T_p\Sigma(\sigma^a)$. We can see this by considering a position vector $t^a \in T_p\Sigma_t(\sigma^a)$ that satisfies $\dot{t}^a t_a = \dot{t}_a t^a = 0$. Let $\dot{t}^a \in N_p\Sigma_t(\sigma^a)$, where $N_p\Sigma_t(\sigma^a)$ is the normal space to $\Sigma_t(\sigma^a) = \{t\} \times \Sigma(\sigma^a)$, and $\dot{t}^a = \nabla_t t^a$ is a covariant derivative of t^a with respect to the time $t \in R \subset R \times \Sigma(\sigma^a)$. From $t_a = g_{ab}t^b$, we obtain $\dot{t}_a = \dot{g}_{ab}t^b + g_{ab}\dot{t}^b$, so $t^a\dot{t}_a = t^a\dot{g}_{ab}t^b + t^a g_{ab}\dot{t}^b$. Since $\dot{t}^a \in N_p\Sigma_t(\sigma^a)$, we have $t^a g_{ab}\dot{t}^b = 0$, thus we get $\dot{g}_{ab}t^a t^b = t^a t^b\nabla_t g_{ab} = 0$ which means that the points of M do not expand nor contract covariantly in the space-like $T_p\Sigma_t(\sigma^a)$, but $t^a t^b\partial_t g_{ab} \neq 0$ is possible.

Therefore, for a diffeomorphism map of $\Sigma(\sigma^a)$ into another space-like surface (consider this as time evolution), it must be

$$\nabla_t(g_{ab}) \notin \Gamma\left(T^*\Sigma_t(\sigma^a) \times T^*\Sigma_t(\sigma^a)\right),$$

and

$$\nabla_t(g_{ab}) \in \Gamma(N^*\Sigma_t(\sigma^a) \times T^*\Sigma_t(\sigma^a)) \oplus \Gamma(N^*\Sigma_t(\sigma^a) \times N^*\Sigma_t(\sigma^a)).$$

This means for each $x, x' \in \Sigma_t(\sigma^a)$, the covariant propagation $g_{ab}(x) \to g_{ab}(x') \equiv g_{ab}(x) + \xi^\mu \nabla_\mu g_{ab}(x)$ does not occur in $T^*_p\Sigma_t(\sigma^a) \times T^*_p\Sigma_t(\sigma^a)$, but it occurs in

$$N^*_p\Sigma_t(\sigma^a) \times T^*_p\Sigma_t(\sigma^a) \oplus N^*_p\Sigma_t(\sigma^a) \times N^*_p\Sigma_t(\sigma^a).$$

This is a changing in the embedding of $\Sigma(\sigma^a)$ in M. Let $t_a \in T^*_p\Sigma_t(\sigma^a)$ satisfy $(\dot{t}_a) = n$, where $(\cdot)$ is matrix notation and n is the normal to $\Sigma_t(\sigma^a)$. This normal is in the direction of the time, so it carries one index; $n = n^0$, thus $(\dot{t}_a) = n_0$. It must also satisfy $(\dot{t}^a) \notin T_p\Sigma_t(\sigma^a)$, so $(g_{ab})(\dot{t}^b) = 0$. Using this in $(\dot{t}_a) = (\dot{g}_{ab})(t^b) + (g_{ab})(\dot{t}^b)$, we obtain $(\dot{t}_a) = (\dot{g}_{ab})(t^b)$, so we get $n_0 = (\dot{g}_{ab})_{0b}t^b$. Thus we get $n_0 n^0 = n^0(\dot{g}_{ab})_{0b}t^b$, and by $n_0 n^0 = 1$, we obtain $n^0(\dot{g}_{ab})_{0b}t^b = 1$. This formula is for determining $(\dot{g}_{ab}) = \nabla_t(g_{ab}) \in \Gamma(N^*\Sigma_t(\sigma^a) \times T^*\Sigma_t(\sigma^a))$.

Let us suggest a formula for determining $(\dot{g}_{ab}) \in \Gamma(N^*\Sigma_t(\sigma^a) \times T^*\Sigma_t(\sigma^a))$ like

$$\nabla_t g_{ab} = f^c_{ab} g_{0c}, \quad f^c_{ab} = f^c_{ba}, \quad t^a t^b f^c_{ab} = 0,$$

where $g_{0c} \in \Gamma(N^*\Sigma_t(\sigma^a) \times T^*\Sigma_t(\sigma^a))$, and $g_{\mu\nu} = (g_{00}, g_{0a}, g_{ab})$ is the full metric. Let us write f^a_{bc} using matrix notation f^a, its elements are $(f^a)_{bc} = f^a_{bc}$. Let $\tilde{f}_a$ satisfy $(\tilde{f}_a)^{bc}(f^{a'})_{bc} = \delta^{a'}_a$. Thus we get the inversion

$$(\tilde{f}_c)^{ab}\nabla_t g_{ab} = (\tilde{f}_c)^{ab}(f^{c'})_{ab} g_{0c'} = \delta^{c'}_c g_{0c'} = g_{0c}, \tag{6}$$

this is obtaining the metric component g_{0a} from $\nabla_t g_{ab}$; the changing of the metric g_{ab} with respect to the time. Thus if we fix the metric component g_{00}, like $g_{00} = -1$, we obtain the map $g_{ab}(t) \to g_{\mu\nu} = (-1, g_{0a}, g_{ab})$. So we have an immersion $\Gamma(T^*\Sigma(\sigma^a)) \to \Gamma(N^*\Sigma_t(\sigma^a) \times T^*\Sigma_t(\sigma^a)) \subset \Gamma(T^*M)$. In another words, for an immersion $M^{(n)} \to M^{(n+1)}$, and under some hypotheses, we can construct a metric on $M^{(n+1)}$ using the metric on M^n. Our hypothesis here is $t^a t^b \nabla_t g_{ab} = 0$.

First we find the matrices f^a then $\tilde{f}_a$. The matrices f^a are symmetric and satisfy $t^b t^c (f^a)_{bc} = t^\top (f^a) t = 0$, where the vector $t^\top = (t_1, t_2, t_3)_p$ is the unit vector in $T_p\Sigma_t(\sigma^a))$, we can write them in a simple form like

$$f^1 = \begin{pmatrix} 0 & 0 & 0 \\ 0 & \cos(\theta_1) & \sin(\theta_1) \\ 0 & \sin(\theta_1) & -\cos(\theta_1) \end{pmatrix}, \quad f^2 = \begin{pmatrix} \cos(\theta_2) & 0 & \sin(\theta_2) \\ 0 & 0 & 0 \\ \sin(\theta_2) & 0 & -\cos(\theta_2) \end{pmatrix},$$

$$f^3 = \begin{pmatrix} \cos(\theta_3) & \sin(\theta_3) & 0 \\ \sin(\theta_3) & -\cos(\theta_3) & 0 \\ 0 & 0 & 0 \end{pmatrix},$$

the angles θ_1, θ_2 and θ_3 can be determined to satisfy $t^a t^b (f^c)_{ab} = 0$, thus we obtain

$$\tan(\theta_1) = \frac{t_3^2 - t_2^2}{2t_3 t_2}, \quad \tan(\theta_2) = \frac{t_3^2 - t_1^2}{2t_3 t_1}, \quad \tan(\theta_3) = \frac{t_2^2 - t_1^2}{2t_2 t_1}.$$

Therefore the matrices $\tilde{f}_a$ can be written in the form

$$\tilde{f}_1 = \begin{pmatrix} 0 & c_1\cos(\theta_3) & 0 \\ c_1\cos(\theta_3) & 2c_1\sin(\theta_3) & \sin(\theta_1) \\ 0 & \sin(\theta_1) & 0 \end{pmatrix}, \quad \tilde{f}_2 = \begin{pmatrix} -2c_1\sin(\theta_3) & c_1\cos(\theta_3) & \cos(\theta_3) \\ c_1\cos(\theta_3) & 0 & c_2\cos(\theta_1) \\ \cos(\theta_3) & c_2\cos(\theta_1) & 2c_2\sin(\theta_1) \end{pmatrix}$$

$$\tilde{f}_3 = \begin{pmatrix} 0 & \cos(\theta_3) & 0 \\ \cos(\theta_3) & -2c_3\sin(\theta_1) & c_3\cos(\theta_1) \\ 0 & c_3\cos(\theta_1) & 0 \end{pmatrix},$$

the constants c_1, c_2 and c_3 are determined to satisfy

$$(\tilde{f}_1)^{ab}(f^1)_{ab} = (\tilde{f}_2)^{ab}(f^2)_{ab} = (\tilde{f}_3)^{ab}(f^2)_{ab} = 1.$$

We can consider that as a continuous changing in the embedding of $\Sigma_t(\sigma^a)$ in M;which is a diffeomorphism map. Since the Lagrangian L_1 depends only on g_{ab} while the term L_2 depends on (g_{00}, g_{0a}, g_{ab}), thus by the previous discussing we have a map $L_1(g_{ab}) \to L_2(g_{00}, g_{0a}, g_{ab})$ (time evolution), with $g_{00} = -1, g_{0a} = (\tilde{f}_a)^{bc}\nabla_t g_{bc}$. As illustrated in the following Figure 1.

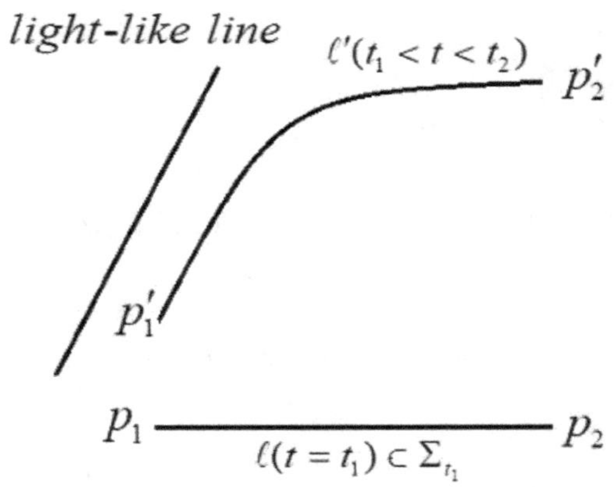

Figure 1. Propagation of space-like surface.

Let $\ell \subset \Sigma_{t_1}$ which propagated to $\ell' \subset \Sigma_{t_1<t<t_2}$, both surfaces are space-like, but the trajectories of the propagation $p_1 \to p'_1$ and $p_2 \to p'_2$ lie in the time-like region. The length of ℓ is also increased. Thus, this means that the 3D metric g_{ab} mapped to 4D metric (g_{00}, g_{0a}, g_{ab}) by Equation (6), which means the length ℓ' is given by

$$\ell'^2 = \int_{p'_1}^{p'_2} \left(-dtdt + g_{0a}dtd\sigma^a + g_{ab}d\sigma^a d\sigma^b\right),$$

which again can be mapped into a space-like surface Σ_{t_2} by an isometric map such $-dtdt + g_{0a}dtd\sigma^a + g_{ab}d\sigma^a d\sigma^b = g'_{a'b'}d\sigma^{a'}d\sigma^{b'}$. which results in Figure 2:

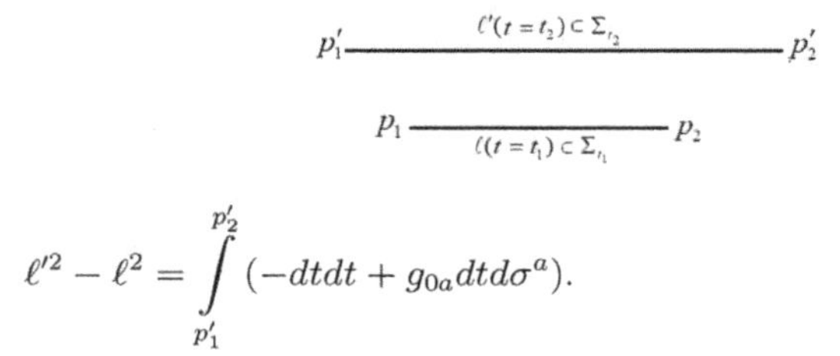

Figure 2. Changing distance between two space-like points after propagation.

$$\ell'^2-\ell^2=\int_{p_1'}^{p_2'}(-dtdt+g_{0a}dtd\sigma^a).$$

The limit $\ell'^2-\ell^2=0$ corresponds to $dt=0$ and $d\sigma^a=0$, not to $-dtdt+g_{0a}dtd\sigma^a=0$ (with $dt\neq 0$ and $d\sigma^a\neq 0$). For example, if two particles P_1 and P_2 exchange photons with wavelength $\lambda=\ell\approx\sigma_2^a-\sigma_1^a$ in $t=t_1$, then in $t=t_2$ they measure a different wavelength, namely $\lambda+\Delta\lambda=\ell'$. The difference is given by $2\lambda\Delta\lambda=-\lambda^2+g_{0a}\lambda^2$, where we set $dt=d\sigma^a=\lambda$. Thus the two particles P_1 and P_2 measure after a time $dt=\lambda$ the difference $2\Delta\lambda/\lambda=-1+g_{0a}$, where g_{0a} is given in Equation (6).

Thus we study the embedding of $3D$ surface Σ_t and its changing in $4D$ manifold M using the $3D$ metric g_{ab} on Σ_t and its derivative with respect to the time.

The Lagrangian $L_1(g_{ab})$ is a function on the space-like space $T_p^*\Sigma_t(\sigma^a)$, therefore, it is independent of time, $\frac{d}{dt}L_1(g_{ab})=0$ on $T_p^*\Sigma_t(\sigma^a)$. We can see this by using the fact

$$\nabla_t(g_{ab})\in\Gamma(N^*\Sigma_t(\sigma^a)\times T^*\Sigma_t(\sigma^a))\oplus\Gamma(N^*\Sigma_t(\sigma^a)\times N^*\Sigma_t(\sigma^a)),$$

we have

$$L_1(g_{ab}(t+dt))-L_1(g_{ab}(t))=\frac{d}{dt}L_1(g_{ab}(t))dt=\frac{\partial L_1(g_{ab})}{\partial g_{ab}}\nabla_t(g_{ab})dt,$$

which is a function on $N_p^*\Sigma_t(\sigma^a)\times T_p^*\Sigma_t(\sigma^a)\oplus N_p^*\Sigma_t(\sigma^a)\times N_p^*\Sigma_t(\sigma^a)$, not on $T_p^*\Sigma_t(\sigma^a)$; which means there are no dynamics on the space-like region $(\Delta x)^2>0$.

Let us write $dt\partial_t L_1(g_{ab})=(\Sigma(\sigma^a),d\theta)$, where (Σ,V) is a projection of $V\in\wedge^4T_p^*M$ onto a surface Σ, the inner product of V with tangent basis in $T_p\Sigma$, defined below in Equation (17), and θ is three-form in the phase space (E_i,ω^{ij}) on $\Sigma(\sigma^a)$. Thus we write

$$\left(\Sigma(\sigma^a),dt\wedge\frac{d}{dt}\theta\right)=dt\frac{d}{dt}L_1(g_{ab})=0.$$

If we write θ as

$$\theta(E,\omega,\Sigma_t(\sigma^a))=(16\pi G)^{-1/2}\varepsilon_{ijk}E^k\wedge R^{ij},\tag{7}$$

with $R^{ij}=(R_{ab})^{ij}d\sigma^a\wedge d\sigma^b$ and $E^i=E_a^i d\sigma^a$, we obtain

$$\left(\Sigma(\sigma^a),dt\wedge\frac{d}{dt}\theta\right)=(16\pi G)^{-1/2}dt\frac{d}{dt}(E_c^k\varepsilon^{abc}\varepsilon_{ijk}(R_{ab})^{ij}e)$$
$$=(16\pi G)^{1/2}dt\frac{d}{dt}L_1=0.$$

Since θ is three-form on $\Sigma(\sigma^a)$, so $d\sigma^a \wedge \frac{\partial}{\partial\sigma^a}\theta = 0$, if we add it to the last formula, we get

$$\left(\Sigma(\sigma^a), dt \wedge \frac{\partial}{\partial t}\theta + d\sigma^a \wedge \frac{\partial}{\partial\sigma^a}\theta\right) = (\Sigma(\sigma^a), d\theta) = 0.$$

This relates the Lagrangian L_1 and θ with this surface. Under arbitrary transformation $(t, \sigma^a) \rightarrow x^\mu$, the two Lagrangian parts L_1 and L_2 will mix. Thus $d\sigma^a = \frac{\partial\sigma^a}{\partial x^\mu}dx^\mu$ and the basis on Σ transforms as

$$\partial_a \wedge \partial_b \wedge \partial_c \rightarrow \frac{\partial x^\mu}{\partial\sigma^a}\frac{\partial x^\nu}{\partial\sigma^b}\frac{\partial x^\rho}{\partial\sigma^c}\partial_\mu \wedge \partial_\nu \wedge \partial_\rho. \tag{8}$$

Therefore, the components of three-form θ transforms as

$$\theta_{abc} \rightarrow \theta_{\mu\nu\rho} = \theta_{abc}\frac{\partial\sigma^a}{\partial x^\mu}\frac{\partial\sigma^b}{\partial x^\nu}\frac{\partial\sigma^c}{\partial x^\rho}.$$

To keep the invariance under this transformation, that is $(\Sigma, d\theta) = 0$ still holds, we write the three-form θ in the phase space (E_i, ω^{ij}) on M, and let $(\Sigma, d\theta)$ be its projection onto Σ. Therefore we write

$$\theta(E, \omega) = (16\pi G)^{-1/2}\varepsilon_{ijk}E^k \wedge R^{ij}, \tag{9}$$

thus we get three-form θ in the phase space (E_i, ω^{ij}) on M, it has internal $SO(3)$ symmetry. Its projection onto $\Sigma_t(\sigma^a)$ is

$$(\Sigma_t(\sigma^a), \theta(E, \omega)) = (16\pi G)^{-1/2}E^k_c\varepsilon^{abc}\varepsilon_{ijk}R^{ij}_{ab}.$$

We write θ using self-dual formalism (Plebanski formalism) [11] in which the connection A^i is a three-complex one-form given by

$$\omega^{IJ} \rightarrow A^i = P^i{}_{IJ}\omega^{IJ}, \tag{10}$$

where P^i_{IJ} is self-dual projector given by Equation (3). The curvature which associates with this connection is

$$F^i = dA^i + \varepsilon^i{}_{jk}A^j \wedge A^k.$$

On the surface $\Sigma_t(\sigma^a)$, it is

$$F^i_{ab} = \frac{1}{2}\left(\partial_b A^i_b - \partial_b A^i_a + \varepsilon^i{}_{jk}A^j_a A^k_b\right).$$

Also

$$e^{[a}_I e^{b]}_J \rightarrow P^{IJ}_i e^{[a}_I e^{b]}_J = E^{ab}_i = \varepsilon^{abc}E_{ic}. \tag{11}$$

Thus we have self-dual plus anti-self-dual projection:

$$e^a_I e^b_J R(\omega)^{IJ}_{ab} \rightarrow \varepsilon^{abc}E_{ia}F^i_{bc} + \varepsilon^{abc}\bar{E}_{ia}\bar{F}^i_{bc}, \tag{12}$$

where $\bar{E}_{ia}$ and $\bar{F}^i_{bc}$ are the Hermitian conjugate of E_{ia} and F^i_{bc}. This projection relates to the decomposition of Lie algebra of the Lorentz group $SO(1,3)$ into two copies of Lie algebra of $SL(2,R)$ [12].

We write the components of the curvature as

$$F^i_{ab} = \frac{1}{2}\left(D_a A^i_b - D_b A^i_a\right) = \frac{1}{2}\left(\partial_a A^i_b - \partial_b A^i_a + \varepsilon^i{}_{jk}A^j_a A^k_b\right), \tag{13}$$

where

$$D_a A^i_b = \partial_a A^i_b + \frac{1}{2}\varepsilon^i{}_{jk} A^j_a A^k_b,$$

which motivates introducing a notation of the covariant derivative like [7]

$$DV^i = dV^i + \frac{1}{2}\varepsilon^i{}_{jk} A^j \wedge V^k,$$

or

$$D_\mu V^i_\nu = \partial_\mu V^i_\nu + \frac{1}{2}\varepsilon^i{}_{jk} A^j_\mu V^k_\nu = \partial_\mu V^i_\nu - \frac{i}{2} A^j_\mu (T^j_A)^{ik} V^k_\nu,$$

so

$$D_\mu = \partial_\mu - \frac{i}{2} A^j_\mu (T^j_A),$$

where the matrix elements $(T^j_A)^{ik} = -i\varepsilon^{jik}$ are the elements of the generators T^j_A in the adjoint representation of the group $SU(2)$ [13]. The coupling constant here is $g = 1$. In general we write this covariant derivative as

$$D_\mu = \partial_\mu - \frac{i}{2} g A^j_\mu (T^j_A). \tag{14}$$

Using the projection in Equation (12), we rewrite θ in the complex phase space (E^a_i, A^i_a) as

$$\theta(E, \omega, \Sigma(\sigma^a)) = (16\pi G)^{-1/2} \varepsilon^{abc} E_{ic} F^i_{ab}(A) e d^3\sigma,$$

where we take in consideration only the first part, the second is obtained by taking the Hermitian conjugate. We can write it as three-form on M as done in Equation (9), we obtain

$$\theta(E, A) = (16\pi G)^{-1/2} E_i \wedge DA^i. \tag{15}$$

Its projection onto $\Sigma_t(\sigma^a)$ is

$$(\Sigma_t(\sigma^a), \theta) = (16\pi G)^{-1/2} \varepsilon^{abc} E_{ia} F^i_{bc}(A),$$

where A^i is complex $SO(3)$ connection, and E_i is a complex one-form as defined in Equation (11).

3. Equation of Continuity on the Hypersurface $\Sigma_t(\sigma^a)$

We have showed that the Lagrangian $L_1(g_{ab})$ is independent of time, $\frac{d}{dt}L_1(g_{ab}) = 0$ on $\Sigma(\sigma^a)$ since $\nabla_t(g_{ab}) \notin \Gamma(T^*\Sigma_t(\sigma^a) \times T^*\Sigma_t(\sigma^a))$. This relates to the fact that there are no dynamics in the space-like $T\Sigma_t(\sigma^a))$; the points of M do not expand nor contract covariantly in this region, $\nabla_\mu t^a \neq 0$. Note that although $\nabla_\mu t^a = 0$, but it may be $\partial_\mu t^a \neq 0$. We had

$$\nabla_t(g_{ab}) \in \Gamma(N^*\Sigma_t(\sigma^a) \times T^*\Sigma_t(\sigma^a)) \oplus \Gamma(N^*\Sigma_t(\sigma^a) \times N^*\Sigma_t(\sigma^a)),$$

which shows that the two parts of the Lagrangian L_1 and L_2 mix by time evolution. Then we wrote $dt\frac{d}{dt}L_1(g_{ab}) = 0$ as $(\Sigma(\sigma^a), d\theta) = 0$, where θ is three-form in the phase space (E_i, ω^{ij}) on M (Equation (7)), its projection onto $\Sigma_t(\sigma^a)$ is Equation (17)

$$(\Sigma_t(\sigma^a), \theta(E, \omega)) = (16\pi G)^{-1/2} \varepsilon_{ijk} \varepsilon^{abc} E^i_a R^{jk}_{bc}(\omega).$$

In self-dual formalism, we obtained $\theta(E,A) = (16\pi G)^{-1/2} E_i \wedge DA^i$, with

$$(\Sigma_t(\sigma^a), \theta(E,A)) = (16\pi G)^{-1/2} \varepsilon^{abc} E_{ia} F^i_{bc}(A).$$

Our condition $(\Sigma(\sigma^a), d\theta) = 0$ makes sense here because of the decomposition $R \times \Sigma$ and fixing a coordinate system σ^a on the hypersurface Σ, this yields to an equation of continuity on this surface. For this purpose, we take the inner product of the four-form $d\theta$ with a tangent basis on the surface $\Sigma_t(\sigma^a)$ at an arbitrary point, we get a one-form co-vector $(\Sigma(\sigma^a), d\theta)$ in the direction of the normal to this surface at that point. Then we set $(\Sigma(\sigma^a), d\theta) = 0$, we obtain an equation of continuity on $\Sigma_t(\sigma^a)$.

Now taking the exterior derivative of Equation (15), we obtain

$$(16\pi G)^{1/2} d\theta = d\left(E_i \wedge DA^i\right) = (DE_i) \wedge DA^i - E_i \wedge DDA^i, \tag{16}$$

where $E_{0i} = 0$. The tri-tangent basic on $\Sigma_t(\sigma^a)$ is $\partial_a \wedge \partial_b \wedge \partial_c$, we rewrite it as $(1/3!)\varepsilon^{abc}\partial_a\partial_b\partial_c$. The projection of $d\theta$ onto this basic is

$$(16\pi G)^{1/2}\left(d\theta, \varepsilon^{abc}\partial_a\partial_b\partial_c\right) = \left((DE_i) \wedge DA^i, \varepsilon^{abc}\partial_a\partial_b\partial_c\right) - \left(E_i \wedge DDA^i, \varepsilon^{abc}\partial_a\partial_b\partial_c\right)$$

where $(\cdot,\cdot)$ is contraction pairing defined by

$$\left(V_{\mu\nu\rho\sigma} dx^\mu \wedge dx^\nu \wedge dx^\rho \wedge dx^\sigma, \varepsilon^{abc}\partial_a\partial_b\partial_c\right) = \varepsilon^{abc} V_{\mu\nu\rho\sigma} dx^{[\mu} \delta^\nu_c \delta^\rho_b \delta^{\sigma]}_a, \tag{17}$$

where the bracket [....] is anti-symmetrization of the indices. Although DDA^i is zero in $4D$ manifold M, but the contraction pairing of $E_i \wedge DDA^i$ with $3D$ basis $\varepsilon^{abc}\partial_a\partial_b\partial_c$ is not zero as we will see, since we do not sum over the time index $\mu = 0$ as we sum over the spatial indices $\mu = 1,2,3$, because we regard dx^0 as normal to the surface $\Sigma_t(\sigma^a)$.

For cotangent basis $\{dx^\mu\}$ and tangent basis $\{\partial_a\}$, this pairing can be defined simply by using inner product like [14]

$$(dx^\mu, \partial_a) = \delta^\mu_a,$$

in which we consider $dx^a = d\sigma^a$ for $a = 1,2,3$, so $E_{0i} = 0$ regarding to our gauge.

Starting with the first term

$$\left((DE_i) \wedge DA^i, \varepsilon^{abc}\partial_a\partial_b\partial_c\right) = \varepsilon^{abc} D_\mu E_{\nu i} D_\rho A^i_\sigma \left(dx^\mu \wedge dx^\nu \wedge dx^\rho \wedge dx^\sigma, \partial_a\partial_b\partial_c\right)$$

we get

$$\begin{aligned} &4\left((DE_i) \wedge DA^i, \varepsilon^{abc}\partial_a\partial_b\partial_c\right) = \\ &-\varepsilon^{abc} D_a E_{bi} D_c A^i_\mu dx^\mu + \varepsilon^{abc} D_a E_{bi} D_\mu A^i_c dx^\mu - \varepsilon^{abc} D_a E_{\mu i} D_b A^i_c dx^\mu + \varepsilon^{abc} D_\mu E_{ai} D_b A^i_c dx^\mu. \end{aligned}$$

The second term is

$$\left(E_i \wedge DDA^i, \varepsilon^{abc}\partial_a\partial_b\partial_c\right) = \varepsilon^{abc} E_{\mu i} D_\nu D_\rho A^i_\sigma \left(dx^\mu \wedge dx^\nu \wedge dx^\rho \wedge dx^\sigma, \partial_a\partial_b\partial_c\right).$$

Doing the same thing, we get

$$4\left(E_i \wedge DDA^i, \varepsilon^{abc}\partial_a\partial_b\partial_c\right) =$$
$$-\varepsilon^{abc}E_{ai}D_bD_cA^i_\mu dx^\mu + \varepsilon^{abc}E_{ai}D_bD_\mu A^i_c dx^\mu - \varepsilon^{abc}E_{ai}D_\mu D_b A^i_c dx^\mu + \varepsilon^{abc}E_{\mu i}D_aD_bA^i_c dx^\mu.$$

Adding the two terms, we obtain

$$4(16\pi G)^{1/2}\left(d\theta, \varepsilon^{abc}\partial_a\partial_b\partial_c\right) =$$
$$-\varepsilon^{abc}D_aE_{bi}D_cA^i_\mu dx^\mu + \varepsilon^{abc}D_aE_{bi}D_\mu A^i_c dx^\mu - \varepsilon^{abc}D_aE_{\mu i}D_bA^i_c dx^\mu + \varepsilon^{abc}D_\mu E_{ai}D_bA^i_c dx^\mu$$
$$+\varepsilon^{abc}E_{ai}D_bD_cA^i_\mu dx^\mu - \varepsilon^{abc}E_{ai}D_bD_\mu A^i_c dx^\mu + \varepsilon^{abc}E_{ai}D_\mu D_bA^i_c dx^\mu - \varepsilon^{abc}E_{\mu i}D_aD_bA^i_c dx^\mu.$$

We define the curvature by using the covariant derivative from Equation (13) as

$$F^i_{\mu c} = \frac{1}{2}\left(D_\mu A^i_c - D_c A^i_\mu\right) = \frac{1}{2}\left(\partial_\mu A^i_c - \partial_c A^i_\mu + g\varepsilon^i{}_{jk}A^j_\mu A^k_c\right), \tag{18}$$

therefore

$$F^{\rho\nu i} = g^{\rho\mu}g^{\nu c}F^i_{\mu c} = \frac{1}{2}\left(D^\rho A^{\nu i} - D^\nu A^{\rho i}\right);\ D^\rho g^{\mu\nu} = 0.$$

Its Hodge dual on the surface Σ with respect to the coordinates (σ^a) is

$$F^{ai} = \varepsilon^{abc}F^i_{bc}.$$

Also we define the complex two-form field from Equation (3) as

$$\Sigma^{bc}_i = E^{bc}_i = \varepsilon^{bca}E_{ai} = \varepsilon^{abc}E_{ai}.$$

Using them in the last formula, we get

$$4(16\pi G)^{1/2}\left(d\theta, \varepsilon^{abc}\partial_a\partial_b\partial_c\right) = 2\varepsilon^{abc}(D_aE_{bi})F^i_{\mu c}dx^\mu - 2(D_aE_{\mu i})F^{ai}dx^\mu$$
$$+2(D_\mu E_{ai})F^{ai}dx^\mu + 2E^{bc}_iD_bF^i_{c\mu}dx^\mu + 2E_{ai}D_\mu F^{ai}dx^\mu - 2E_{\mu i}D_aF^{ai}dx^\mu.$$

And using

$$\varepsilon^{abc}(D_aE_{bi})F^i_{\mu c}dx^\mu = -\varepsilon^{acb}(D_aE_{bi})F^i_{\mu c}dx^\mu = -(D_aE^{ac}_i)F^i_{\mu c}dx^\mu,$$

we obtain

$$2(16\pi G)^{1/2}\left(d\theta, \varepsilon^{abc}\partial_a\partial_b\partial_c\right) =$$
$$-(D_aE^{ac}_i)F^i_{\mu c}dx^\mu - (D_aE_{\mu i})F^{ai}dx^\mu + (D_\mu E_{ai})F^{ai}dx^\mu - E^{bc}_iD_bF^i_{\mu c}dx^\mu$$
$$+E_{ai}D_\mu F^{ai}dx^\mu - E_{\mu i}D_aF^{ai}dx^\mu.$$

With

$$-(D_aE^{ac}_i)F^i_{\mu c}dx^\mu - E^{bc}_iD_bF^i_{\mu c}dx^\mu = -D_a\left(E^{ac}_iF^i_{\mu c}dx^\mu\right),$$

it becomes

$$2(16\pi G)^{1/2}\left(d\theta, \varepsilon^{abc}\partial_a\partial_b\partial_c\right) = -D_a(E^{ac}_iF^i_{\mu c})dx^\mu - D_a(E_{\mu i}F^{ai})dx^\mu + D_\mu(E_{ai}F^{ai})dx^\mu.$$

As we suggested before, we let the normal of the surface $\Sigma_t(\sigma^a)$ be in direction of the time dx^0, so $\left(d\theta, \varepsilon^{abc}\partial_a\partial_b\partial_c\right)$ is in direction of the time. Therefore we set $\mu = 0$, thus we get

$$2(16\pi G)^{1/2}\left(d\theta, \varepsilon^{abc}\partial_a\partial_b\partial_c\right) = -D_a(E_i^{ac}F_{0c}^i)dx^0 - 2D_a(E_{0i}F^{ai})dx^0 + D_0(E_{ai}F^{ai})dx^0.$$

The vector $\left(d\theta, \varepsilon^{abc}\partial_a\partial_b\partial_c\right)$ is one-form in the direction of the normal to the surface $\Sigma_t(\sigma^a)$. It is zero as we mentioned before, thus we get

$$-D_a(E_i^{ac}F_{0c}^i)dx^0 - D_a(E_{0i}F^{ai})dx^0 + D_0(E_{ai}F^{ai})dx^0 = 0,$$

or

$$-D_a(E_i^{ac}F_{0c}^i) - D_a(E_{0i}F^{ai}) + D_0(E_{ai}F^{ai}) = 0.$$

We write it as

$$D_a(E_i^{ac}F_{c0}^i) - D_a(E_{0i}F^{ai}) + D_0(E_{ai}F^{ai}) = 0.$$

The term $(E_{ai}F^{ai})$ is scalar, so $D_0(E_{ai}F^{ai}) = \partial_0(E_{ai}F^{ai})$, the vector $E_i^{ab}F_{b0}^i$ is a usual vector field on Σ, it does not carry a Lorentz index, so $D_a(E_i^{ab}F_{b0}^i) = \partial_a(E_i^{ab}F_{b0}^i)$ and $E_{0i} = 0$, thus we get

$$\partial_a(E_i^{ab}F_{b0}^i) + \partial_0(E_{ai}F^{ai}) = 0,$$

using $E_{ai}F^{ai} = \frac{1}{2}E_{abi}F^{abi} = \frac{1}{2}\Sigma_{abi}F^{abi}$

$$\partial_a\left(\Sigma_i^{ab}F_{b0}^i\right) + \partial_0\left(\frac{1}{2}\Sigma_{abi}F^{abi}\right) = 0. \tag{19}$$

This equation shows that there is a relation between Σ^i and F^i in the space (Σ^i, F^i) on $3+1$ manifold $R \times \Sigma$. Usually this relation is written as $F^i = \psi^i{}_j\Sigma^j + \bar{\psi}^i{}_j\bar{\Sigma}^j$. We can find that relation by regarding this equation as an equation of continuity with respect to a Lagrangian like $L(F^{0ai}, DA^i)$ that satisfies the action principle $\delta S(DA^i) = 0$ and the invariance under continuous symmetries of GR. Therefore we regard $\frac{1}{2}c\Sigma_{abi}F^{abi}$ as energy density T^{00}, and $c\Sigma_i^{ab}F_{b0}^i$ as momentum density T^{0a}, where c is constant for satisfying the units. Then we search for a suitable Lagrangian and Hamiltonian with canonical relations that correspond to the same continuity equation according to the quantum fields theory, we do this at the flat-space–time limit and generalize it to an arbitrary curved space–time.

In scalar field ϕ theory, the Lagrangian is [15]

$$L(\phi, \partial\phi) = \pi\partial_0\phi - H(\pi, \phi),$$

the conjugate momentum is $\pi = \partial_0\phi = -\partial^0\phi$. The conserved momentum–energy tensor is

$$T^{00} = H(\pi, \phi) = \pi\partial_0\phi - L(\phi, \partial_\mu\phi) \text{ and } T^{0a} = P^a = \partial^0\phi\partial^a\phi = -\pi\partial^a\phi.$$

In the flat limit of the space–time, our momentum–energy tensor is

$$T^{0a} = c\Sigma^a{}_{bi}F^{0bi} = -c\Sigma_i^{ab}F_{0b}^i = c\Sigma_i^{ab}F_{b0}^i,$$

and

$$T^{00} = \frac{1}{2}c\Sigma_i^{ab}F_{ab}^i = \frac{1}{4}ce_i^a e_j^b F_{ab}^{ij}, \tag{20}$$

comparing it with the momentum $\partial^0\phi\partial^a\phi = -\pi\partial^a\phi$, we conclude that our conjugate momentum is $\pi^{bi} \sim -F^{0bi}$ [16–19].

By considering a Lagrangian of the form $L(F^{0ai}, DA^i)$ with corresponding Hamiltonian like $H(\pi_{ai}, DA^i)$ and using the action principle $\delta S(DA^i) = 0$ and the diffeomorphism invariance on the surface $\Sigma_t(\sigma^a)$, we obtain the momentum–energy tensor like [16–18]

$$T^{00} = \pi_{ai}F^{0ai} - L(F^{0ai}, DA^i) \text{ and } T^{0a} = c\pi_{bi}F^{abi}.$$

Comparing them with our momentum–energy tensor $T^{00} = \frac{1}{2}c\Sigma_{abi}F^{abi}$ and $T^{0a} = c\Sigma^a{}_{bi}F^{0bi}$, we conclude

$$T^{0a} = \pi_{bi}F^{abi} = c\Sigma^a{}_{bi}F^{0bi}. \tag{21}$$

In general, the curvature $F^i_{\mu\nu}$ can be written as [20,21]

$$F^i_{\mu\nu} = \psi^i{}_j\Sigma^j_{\mu\nu} + \psi'^i{}_j\bar{\Sigma}^j_{\mu\nu}, \tag{22}$$

so $\Sigma^{\mu\nu}_i F^i_{\mu\nu} = \psi^i{}_i$ since $\Sigma^{\mu\nu}_i\Sigma^j_{\mu\nu} = \delta^j_i$ and $\Sigma^{\mu\nu}_i\bar{\Sigma}^j_{\mu\nu} = 0$.

Using Equation (22) in Equation (21), we get

$$T^{0a} = \pi_{bi}\psi^i{}_j\Sigma^{abj} + \pi_{bi}\psi'^i{}_j\bar{\Sigma}^{abj} = c\Sigma^a{}_{bi}F^{0bi},$$

therefore we set $\psi'^i{}_j = 0$, so $F^i_{\mu\nu} = \psi^i{}_j\Sigma^j_{\mu\nu}$, and $\Sigma^i_{\mu\nu} = (\psi^{-1})^i{}_jF^j_{\mu\nu}$, we obtain

$$\pi_{bi}\psi^i{}_j\Sigma^{abj} = -c\Sigma^{abi}F_{0bi} \text{ so } \pi^{bi} = c(\psi^{-1})^i{}_jF^{0bj} = c\Sigma^{0bi}, \tag{23}$$

which means that Σ^{0bi} is conjugate momentum of A^i_a. Using it in the Hamiltonian $H = T^{00}$:

$$H = \pi_{ai}F^{0ai} - L(F^{0ai}, DA^i) = -c\Sigma_{0ai}F^{0ai} - L(F^{0ai}, DA^i),$$

then using our energy density $\frac{1}{2}c\Sigma_{abi}F^{abi}$, we get the Lagrangian

$$L(E, F^{0ai}, DA^i) = \pi_{ai}F^{0ai} - H = -c\Sigma_{0ai}F^{0ai} - \frac{1}{2}c\Sigma_{abi}F^{abi}. \tag{24}$$

Therefore

$$L(E, F) = -\frac{1}{2}c(\Sigma_{0ai}F^{0ai} + \Sigma_{0ai}F^{a0i}) - \frac{1}{2}c\Sigma_{abi}F^{abi},$$

so

$$L(E, A) = -\frac{1}{2}c\Sigma_{\mu\nu i}F^{\mu\nu i}, \text{ or } L(E, A) = -\frac{1}{2}c\Sigma_{\mu\nu i}F^{\mu\nu i}\sqrt{-g}d^4x,$$

where $\Sigma^i = E^i$ and F^i are defined in Equations (3) and (18). This Lagrangian has a symmetry of the complex group $SO(3)$ and the self-dual of Lorentz group $SO(3,1)$. The contraction is defined by using the metric $g_{\mu\nu} = \eta_{IJ}e^I_\mu e^J_\nu$. This Lagrangian corresponds to self-dual part of Equation (12). To get the total Lagrangian, we add the Hermitian conjugate, we obtain $L_{dual\ GR}(E, A, \bar{E}, \bar{A}) = L_{Left\ GR}(E, A) + L_{Right\ GR}(\bar{E}, \bar{A})$:

$$L_{dual\ GR}(\Sigma, A, \bar{\Sigma}, \bar{A}) = \frac{-c}{2}\Sigma^{\mu\nu}_i F^i_{\mu\nu}e + \frac{-c}{2}\bar{\Sigma}^{\mu\nu}_i\bar{F}^i_{\mu\nu}e, \tag{25}$$

the curvature $\bar{F}^i$ is the Hermitian conjugate $F^i = P^i_{IJ}R^{IJ}$ (10). The Lorentz group is $SO(3,1)$, its Lie algebra is reducible and can be decomposed into two copies of the Lie algebra of $SU(2)$:

$SO(3,1,C) \cong SO(3,C)_{Left} \times SO(3,C)_{Right}$.

The complex connection $A^i = P^i_{IJ}\omega^{IJ}$ takes values in Lie algebra of $SO(3,C)_{Left}$, while its Hermitian conjugate $\bar{A}^i = \bar{P}^i_{IJ}R^{IJ}$ takes values in Lie algebra of $SO(3,C)_{Right}$. Thus the Lorentz invariance is satisfied by the uni-variance under the two groups $SL(2,R)_{Left}$ and $SL(2,R)_{Right}$ [11,12].

To determine the constant c, we write this Lagrangian in the form

$$(16\pi G)^{-1} e^{\mu}_{I} e^{\nu}_{J} (R_{\mu\nu})^{IJ} e,$$

by using properties of the projection P^i_{IJ}

$$P^{IJ}_i P^i_{KL} + \bar{P}^{IJ}_i \bar{P}^i_{KL} = \frac{1}{2}(\delta^I_K \delta^J_L - \delta^I_L \delta^J_K), \text{ and } P^i_{IJ}\bar{P}^{IJ}_k = 0, \tag{26}$$

the Lagrangian (Equation (25)) becomes $(-c/2)e^{\mu}_{I} e^{\nu}_{J} (R_{\mu\nu})^{IJ} e$, thus $-c/2 = (16\pi G)^{-1}$. Therefore

$$L_{dual\ GR}(A,\Sigma,\bar{\Sigma},\bar{A}) = \frac{1}{16\pi G}\Sigma^{\mu\nu}_i F^i_{\mu\nu} e + \frac{1}{16\pi G}\bar{\Sigma}^{\mu\nu}_i \bar{F}^i_{\mu\nu} e, \tag{27}$$

this Lagrangian is similar to the Plebanisky Lagrangian, but it is not multiplied by the imaginary number i and does not include the cosmological constant term.

4. Yang–Mills Theory of Gravity

By regarding the local Lorentz symmetry as a gauge symmetry with spin connection ω^{IJ} (or A^i) as gauge fields, we recognize Yang–Mills theory in gravity. But not full gravity, since in the Yang–Mills theory, the variables are connections and conserved currents, while in the gravity the metric is also variable. The local Lorentz symmetry generates locally conserved currents, and those currents are coupled to spin connection ω^{IJ}. This makes the local Lorentz symmetry a gauge symmetry with the Lorentz group as a gauge group. Also, these currents must be conserved and vanish in the vacuum.

From the formula $F^i = \psi^i{}_j\Sigma^j + \bar{\psi}^i{}_j\bar{\Sigma}^j$, we can get the inversion $\Sigma^i = \xi^i{}_jF^j + \bar{\xi}^i{}_j\bar{F}^j$ by inserting it back, we obtain

$$\psi^i{}_j\xi^j{}_k + \bar{\psi}^i{}_j\bar{\xi}^j{}_k = \delta^j_k \text{ and } \psi^i{}_j\bar{\xi}^j{}_k + \bar{\psi}^i{}_j\xi^j{}_k = 0.$$

We can get the equation of motion $\Sigma^i = \xi^i{}_jF^j + \bar{\xi}^i{}_j\bar{F}^j$ from this Lagrangian (Equation (27)) by adding terms like

$$16\pi G L_{dual\ GR}(A,\Sigma,\bar{\Sigma},\bar{A}) = \Sigma^{\mu\nu}_i F^i_{\mu\nu} e - \psi_{ij}\Sigma^{\mu\nu i}\Sigma^j_{\mu\nu} e - \psi'_{ij}\bar{\Sigma}^{\mu\nu i}\Sigma^j_{\mu\nu} e + \text{C.C},$$

therefore the $\delta L/\Sigma^{\mu\nu}_i = 0$ and $\delta L/\delta F^i_{\mu\nu} = 0$ yields $F^i = \psi^i{}_j\Sigma^j + \bar{\psi}^i{}_j\bar{\Sigma}^j$ and $D\Sigma^i = 0$. But by using properties of the self-dual projection from Equation (26), we obtain $\bar{\Sigma}^{\mu\nu i}\Sigma^j_{\mu\nu} = 0$ and $\Sigma^{\mu\nu i}\Sigma^j_{\mu\nu} = \delta^{ij}$, but this does not change the equations of motion.

But as we will see, if there is a Lorentz current, like the spin current of the spinor field, then $\delta L/\delta F^i_{\mu\nu} \neq 0$, therefore to keep $D\Sigma^i = 0$, and to also keep $De^I = 0$, we add a term like $F_{\mu\nu i}F^{\mu\nu i}e$. This is done in order to insert back $\Sigma^i \sim F^i$ in $\psi_{ij}\Sigma^{\mu\nu i}\Sigma^j_{\mu\nu}$ into the Lagrangian. Therefore we write

$$L_{dual\ GR}(E,\Sigma,\bar{\Sigma},\bar{A}) = (16\pi G)^{-1}(\Sigma^{\mu\nu}_i F^i_{\mu\nu} e - \psi_{ij}\Sigma^{\mu\nu i}\Sigma^j_{\mu\nu})e + kF^{\mu\nu}_i F^i_{\mu\nu} e + \ldots + \text{C.C},$$

where k is a constant can relate to a coupling constant of Lorentz current with the spin connection A^i.

This Lagrangian includes $F^{\mu\nu i}F^{i}_{\mu\nu}e$, F^{i} is given in Equation (18), similarly to Lagrangian in Yang–Mills theory with gauge group $SO(3)$, and the self-dual of Lorentz group $SO(3,1)$. It depends only on the connection A^{i}, thus it describes the changing of the local Lorentz basis. Therefore $F^{\mu\nu i}F^{i}_{\mu\nu}e$ reads the invariance only under the local Lorentz transformations. Also, it is a topological invariance, so it allows the free propagation of spin connection A^{i}. But by relating A^{i} with the triad e^{I}, and relating e^{I} with the metric $g_{\mu\nu}$ by $g_{\mu\nu} = \eta_{IJ}e^{I}_{\mu}e^{J}_{\nu}$, this breaks the free propagation of spin connection A^{i} as free waves, except in the background approximation of the metric, the result is gravitational waves. Similarly to the free electromagnetic field.

By using the properties of self-dual projection, this Lagrangian can be written using the Riemannian tensor $R(\omega)$ as

$$L = \frac{1}{16\pi G}Re + kR^{2}e + ...$$

We find the role of the term $F^{\mu\nu i}F^{i}_{\mu\nu}e$ by including the interaction of mass-less spinor particles with gravity. Since they are massless, their energies are small so that their interaction with the gravitational field e^{I} is weak, but their interaction with spin connection takes place. The interaction term is

$$\omega^{IJ}_{\mu}e^{\mu}_{K}\bar{\psi}\gamma^{K}S_{IJ}\psi e = \omega^{IJ}_{\mu}J^{\mu}_{IJ}e, \tag{28}$$

where $J^{\mu}_{IJ} = e^{\mu}_{K}J^{K}_{IJ} = e^{\mu}_{K}\bar{\psi}\gamma^{K}S_{IJ}\psi$ is Lorentz current. If we add this term to the Lagrangian

$$(16\pi G)^{-1}e^{\mu}_{I}e^{\nu}_{J}(R_{\mu\nu})^{IJ}e,$$

we get

$$(16\pi G)^{-1}e^{\mu}_{I}e^{\nu}_{J}(R_{\mu\nu})^{IJ}e + \omega^{IJ}_{\mu}J^{\mu}_{IJ}e.$$

So the equation of motion for ω^{IJ}_{μ} is

$$-(16\pi G)^{-1}D_{\mu}(e^{\mu}_{[I}e^{\nu}_{J]}e) + J^{\mu}_{IJ} = 0.$$

In self-dual formalism, this equation becomes

$$-(16\pi G)^{-1}D_{\mu}\Sigma^{\mu\nu i} + J^{\nu i} = 0.$$

These two equations say that the Lorentz current is the source for the gravitational field e^{I}, but this is not right since the energy is the source for it, also we choose $De^{I} = 0$. Thus, these equations do not hold. But if we use the Lagrangian

$$L = (16\pi G)^{-1}\Sigma^{\mu\nu}_{i}F^{i}_{\mu\nu}e + kF^{\mu\nu}_{i}F^{i}_{\mu\nu}e + A^{i}_{\mu}J^{\mu}_{i}e + C.C, \tag{29}$$

the equation of motion for A^{i} becomes

$$-(16\pi G)^{-1}D_{\mu}(\Sigma^{\mu\nu i}e) - \frac{1}{2}kD_{\mu}F^{\mu\nu i}e + J^{\nu i} = 0.$$

Thus we choose $D_{\mu}(\Sigma^{\mu\nu i}e) = 0$ and

$$-\frac{1}{2}kD_{\mu}F^{\mu\nu i} + J^{\nu i} = 0. \tag{30}$$

The term $D_\mu F^{\mu\nu i}$ includes only the spin connection A^i, so the Lorentz current $e^\nu_I \bar{\psi}\gamma^I S^i \psi$ contributes as a source for the spin connection A^i, and so interacts with it. This relate to the fact that we can regard the local Lorentz symmetry as a gauge group with spin connection ω^{IJ} (or A^i) as gauge fields. The Lorentz current is conserved since

$$D_\nu D_\mu F^{\mu\nu i} = \frac{1}{2}[D_\nu, D_\mu]F^{\mu\nu i} = -\frac{1}{2}\varepsilon^i{}_{jk}F^j_{\mu\nu}F^{\mu\nu k} = 0 \rightarrow D_\nu J^{\nu i} = 0.$$

Furthermore, Equation (30) allows us to calculate a Lorentz current for a given curvature $F^i_{\mu\nu}$, although the curvatures $R^{IJ}_{\mu\nu}$ and $F^i_{\mu\nu}$ are calculated using only the energy–momentum tensor. Although there are Lorentz currents associated with matter, those currents relate to the local Lorentz symmetry.

We need to prove $\nabla^\mu R_{\mu\nu\rho\sigma} = 0$ in the vacuum, where $R_{\mu\nu\rho\sigma}$ is Riemann curvature tensor, it satisfies $R_{\mu\nu\rho\sigma} = -R_{\nu\mu\rho\sigma}$, $R_{\mu\nu\rho\sigma} = -R_{\mu\nu\sigma\rho}$ and $R_{\mu\nu\rho\sigma} = R_{\rho\sigma\mu\nu}$. In the vacuum, we have the equality

$$R_{\mu\nu} = constant \times Rg_{\mu\nu},$$

where $R_{\mu\nu}$ is a Ricci tensor and $R = g^{\mu\nu}R_{\mu\nu}$. This equality is equivalent to another equality [11]:

$$R_{\mu\nu} = constant \times Rg_{\mu\nu} \Leftrightarrow {}^*R_{\mu\nu\rho\sigma} = R^*{}_{\mu\nu\rho\sigma},$$

with Hodge operator

$${}^*R_{\mu\nu\rho\sigma} = \frac{1}{2}\epsilon_{\mu\nu}{}^{\mu'\nu'}R_{\mu'\nu'\rho\sigma},\ R^*{}_{\mu\nu\rho\sigma} = \frac{1}{2}R_{\mu\nu\rho'\sigma'}\epsilon^{\rho'\sigma'}{}_{\rho\sigma},$$

the anti-symmetric tensor $\epsilon^{\mu\nu\rho\sigma}$ is the volume four-form for metric $g_{\mu\nu}$. Therefore in the vacuum, we have

$$\epsilon_{\mu\nu}{}^{\mu'\nu'}R_{\mu'\nu'\rho\sigma} = R_{\mu\nu\rho'\sigma'}\epsilon^{\rho'\sigma'}{}_{\rho\sigma},\ \text{so}\ \epsilon_{\mu\nu\mu'\nu'}R^{\mu'\nu'}{}_{\rho\sigma} = R_{\mu\nu}{}^{\rho'\sigma'}\epsilon_{\rho'\sigma'\rho\sigma},$$

acting by ∇_γ on both sides, with $\nabla_\gamma \epsilon^{\mu\nu\rho\sigma} = 0$ we get

$$\epsilon_{\mu\nu\mu'\nu'}\nabla_\gamma R^{\mu'\nu'}{}_{\rho\sigma} = \nabla_\gamma R_{\mu\nu}{}^{\rho'\sigma'}\epsilon_{\rho'\sigma'\rho\sigma}.$$

The summing here is over μ', ν', ρ' and σ', while μ, ν, ρ, σ and γ are fixed. Then multiplying both sides by $\epsilon^{\gamma\mu\nu\alpha}$:

$$\epsilon^{\gamma\mu\nu\alpha}\epsilon_{\mu\nu\mu'\nu'}\nabla_\gamma R^{\mu'\nu'}{}_{\rho\sigma} = \epsilon^{\gamma\mu\nu\alpha}\nabla_\gamma R_{\mu\nu}{}^{\rho'\sigma'}\epsilon_{\rho'\sigma'\rho\sigma}.$$

We note that by summing over γ, μ, ν in $\epsilon^{\gamma\mu\nu\alpha}\nabla_\gamma R_{\mu\nu}{}^{\rho'\sigma'}$ for each fixed ρ, σ, α, we obtain the Bianchi identity $\epsilon^{\gamma\mu\nu\alpha}\nabla_\gamma R_{\mu\nu}{}^{\rho'\sigma'} = 0$, therefore

$$\epsilon^{\gamma\mu\nu\alpha}\epsilon_{\mu\nu\mu'\nu'}\nabla_\gamma R^{\mu'\nu'}{}_{\rho\sigma} = 0.$$

It becomes

$$2(\delta^\alpha_{\mu'}\delta^\gamma_{\nu'} - \delta^\gamma_{\mu'}\delta^\alpha_{\nu'})\nabla_\gamma R^{\mu'\nu'}{}_{\rho\sigma} = 0 \rightarrow 2\nabla_\gamma R^{\alpha\gamma}{}_{\rho\sigma} - 2\nabla_\gamma R^{\gamma\alpha}{}_{\rho\sigma} = -4\nabla_\gamma R^{\gamma\alpha}{}_{\rho\sigma} = 0,$$

so $\nabla^\mu R_{\mu\nu\rho\sigma} = 0$ in the vacuum. Therefore the Lorentz current $J^{IJ}_\nu = \nabla^\mu R^{IJ}_{\mu\nu} = e^{I\rho}e^{J\sigma}\nabla^\mu R_{\mu\nu\rho\sigma}$ vanish in the vacuum. Using the self-dual projection, we find

$$J^i_\nu = \nabla^\mu F^i_{\mu\nu} = P^i_{IJ}\nabla^\mu R^{IJ}_{\mu\nu}$$

also vanish in the vacuum. Therefore this current associates only with matter. We can also prove this by using the formula $F^i = \psi^i{}_j\Sigma^j + \bar{\psi}^i{}_j\bar{\Sigma}^j$, and by setting $\bar{\psi}^i{}_j = 0$ in the vacuum [11]. Using

$$(*d*F^i)_\mu = \nabla^\nu F^i_{\nu\mu},$$

with $*F^i = \psi^i{}_j(*\Sigma^j) + \bar{\psi}^i{}_j(*\bar{\Sigma}^j)$, and

$$*\Sigma^j = -i\Sigma^j, \;\; *\bar{\Sigma}^j = i\bar{\Sigma}^j, \tag{31}$$

Hodge operator here is with respect to the metric $g_{\mu\nu}$, we get

$$\nabla^\nu F^i_{\nu\mu} = i(*(-d\psi^i{}_j\Sigma^j + d\bar{\psi}^i{}_j\bar{\Sigma}^j))_\mu.$$

The first term becomes

$$d(\psi^i{}_j\Sigma^j) = d(\psi^i{}_j\Sigma^j + \bar{\psi}^i{}_j\bar{\Sigma}^j) - d(\bar{\psi}^i{}_j\bar{\Sigma}^j), \text{ so } d(\psi^i{}_j\Sigma^j) = dF^i - d(\bar{\psi}^i{}_j\bar{\Sigma}^j),$$

and by Bianchi identity $dF^i = 0$, we obtain $d(\psi^i{}_j\Sigma^j) = -d(\bar{\psi}^i{}_j\bar{\Sigma}^j)$. Therefore

$$\nabla^\nu F^i_{\nu\mu} = i(*(d(\bar{\psi}^i{}_j\bar{\Sigma}^j) + d\bar{\psi}^i{}_j\bar{\Sigma}^j))_\mu = 2i(*d(\bar{\psi}^i{}_j\bar{\Sigma}^j))_\mu = -2i(*d(\psi^i{}_j\Sigma^j))_\mu. \tag{32}$$

Since $\bar{\psi}^i{}_j = 0$ in the vacuum, we get $\nabla^\nu F^i_{\nu\mu} = 0$.

Therefore for a Lorentz J^i_μ current that associates with matter, we get

$$\nabla^\nu F^i_{\nu\mu} = 2i(*d(\bar{\psi}^i{}_j)\bar{\Sigma}^j)_\mu = J^i_\mu, \text{ so } -2id\bar{\psi}^i{}_j \wedge \bar{\Sigma}^j = *J^i,$$

Where we used $d\bar{\Sigma}^j = 0$ and $J^i = J^i_\mu dx^\mu$, with property of Hodge dual twice operation on p-form V in n dimensions: $**V = (-1)^{pq+t}V$, where $q = n - p$, and t is the number of negative eigenvalues of the metric tensor [14]. In our case we have $n = 4,\; p = 3,\; t = 1$.

Same thing we get for $\psi^i{}_j$, (Equation (32)):

$$\nabla^\nu F^i_{\nu\mu} = -2i(*(d\psi^i{}_j \wedge \Sigma^j))_\mu = J^i_\mu, \text{ so } -2id\psi^i{}_j \wedge \Sigma^j = *J^i.$$

Therefore

$$-2i(D_\mu\psi^i{}_j)\Sigma^j_{\nu\rho}dx^\mu \wedge dx^\nu \wedge dx^\rho = J^{i\sigma}\epsilon_{\mu\nu\rho\sigma}dx^\mu \wedge dx^\nu \wedge dx^\rho/3!,$$

so

$$2i(D_\mu\psi^i{}_j)\Sigma^j_{\nu\rho}\epsilon^{\mu\nu\rho\sigma} = J^{i\sigma}.$$

But

$$\Sigma^j_{\nu\rho}\epsilon^{\nu\rho\mu\sigma}/3! = (*\Sigma^j)^{\mu\sigma} = (-i\Sigma^j)^{\mu\sigma},$$

(we used self-dual properties Equation (31)), we obtain

$$(D_\mu\psi^i{}_j)\Sigma^{j\mu\nu} = \frac{J^{i\nu}}{2\times 3!}. \tag{33}$$

The quantity $2(D_\mu \psi^i{}_j)\Sigma^{j\mu\nu}$ depends on the triads e^I, and on spin connection ω^{IJ}, while $J^{i\nu}$ relates matter. Therefore by solving this equation for a given Lorentz current $J^{i\sigma}$, we get solutions for e^I and ω^{IJ}. But this formula reads only the contributing of Lorentz current in $\psi^i{}_j$, there is a contributing of cosmological constant and matter in symmetric part of ψ; $tr(\psi) = -\Lambda - 2\pi GT$, where T is trace of energy-momentum tensor $T_{\mu\nu}$ [12,22]. Since Lorentz current generates local Lorentz transformation, we expect that $D_\mu \psi^i{}_j$ in Equation (33) is anti-symmetric, this distinguishes contributing of $J^{i\sigma}$ from those of cosmological constant and matter. So we write

$$2(D_\mu \psi_k)\varepsilon^{ki}{}_j \Sigma^{j\mu\nu} = J^{i\nu}, \tag{34}$$

thus $J^{i\nu}$ contributes in anti-symmetric part of $\psi^i{}_j := \psi_k \varepsilon^{ki}{}_j$, where ψ^i is vector field in local Lorentz frame.

Let us write $D_\mu \psi^i = P^i_{IK} J^{IK}_\mu$, with Lorentz current $J^{IK}_\mu = J^{IK}_L e^L_\mu$ defined in Equation (28). The relationship (linear) between J^{IK}_μ and $J^{i\nu} = J^i_I e^{I\nu}$ can be determined by inserting $D_\mu \psi^i = P^i_{IK} J^{IK}_\mu$ in Equation (34). It is easier to solve

$$D_\mu \psi^i = P^i_{IK} J^{IK}_\mu$$

in region away from matter where $J^{IK}_\mu = 0$, so we can solve it in background approximation, $g \approx \eta$, thus $\psi^i \to (\psi^r, \psi^\theta, \psi^\phi)$ in spherically coordinates. If we assume that the vector field ψ^r depends only on the radius r, we get

$$\frac{1}{r^2}\partial_r(r^2 \psi^r(r)) = 0 \to \psi^r(r) = \frac{a^r}{r^2},$$

where a^r is constant vector. Therefore the contributing of Lorentz current $J^{i\nu}$ in the curvature $F^i_{\mu\nu}$ is

$$F^i_{\mu\nu} := \frac{1}{r^2} a_k \varepsilon^{ki}{}_j \Sigma^j_{\mu\nu},$$

therefore the contributing of $J^{i\nu}$ in F^2 is

$$F^i_{\mu\nu} F^{\mu\nu}_i := \frac{2}{r^4} a^2, \text{ with } \Sigma^{\mu\nu}_i \Sigma^j_{\mu\nu} = \delta^j_i.$$

This formula satisfies $\oint_S \left\| F^i_{\mu\nu} \right\| dS = 4\pi\sqrt{2a^2} = constant$, it is similar to electric field $\oint_S \vec{E} \cdot d\vec{S} = Q/\varepsilon_0$. This is similarity between GR and Yang-Mills theory.

The Lorentz current is not associated only with spinor particles, the Lorentz symmetry for arbitrary field produces a global conserved Lorentz current like ([15], section 22)

$$M^{IJK} = x^J T^{IK} - x^K T^{IJ}.$$

T^{IK} is energy-momentum tensor in flat space–time. Locally we write this as

$$M^{\mu JK} = a^\nu (e^J_\nu T^{\mu K} - e^K_\nu T^{\mu J}),$$

where a^ν is constant, therefore

$$\nabla_\mu M^{\mu JK} = 0; \ \nabla_\mu e^J_\nu = 0, \ \nabla_\mu T^{\mu K} = 0, \ T^{\mu K} = e^K_\nu T^{\mu\nu}.$$

The existence of a conserved spin current J^i_μ that couples to A^i lets us to believe in local Lorentz symmetry as a gauge group with spin connection ω^{IJ} (or A^i) as gauge fields similarly to Yang–Mills theory.

5. Beta Function

We assume that the interaction of mass-less particles with the gravity is dominated by interaction of their spin current (Lorentz current) with the connection A^i_μ. We can use the Lagrangian $F_{\mu\nu i}F^{\mu\nu i}$ to describe the interaction of left-handed fermions with the connection A^i_μ. We choose a representation of $SL(2,C)$ in which we have $K^i = iJ^i$, where K^i are boost generators and J^i are rotation generators [23]. Our connection A^i is a one-form complex given by the self-dual projection $A^i_\mu = P^i_{IJ}\omega^{IJ}_\mu$ of the spin connection ω^{IJ}_μ according to the decomposition $so(3,1:C) = so(3:C) \oplus so(3:C)$ [24]. Therefore

$$A^iJ^i = Re(A^i)J^i + iIm(A^i)J^i = Re(A^i)J^i + Im(A^i)K^i,$$

or $A^iJ^i + A'^iK^i$, where A^i and A'^i are real.

For simplicity let us choose $A'^i = \gamma A^i$, with a constant $\gamma \in R$. Thus the connection becomes

$$A^iJ^i + A'^iK^i = A^iJ^i + i\gamma A^iJ^i = A^i\left(J^i + i\gamma J^i\right) = A^iT^i \in \Omega^1(M, sl(2,C)),$$

with new generators $T^i = J^i + i\gamma J^i = (1+i\gamma)\,J^i \in sl(2,C)$.

Therefore the coupling of the connection A^i with left-handed spinor field is $\frac{1}{2}gA^i_\mu e^\mu_I \psi^+\bar{\sigma}^I T^i\psi e$, where g is coupling constants comes from using the covariant derivative seen in Equation (14). So we write the fermion-gravity Lagrangian as

$$L(A,\psi) = ie^\mu_I\psi^+\bar{\sigma}^I\partial_\mu\psi e + \frac{1}{4}F_{\mu\nu i}F^{\mu\nu i}e + \frac{1}{2}gA^i_\mu e^\mu_I\psi^+\bar{\sigma}^I T^i\psi e.$$

Using the metric $g_{\mu\nu} = \eta_{IJ}e^I_\mu e^J_\nu$, we obtain

$$F_{\mu\nu i}F^{\mu\nu i} = g_{\mu\mu'}g_{\nu\nu'}F^{\mu'\nu' i}F^{\mu\nu i} = e^I_\mu e_{I\mu'}e^J_\nu e_{J\nu'}F^{\mu'\nu' i}F^{\mu\nu i}.$$

In background space–time, we have $e^i_\mu(x) = \delta^i_\mu + h^i_\mu(x)$, so $g_{\mu\nu}(x) = \eta_{IJ}\delta^I_\mu\delta^J_\nu + ...$, where η is the Minkowski metric. Therefore the gravity and spinor Lagrangian approximates to

$$L(A,\psi) = i\psi^+\bar{\sigma}^\mu\partial_\mu\psi + \frac{1}{4}\eta_{\mu'\mu}\eta_{\nu'\nu}F^{\mu'\nu' i}F^{\mu\nu i} + \frac{1}{2}gA^i_\mu\psi^+\bar{\sigma}^\mu T^i\psi + ...$$

The remaining term includes the interaction with the fluctuated gravitational field $h^i_\mu(x)$, this interaction relates with local invariance under diffeomorphism of M. Let us consider only the part

$$L(A,\psi) = i\psi^+\bar{\sigma}^\mu\partial_\mu\psi + \frac{1}{4}\eta_{\mu'\mu}\eta_{\nu'\nu}F^{\mu'\nu' i}F^{\mu\nu i} + \frac{1}{2}gA^i_\mu\psi^+\bar{\sigma}^\mu T^i\psi,$$

which is invariant under local Lorentz transformation, the compatible currents take values in $sl(2,C)$, thus A^i is coupled to Lorentz currents. By that we have included only the invariance under local Lorentz transformation and excluded the local invariance under diffeomorphism of M. Thus we describe the GR using Lorentz frames as vector bundle over basis space M with connection $A^i \in T^*_pM \times sl(2,C)$.

It is similar to the Lagrangian of Yang–Mills theory for spinor field, but with the generators $\frac{1}{2}T^i = \frac{1}{2}(1+i\gamma)\,J^i$, so to get results from the usual theory, we just replace the generators J^i with $\frac{1}{2}(1+i\gamma)\,J^i$.

For example, to get the beta function $\beta(g) = \partial g/\partial ln(M)$ for our Lagrangian, we use the beta function of Yang–Mills theory for spinor field with symmetry group like $SU(N)$, it is given by [15,25]

$$\beta(g) = -\left[\frac{11}{3}T(A) - \frac{4}{3}n_f T(R)\right]\frac{g^3}{16\pi^2} + O(g^5).$$

The numbers $T(A)$ and $T(R)$, are given in

$$Tr(J_R^a J_R^b) = T(R)\delta^{ab} \; and \; Tr(J_A^a J_A^b) = T(A)\delta^{ab},$$

where J_R^a are generators for the fundamental representation of $SU(N)$ and J_A^a are generators for the adjoint representation. To get them for our Lagrangian, we have to start with the commutation relation $[J^a, J^b] = if^{abc}J^c$ and note that we can multiply both sides by $\frac{1}{4}(1+i\gamma)^2$ to get

$$\left[\frac{1}{2}(1+i\gamma)J^a, \frac{1}{2}(1+i\gamma)J^b\right] = i\frac{1}{2}(1+i\gamma)f^{abc}\frac{1}{2}(1+i\gamma)J^c,$$

thus we get new anti-symmetric structure constants $\frac{1}{2}(1+i\gamma)f^{abc}$, although the new generators are not hermitian, but this does not violates the methods of deriving the beta function, the necessary thing in deriving it is keeping $T(A)$, $T(R)$ and f^{abc} constants [15,26,27]. Anyway, we will absorb the modification factor $\frac{1}{2}(1+i\gamma)$ into the coupling constant g, so we have $SU(2)$ gauge group with complex coupling constant like $\frac{1}{2}(1+i\gamma)g$. Therefore we obtain

$$Tr[J_R^a J_R^b] = T(R)\delta^{ab} \rightarrow Tr\left[\frac{1}{2}(1+i\gamma)J_R^a\frac{1}{2}(1+i\gamma)J_R^b\right] = \frac{1}{4}(1+i\gamma)^2 T(R)\delta^{ab}$$

and

$$Tr[J_A^a J_A^b] = T(A)\delta^{ab} \rightarrow Tr\left[\frac{1}{2}(1+i\gamma)J_A^a\frac{1}{2}(1+i\gamma)J_A^b\right] = \frac{1}{4}(1+i\gamma)^2 T(A)\delta^{ab}.$$

Thus to get beta function for our Lagrangian, we replace $T(A)$ with $\frac{1}{4}(1+i\gamma)^2 T(A)$ and $T(R)$ with $\frac{1}{4}(1+i\gamma)^2 T(R)$, so using this in the beta function

$$\beta(g) = -\left[\frac{11}{3}T(A) - \frac{4}{3}n_f T(R)\right]\frac{g^3}{16\pi^2} + O(g^5),$$

we get

$$\beta(g) = -\frac{1}{4}(1+i\gamma)^2\left[\frac{11}{3}T(A) - \frac{4}{3}n_f T(R)\right]\frac{g^3}{16\pi^2} + O(g^5).$$

For the group $SU(2)$, we have $T(A) = 2$ and $T(R) = \frac{1}{2}$, so for $n_f = 1$, we obtain beta function like

$$\beta(g) = -\frac{5}{3}(1+i\gamma)^2\frac{g^3}{16\pi^2} + O(g^5).$$

Taking in consideration the first statement we obtain

$$\frac{\partial g}{\partial ln(M)} = -\frac{5}{3}(1+i\gamma)^2\frac{g^3}{16\pi^2},$$

which can be solved for energy scale M as

$$g^{-2}(M) = \frac{10}{3 \cdot 16\pi^2}(1+i\gamma)^2 ln\left(\frac{M}{M_0}\right) + g^{-2}(M_0).$$

Usually the coupling constant is real, so for our one, we consider that the interaction strength is governed by the real part of the coupling constant g, this is

$$g_{int}^{-2}(M) = \frac{10}{3 \cdot 16\pi^2}\left(1-\gamma^2\right) ln\left(\frac{M}{M_0}\right) + g_{int}^{-2}(M_0),$$

thus the behavior of interaction of the left-fermions with the spin connection A^i according to our Lagrangian depends on γ, if $\gamma > 1$, then $\beta(g) > 0$, which means that this interaction becomes stronger as the energy increases, until breaking the perturbation at some energy scale. It is natural to consider $\gamma > 1$ since we expect $A'^i = A^i_{boost} > A^i_{rotation} = A^i$ in nature; the gravity effects the particles by changing their energies (like accelerating a particle via a gravitational field) not by changing their angular momentums. So from $A' = \gamma A$, we have $\gamma > 1$ for this case.

But when does the case $\gamma > 1; \beta(g) < 0$ appear? it appears when $A^i_{rotation} > A^i_{boost}$, in this case, the gravity induces a rotation of the inertial frame (the Lorentz frame moves with a particle, or the Lorentz frame in which the particle has a constant speed) more than changing the energy of that particle (accelerating). This occurs when there is the smallest distance between two particles which interact by their gravitational field. At this distance, the velocities of the two particles are constant, so the interaction by the gravity is dominated by changing their angular momentum, thus $A' > A$ so $\gamma > 1$. This situation appears in the back holes; which is a confinement of $\beta(g) < 0$.

6. Conclusions

We have considered the spacial Lorentz orthonormal basis (e^1, e^2, e^3) as an element in a vector bundle with real spin connection ω^{ij}, which takes values in the Lie algebra of group $SO(3)$ or $SU(2)$. We considered this vector bundle as a tangent vector bundle on the $3D$ hypersurface of constant time $\Sigma_t(\sigma^a)$, this allowed us to define three-form $\theta = \text{constant} \times \varepsilon_{ijk} E^i \wedge R^{jk}$ in the phase space (E^a_i, ω^{ij}_a) on this surface. By arbitrary transformation of x^μ, the three-form θ becomes on M, but $(\Sigma(\sigma^a), d\theta) = 0$ is always satisfied. By doing the same thing, we obtained the equation $(\Sigma(\sigma^a), d\theta) = 0$ using self-dual and anti-self-dual formalism. This equation produces an equation of continuity on the hypersurface $\Sigma_t(\sigma^a)$. We found that Σ^{0a}_i is a conjugate momentum of A^i_a where $\Sigma^{ab}_i F^i_{ab}$ is its energy density. We saw that we have to include the term $L(DA^i)ed^4x = (1/4)F_{\mu\nu i}F^{\mu\nu i}ed^4x$ in the GR Lagrangian, since there is a conserved spin current that couples to A^i. This is the similarity between GR and Yang–Mills theory of gauge fields. If we can solve the GR equations using only spin current, we may consider GR as Yang–Mills theory of gauge fields on a curved space–time manifold with spin connection A^i as a gauge field.

References

1. Montesinos, M.; González, D.; Celada, M.; íaz, B. Reformulation of the symmetries of first-order general relativity. *Class. Quant. Grav.* **2017**, *34*, 205002. [CrossRef]
2. Weatherall, J.O. Fiber Bundles, Yang–Mills Theory, and General Relativity. *Synthese* **2016**, *193*, 2389. [CrossRef]
3. Fleischhack, C. On Ashtekar's Formulation of General Relativity. *J. Phys. Conf. Ser.* **2012**, *360*, 012022. [CrossRef]
4. Arnowitt, R.; Deser, S.; Misner, C.W. The Dynamics of General Relativity. *Gen Relat. Grav.* **2008**, *40*, 1997–2027. [CrossRef]
5. Wald, R.M. *General Relativity*; University of Chicago Press: Chicago, IL, USA, 1984; ISBN-13:978-0226870335.
6. Ponomarev, V.N.; Barvinsky, A.O.; Obukhov, Y.N. *Gauge Approach and Quantization Methods in Gravity Theory*; Nauka: Moscow, Russia, 2017; ISBN 978-5-02-040047-4. [CrossRef]

7. Rosas-Rodriguez, R. Alternative variables for the dynamics of general relativity. *Int. J. Mod. Phys. A* **2008**, *23*, 895–908. [CrossRef]
8. Rovelli, C. Ashtekar formulation of general relativity and loop-space nonperturbative quantum gravity: A report. *Class. Quantum Grav.* **1991**, *8*, 1613. [CrossRef]
9. Rovelli, C. Area is the length of Ashtekar's triad field. *Phys. Rev. D* **2013**, *87*, 089902. [CrossRef]
10. Rovelli, C. *Quantum Gravity*; Cambridge University Pres: Cambridge, UK, 2004; ISBN 978-0521715966.
11. Krasnov, K. Plebański formulation of general relativity: a practical introduction. *Gen. Relat. Grav.* **2011**, *43*, 1. [CrossRef]
12. Bennett, D.L.; Laperashvili, L.V.; Nielsen, H.B.; Tureanu, A. Gravity and Mirror Gravity in Plebanski Formulation. *Int. J. Mod. Phys. A* **2013**, *28*, 1350035. [CrossRef]
13. Georgi, H. *Lie Algebras in Particle Physics*; CRC Press, Taylor & Francis Group: Boca Raton, FL, USA, 2000; ISBN 9780429499210. [CrossRef]
14. Pope, C. *Geometry and Group Theory*; Texas A&M University Press: College Station, TX, USA, 2008.
15. Srednicki, M. *Quantum Field Theory*; Cambridge University Press: Cambridge, UK, 2007; ISBN 9780521864497.
16. Struckmeier, J.; Redelbach, A. Covariant Hamiltonian field theory. *Int. J. Mod. Phys. E* **2008**, *17*, 435–491. [CrossRef]
17. Zambrano, G.E.; Pimentel, B.M. Canonical structure of gauge invariance proca's electrodynamics theory. *Momento* **2018**, *56*, 26–44, 2018. [CrossRef]
18. Creutz, M.; Muzinich, I.J.; Tudron, T.N. Gauge fixing and canonical quantization. *Phys. Rev. D* **1979**, *19*, 531. [CrossRef]
19. Utiyama, R. Invariant Theoretical Interpretation of Interaction. *Phys. Rev.* **1956**, *101*, 1597. [CrossRef]
20. Engle, J.; Noui, K.; Perez, A.; Pranzetti, D. Black hole entropy from an SU(2)-invariant formulation of Type I isolated horizons. *Phys. Rev. D* **2010**, *82*, 044050. [CrossRef]
21. Engle, J.; Noui, K.; Perez, A. Black hole entropy and SU(2) Chern-Simons theory. *Phys. Rev. Lett.* **2010**, *105*, 031302. [CrossRef] [PubMed]
22. Tennie, F.; Wohlfarth, M.N.R. Consistent matter couplings for Plebanski gravity. *Phys. Rev. D* **2010** *28*, 104052. [CrossRef]
23. Başkal, S.; Kim, Y.; Noz, M. Loop Representation of Wigner's Little Groups. *Symmetry* **2017**, *9*, 97. [CrossRef]
24. Herfray, Y. Pure Connection Formulation, Twistors and the Chase for a Twistor Action for General Relativity. *J. Math. Phys.* **2017**, *58*, 112505. [CrossRef]
25. Herzog, F.; Ruijl, B.; Ueda, T.; Vermaseren, J.A.M.; Vogt, A. The five-loop beta function of Yang–Mills theory with fermions. *JHEP* **2017**, *1702*, 90. [CrossRef]
26. Cui, J.W.; Wu, Y.L. One-Loop Renormalization of Non-Abelian Gauge Theory and beta Function Based on Loop Regularization Method. *Int. J. Mod. Phys. A* **2008**, *23*, 2861–2913. [CrossRef]
27. Peskin, M.E.; Schroeder, D.V. *An Introduction to Quantum Field Theory*; Westview Press: Boulder, CO, USA, 1995; ISBN 9780201503975.

3

Parameterized and Consistency Tests of Gravity with Gravitational Waves: Current and Future †

Zack Carson and Kent Yagi *

Department of Physics, University of Virginia, Charlottesville, VA 22904, USA
* Correspondence: ky5t@virginia.edu
† Presented at the Recent Progress in Relativistic Astrophysics, Shanghai, China, 6–8 May 2019.

Published: 20 August 2019

Abstract: Gravitational wave observations offer unique opportunities to probe gravity in the strong and dynamical regime, which was difficult to access previously. We here review two theory-agnostic ways to carry out tests of general relativity with gravitational waves, namely (i) parameterized waveform tests and (ii) consistency tests between the inspiral and merger-ringdown portions. For each method, we explain the formalism, followed by results from existing events, and finally we discuss future prospects with upgraded detectors, including the possibility of using multi-band gravitational-wave observations with ground-based and space-borne interferometers. We show that such future observations have the potential to improve upon current bounds on theories beyond general relativity by many orders of magnitude. We conclude by listing several open questions that remain to be addressed.

Keywords: gravitational waves; black holes; general relativity; modified theories of gravity

---

1. Introduction

Einsteins' famous theory of general relativity (GR) has proven to be wildly successful for over 100 years, accurately predicting many astrophysical phenomena observed to this very day. Throughout this period of time, many have attempted to prove the theory incorrect or merely just one piece of a more grand theory of nature with various observational and experimental schemes. All have met with the same result: GR still standing true with absolutely no statistically significant signs of deviation. With such an outstanding history of success, why must we continue to test the theory of GR? The answer is simple: There yet remains a plethora of unanswered questions stemming from mysterious observations seen throughout this time. These open questions include, and are not limited to, the unification of GR and quantum mechanics [1–6], dark matter and its influence on galactic rotation curves [1–4,7], dark energy and the ensuing accelerated expansion of the universe [2,5,6,8], the strange inflationary period seen in the early universe [1–3,6], and the matter-anti-matter asymmetry found in the present universe [1,3]. To date, there have been several proposed theories of gravity, both modifications or extensions to GR, as well as entirely new theories, many of which could be used to explain several of the above-listed open astrophysical/cosmological questions. While these theories could potentially reduce to the GR we know in the weak-field environments typically observed, they could very well become active in the extreme-gravity regime where the gravitational fields are strong, non-linear, and highly-dynamical.

For the last century, many attempts have been made to determine and constrain the various proposed modified theories of gravity found in the literature. When probed in the weak-field and static environments such as the local solar system, observations of photon-deflection, Shapiro time-delay, perihelion advance of Mercury, the Nordtvedt effect, and more [9] have determined no deviations from GR. Similarly, observations concerning the strong-field and static systems of binary pulsar systems [10,11] have also shown to be consistent with GR. Further, large-scale cosmological observations [1,2,6,8,12] have also identified no deviations. More recently, the groundbreaking gravitational wave (GW) observations of coalescing black holes (BHs) [13,14] and neutron stars (NSs) [15] by the Laser Interferometer Gravitational-Wave Observatory (LIGO) and Virgo Collaborations (LVC) have provided us with the unique opportunity to study fascinating extreme-gravity environments. To date, all such confirmed events have similarly found no deviations from GR [16,17].

While the current extreme-gravity tests of GR have yet to discover ground-breaking results, hope is not lost, as the field of gravitational wave astronomy is still in its infancy. While monumental in their engineering design and successful sensitivity, the current LVC Observing Run 2 (O2) infrastructure is limited by noise. Due to the LVC's overwhelming successes on the GW front, several proposed, planned, and even funded gravitational wave detectors are in the works. Several planned upgrades to the current LIGO detectors, aLIGO, A+, and Voyager [18,19], are currently underway with large improvements in the design sensitivity. Furthermore, new "third generation" interferometers Cosmic Explorer [19] (CE), and the Einstein Telescope [20] (ET) with up to $\sim$ 100 times the sensitivity of current detectors are currently in the planning stages. Finally, space-based laboratories, such as the Laser Interferometer Space Antenna (LISA) [21], TianQin [22,23], the Deci-Hertz Interferometer Gravitational wave Observatory (DECIGO) [24], and B-DECIGO [25], are currently in progress with sensitivities to GWs in the sub-Hz frequency bands. For stellar-mass binary BHs, ground-based detectors sensitive to high GW frequencies can largely probe non-GR effects, as they become more active at high relative velocities; while space-based detectors that are operative at low frequencies, are more suited to probing low-velocity effects. With such a promising future of observational GW astrophysics, probes of modified theories of gravity stand a highly increased likelihood of observing possible deviations from GR.

In the following document, we summarize the past, present, and future considerations for testing GR in the extreme gravity environments of merging BHs. In particular, we consider both parameterized tests of GR and consistency tests between the inspiral and merger-ringdown portions of the GW signal. The former tests allow one to map generalized non-GR effects entering the gravitational waveform with a certain velocity dependence, to most proposed modified theories of gravity and their associate theoretical parameters. The latter allows one to test how consistent the obtained signal is with the predictions of GR as a whole, granting a gauge on how much the inspiral and merger-ringdown signals agree. Specifically, we present current and projected bounds on the Einstein dilaton Gauss–Bonnet (EdGB) [26,27], dynamical Chern–Simons (dCS) [28–31], scalar-tensor theories [32,33], noncommutative theories [34,35], time-varying G theories [9,36,37], time-varying BH mass theories [38,39], and massive gravity [40–43]. We discuss the current constraints and progress, followed byy estimated future bounds. The latter is considered from single-band detections on both ground- and space-based GW detectors, as well as the multi-band observations between both detector types. We find that orders-of-magnitude improvements can be made upon using such future considerations, for both parameterized and consistency tests of GR.

The organization of this article is as follows. Section 2 details the parameterized tests of GR, starting off with the formulation and techniques used, followed up with the current status and future predictions of constraints on non-GR effects, finishing up with a discussion of multi-band observations. Section 3 follows suit with the same organization for the inspiral-merger-ringdown consistency tests of GR. Finally, in Section 4, we conclude and discuss the open questions yet remaining in the testing of non-GR effects. Throughout this document, we utilize the geometric units of $G = 1 = c$, unless otherwise stated.

2. Parameterized Tests

2.1. Formulation

Instead of comparing GW data with template waveforms in specific modified theories of gravity one by one, a more efficient approach is to first compare the data with template waveforms that can capture generic non-GR modifications, and then map the information from generic non-GR parameters to that of parameters in each theory. Various formalisms exist for such a theory-agnostic approach [16,17,44–49]. Here we follow the parameterized post-Einsteinian (ppE) formalism [47], in which the frequency-domain waveform is given by

$$\tilde{h}(f) = \tilde{h}_{\rm GR}(1 + \alpha\, u^a)e^{i\delta\Psi}\,, \quad \delta\Psi = \beta u^b\,. \tag{1}$$

Here $\tilde{h}_{\rm GR}$ is the GR waveform (for which we use the IMRPhenomD waveform [50,51] for spin-aligned binary black holes (BBHs) with circular orbits) while $u = (\pi \mathcal{M} f)^{1/3}$ is the effective relative velocity of binary constituents with $\mathcal{M}$ and f representing the chirp mass and GW frequency respectively. (α, a, β, b) are known as the ppE parameters. α and β denote the overall magnitude of the non-GR term in the amplitude and phase respectively, while a and b characterize at which post-Newtonian (PN)[1] order the correction enters the gravitational waveform in the amplitude and phase. The mapping between these ppE parameters and theoretical constants in various modified theories of gravity can be found in Tables I and II of [37].

Let us now prepare the basics of the Fisher analysis methods used frequently in this document for parameter estimation of template parameters θ^a. Commonly used as a less-computationally expensive alternative to a full Bayesian statistical analysis, the Fisher analysis is a good approximation for loud enough events. The signal-to-noise-ratio (SNR) ρ of such events is defined as

$$\rho \equiv \sqrt{(h|h)}, \tag{2}$$

where h is the gravitational waveform template, and the inner product $(a|b)$ is defined to be

$$(a|b) \equiv 2\int_{f_{\rm low}}^{f_{\rm high}} \frac{\tilde{a}^*\tilde{b} + \tilde{b}^*\tilde{a}}{S_n(f)} df. \tag{3}$$

In the above expression, $S_n(f)$ represents the spectral noise density of the given detector, and $f_{\rm high,low}$ are the cutoff frequencies, again dependent on the detector.

Assuming a Gaussian-distributed noise pattern, and Gaussian prior distributions on waveform template parameters, the parameters θ^a assuming a GW signal s can be found to follow [52]

$$p(\theta^a|s) \propto p^{(0)}_{\theta^a} \exp\left[-\frac{1}{2}\Gamma_{ij}\Delta\theta^i\Delta\theta^j\right]. \tag{4}$$

In the above distribution, $\Delta\theta^i \equiv \theta^i - \hat{\theta}^i$ with $\hat{\theta}^i$ representing the maximum likelihood value of θ^i, $p^{(0)}_{\theta^a}$ is the prior probability distribution which we assumed to be Gaussian with root-mean-square errors $\sigma^0_{\theta^a}$, and Γ_{ij} is the Fisher information matrix determined to be

$$\Gamma_{ij} \equiv (\partial_i h|\partial_j h). \tag{5}$$

[1] A term of nPN order in the waveform is proportional to $(u/c)^{2n}$ relative to the leading-order term.

The resulting 1σ root-mean-square errors on template parameters θ^a can be written directly as

$$\Delta\theta_i \approx \sqrt{(\tilde{\Gamma}_{ii})^{-1}}, \tag{6}$$

with the effective Fisher matrix defined by [53–55]

$$\tilde{\Gamma}_{ij} \equiv \Gamma_{ij} + \frac{1}{(\sigma^0_{\theta^a})^2}\delta_{ij}. \tag{7}$$

Finally, if one desires to combine the information from N detectors, the resultant effective Fisher matrix becomes

$$\tilde{\Gamma}^{\text{total}}_{ij} = \sum_{k=1}^{N} \Gamma^{(k)}_{ij} + \frac{1}{(\sigma^0_{\theta^a})^2}\delta_{ij}, \tag{8}$$

where $\Gamma^{(k)}_{ij}$ denotes the Fisher matrix from the k-th detector.

2.2. Current Bounds

We now review bounds on the ppE parameters from the observed GW events to date. Figure 1 presents upper bounds on β as a function of the PN order the leading correction enters, for GW150914 and GW151226. Observe that GW151226 gives stronger bounds than GW150914 due to a larger number of GW cycles and smaller relative velocity of BHs. For comparison, we also show bounds from solar system experiments and binary pulsar observations. Notice that GW observations have an advantage on probing positive PN corrections over binary pulsar observations. The solar system bound at 1PN is much stronger than the GW bounds, though the former can only probe corrections in the conservative sector (modifications to the binding energy and Kepler's law) while the latter can probe both the conservative and dissipative sectors (GW emission).

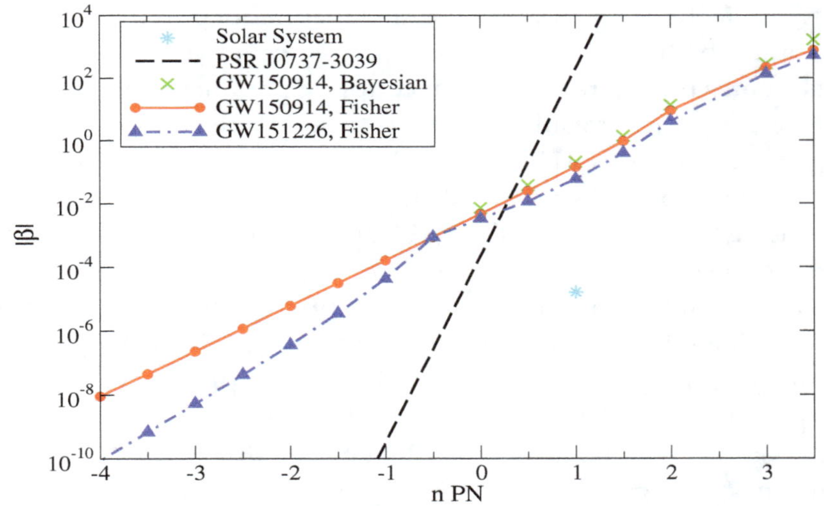

Figure 1. The 90% credible upper bounds on the parameterized post-Einsteinian (ppE) parameter β at each PN order the correction enters, using solar system experiments [56] (cyan star), binary pulsar observations [57] (black dashed), GW150914 with Bayesian [17] (green crosses) and Fisher [58] (red solid) analyses, and GW151226 with a Fisher analysis [58] (blue dotted-dashed). This figure is taken from [39,58].

We next map the bounds on the ppE parameter in Figure 1 to those on example modified theories of gravity as summarized in Table 1. For example, the GW151226 bounds on EdGB are comparable to other existing bounds, while other bounds are typically weaker. However, the GW bounds have meaning as they are the first constraints obtained in the strong/dynamical field regime.

The bounds summarized in Table 1 are derived mainly from corrections in the waveform phase. We showed in [59] that amplitude corrections can give comparable bounds to those from phase corrections for massive binaries like GW150914, though inclusion of the former does not affect the bounds compared to the case where one only includes corrections in the phase, which justifies many previous works, e.g., [16,38,55,58,60–62].

Table 1. Each example theory (1st column) violates certain fundamental aspects of general relativity (GR) (2nd column: the strong equivalence principle (SEP), Lorentz invariance (LI), four-dimensional spacetime (4D), and massless gravitons ($m_g = 0$)) and the leading correction enters in the gravitational waveform at certain PN orders (3rd column). Each representative parameter (4th column) has been constrained from GW150914 (5th column), GW151226 (6th column) and from other observations (7th column). The top (bottom) row within massive graviton corresponds to modifications in the dynamical (propagation/conservative) sector. This table is taken and edited from [39,58].

Theory	GR Pillar	PN	Repr. Parameters	GW150914	GW151226	Other Bounds
EdGB scalar-tensor	SEP	-1	$\sqrt{\lvert\alpha_{\rm EdGB}\rvert}$ [km] $\lvert\dot{\phi}\rvert$ [1/sec]	— —	5.7 [31], 4.3 [59], 3.5 [63] 1.1×10^4 [59]	10^7 [64], 2 [65–67] 10^{-6} [32]
dCS	SEP, LI	$+2$	$\sqrt{\lvert\alpha_{\rm dCS}\rvert}$ [km]	—	—	10^8 [68,69]
Time-Varying M	4D	-4	$\dot{M}$ [$M_\odot$/yr]	$\mathbf{4.2 \times 10^8}$	$\mathbf{5.3 \times 10^6}$	—
Time-Varying G	SEP	-4	$\lvert\dot{G}\rvert$ [10^{-12}/yr]	$\mathbf{5.4 \times 10^{18}}$ [58] $\mathbf{7.2 \times 10^{18}}$ [59]	$\mathbf{1.7 \times 10^{17}}$ [58] $\mathbf{2.2 \times 10^{16}}$ [59]	0.1–1 [70–73]
Massive graviton	$m_g = 0$	-3 $+1$	m_g [eV]	$\mathbf{6.4 \times 10^{-14}}$ $\mathbf{10^{-22}}$ [16,17]	$\mathbf{10^{-14}}$ [74], $\mathbf{3.1 \times 10^{-14}}$ $\mathbf{2.9 \times 10^{-22}}$ [17,77]	10^{-21}–10^{-19} [75,76] 10^{-30}–10^{-23} [78–83]

2.3. Future Bounds

Now that we have discussed the current status of parameterized tests of GR, let us now focus our attention on the future prospects of such tests [54,55,61,84–88]. Chamberlain and Yunes [86] considered the theoretical physics implications on various modified theories of gravity from BBH mergers detected by future GW detectors, which nicely complements [58] reviewed in Section 2.2. In [87,88], we similarly presented estimates on future bounds on coupling parameters for various modified theories of gravity. Here, we sum up these results for the future GW detectors Cosmic Explorer [19] (CE), LISA [21], TianQin [23], B-DECIGO [25], and DECIGO [24]. The first detector considered is a future-planned, third-generation ground-based detector with roughly $\sim$ 100 times the sensitivity of the advanced LIGO design sensitivity (aLIGO) [19], and the last four are future-planned space-based detectors. The former is exceedingly efficient at probing GWs in the high frequency regime (1–10^4 Hz), while the latter have larger arm lengths allowing them to proficiently probe the lower frequency bands (10^{-4}–1 Hz for LISA and TianQin, and 10^{-2}–10^2 Hz for B-DECIGO and DECIGO).

In this document, we summarize the results of [87,88], displaying constraints on the following modified theories of gravity (together with the theoretical constant and the PN order at which the leading correction enters): Einstein–dilaton Gauss–Bonnet (EdGB) gravity [26,27] ($\alpha_{\rm EdGB}$, -1PN order), dynamical Chern–Simons (dCS) gravity [28–31] ($\alpha_{\rm dCS}$, +2PN order), scalar-tensor theories [32,33] ($\dot{\phi}$, -1PN order), noncommutative gravities [34,35] (Λ, +2PN order), time-varying G theories of gravity [9,36,37] ($\dot{G}$, -4PN order), varying BH mass theories of gravity [38,39] ($\dot{M}$, -4PN order), massive graviton via dynamical effects [43,75] (m_g, -3PN), and massive graviton via the modified dispersion relation of the graviton [41] (m_g, $+1$PN order). See Berti et al. [39,60] for a comprehensive summary regarding these theories of gravity, as well as Tahura and Yagi's [37] summary of the ppE expressions used here.

Figure 2 (blue and maroon data points) displays the resulting constraints from GW150914-like events on each modified theory of gravity for future GW detectors CE, LISA, TianQin, B-DECIGO, and DECIGO. Additionally shown are the current observational constraints found in the literature. We observe that bounds on EdGB gravity can be improved upon with all four space-based detectors, dCS gravity can only be improved upon with DECIGO (further, CE, LISA, TianQin, and B-DECIGO do not satisfy the small-coupling approximation used to derive corrections to the waveform and thus no valid bounds can be placed), noncommutative gravities can be improved upon with all five future GW detectors considered here, and massive graviton bounds (dynamical and propagation) can be improved upon only with DECIGO. In general, ground-based (space-based) detectors have the advantage on probing corrections entering at positive (negative) PN orders for GW150914-like events. See also [89] for future prospects on probing EdGB gravity and scalar-tensor theories with a mixed binary consisting of one BH and one NS.

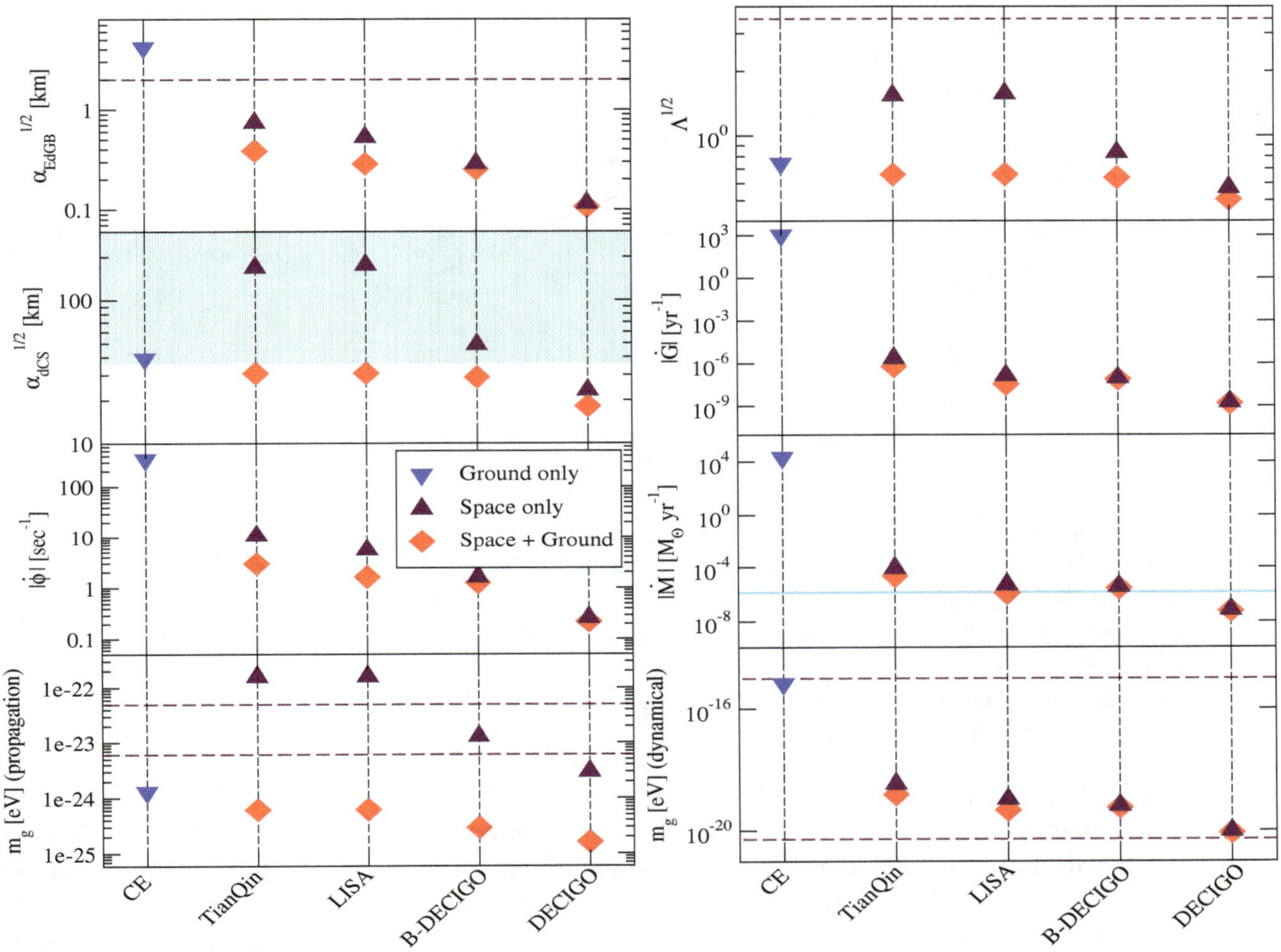

Figure 2. The 90% upper-bound credible level constraints on the parameters representative of the modified theories of gravity considered in [88] for GW150914-like events. Bounds are presented for Einstein dilaton Gauss–Bonnet (EdGB) gravity, dynamical Chern–Simons (dCS) gravity, scalar tensor theories, noncommutative gravity, varying-*G* theories, black hole (BH) mass-varying theories, and massive graviton (dynamical and propagation). For EdGB, dCS and scalar-tensor theories, the bounds are only meaningful outside of the blue shaded region where the small coupling approximations are violated. The dashed maroon lines correspond to the current bounds in the literature. The cyan line in the second-to-last right panel corresponds to the Eddington accretion rate: the maximum rate GW150914-like events can accrete in-falling matter under spherical symmetry. This figure is taken and edited from [88].

2.4. Multi-Band Bounds

In this section, we follow up the previous section by considering the combination of both observations from space and Earth, enabling the so called multi-band observations [87,88,90–93], which allows one to constrain modified theories of gravity entering at all PN orders. Following the observation of GW150914, Sesana [94] showed how joint multi-band observations of GW150914-like events could be made with both LISA and ground-based detectors. These events would first be observed in their early inspiral stage by space-based interferometers before leaving the space-band at 1 Hz ($\sim$ 100 Hz for B-DECIGO and DECIGO) for several months before entering the ground band again to merge at $\sim$ 300 Hz, as can be seen in Figure 3. The multi-band event rates for these objects have been found to be on the order of $\mathcal{O}(1)$ by Gerosa et al. [95], due to various technical details previously unconsidered [94,96]. It was similarly shown in [96,97] that multi-band observations could be made for more massive BBHs, as well as with binary NSs [25].

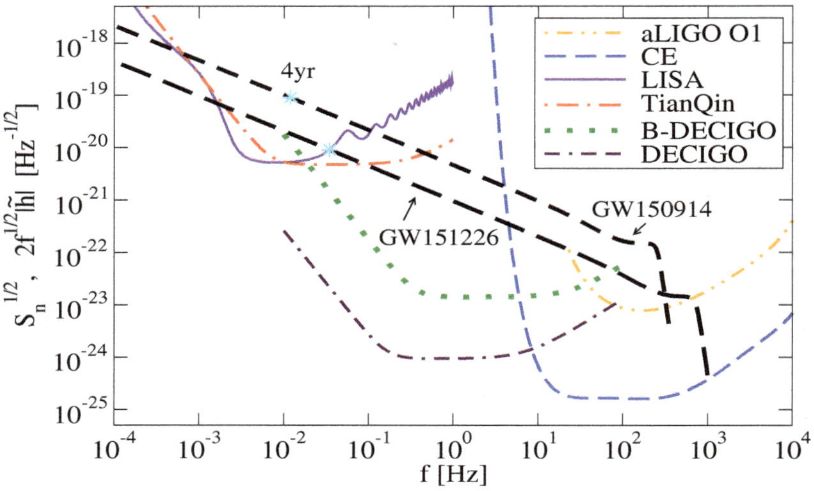

Figure 3. The (square root of) spectral noise densities $\sqrt{S_n(f)}$ of the gravitational-wave interferometers discussed in this document. The characteristic amplitudes $2\sqrt{f}|\tilde{h}(f)|$ for both events GW150914 and GW151226 are also displayed, with four years prior to merger shown as cyan stars. The ratio between $2\sqrt{f}|\tilde{h}(f)|$ and $\sqrt{S_n(f)}$ roughly corresponds to the signal-to-noise-ratio (SNR) of the event. Observe how the early inspiral portions of the BH coalescences are observed by the space-based detectors, while the late inspiral and merger-ringdown portions are observed by the ground-based detectors. This figure is taken and edited from [87].

In addition to providing more effective probes of gravity, multi-band observations have a myriad of other useful applications. Foremost, the early detections of binary coalescences could give alert to both ground-based detectors and electromagnetic telescopes for follow-up observations of the merger-ringdown event [94]. The former will also allow one to optimize ground-based GW detector sensitivities to further improve upon tests of GR [98]. On the other hand, successful observations of merger-ringdown events with ground-based detectors could allow one to revisit old space-based data and recover sub-threshold events [99], lowering the SNR threshold from 15 to 9 [100] for space-based detectors, which can result in an increased total number of detections [96,99,100]. Finally, the multi-band GW observations of coalescence events have been shown to improve upon the measurement accuracy of several binary parameters, in particular the masses, spins and sky-positions [90,92,96,101].

The red data points in Figure 2 summarize the results determined in [87,88], for the constraint of the eight modified theories of gravity considered here. We observe that, regardless of the PN order at which each effect enters the gravitational waveform, the multi-band observation can improve upon bounds obtained from either the space-based or ground-based detections alone. In particular for the case of dCS gravity, we see that the single-band observations with either detector type fails to satisfy the small coupling approximation (with the exception of DECIGO). Only when utilizing multi-band detections will this approximation become valid, allowing for constraints on $\sqrt{\alpha_{\text{dCS}}}$ to be placed, several orders-of-magnitude stronger than the current constraints. We also observe that several other alternative theories of gravity can be constrained stronger than the current bounds found in the literature with multi-band observations. We refer to [88] for a comprehensive list of constraints presented here for both single- and multi-band observations.

3. Inspiral-Merger-Ringdown Consistency Tests

Let us next test the consistency between the inspiral and merger-ringdown parts of GW signals in a theory-agnostic method. This test, aptly named the inspiral-merger-ringdown (IMR) consistency test [16,17,102–104], allows one to independently compare the two parts of the signal, assuming GR is correct. This is accomplished by estimating the remnant BH's mass M_f and spin χ_f from each portion of the waveform, and comparing the two. Any statistically significant inconsistencies between the two could be presented as evidence for deviations from GR.

3.1. Formulation

In this section, we discuss the formulation and techniques used to carry out the IMR consistency test of GR. We make an assumption that GR is the correct theory of gravity and use the GR waveform templates. To begin, the entire IMR GW signal is divided into the inspiral (I) and merger-ringdown (MR) portions. The transitional frequency between the two is defined to be $f_{\rm trans} = 132$ Hz for GW150914-like events [17]. Then, one can estimate the four-dimensional probability distributions $P_{\rm I}(m_1, m_2, \chi_1, \chi_2)$ and $P_{\rm MR}(m_1, m_2, \chi_1, \chi_2)$ between the BH masses m_A and spins χ_A from each portion. Such distributions can be obtained with a comprehensive Bayesian analysis as was done in [16,17,102–104], or approximated with the simpler Fisher analysis techniques [52] discussed in Section 2.1, which will be used here. As a result, the probability distributions will take a Gaussian form, centered at the injected masses and spins. Following this, the numerical relativity (NR) fits obtained in [50] in GR allows one to predict the remnant BH's mass $M_f^{\rm I,MR}(m_1, m_2, \chi_1, \chi_2)$ and spin $\chi_f^{\rm I,MR}(m_1, m_2, \chi_1, \chi_2)$ from each waveform, entirely from the constituent BH masses and spins. A Jacobian transformation matrix constructed out of such NR fits freely transforms the four-dimensional probability distributions obtained previously into the two-dimensional probability distributions $P_{\rm I}(M_f, \chi_f)$ and $P_{\rm MR}(M_f, \chi_f)$. Finally, the consistency between these two distributions provides valuable insight about the gravitational nature of the signal as compared to the assumed theory of GR. Any inconsistencies between the two may point to a modified theory of gravity presenting itself somewhere throughout the entire GW signal.

Typically, the agreement between the two probability distributions above can be measured by once again transforming them together into a joint-probability distribution. We define the new variables ϵ and σ as

$$\epsilon \equiv \frac{\Delta M_f}{\bar{M}_f} \equiv 2\frac{M_f^{\rm I} - M_f^{\rm MR}}{M_f^{\rm I} + M_f^{\rm MR}}, \qquad \sigma \equiv \frac{\Delta \chi_f}{\bar{\chi}_f} \equiv 2\frac{\chi_f^{\rm I} - \chi_f^{\rm MR}}{\chi_f^{\rm I} + \chi_f^{\rm MR}}. \tag{9}$$

Here $\Delta M_f \equiv M_f^{\rm I} - M_f^{\rm MR}$ and $\Delta\chi_f \equiv \chi_f^{\rm I} - \chi_f^{\rm MR}$ describe the differences in the final mass and spin estimates between the inspiral and merger-ringdown signals under the GR assumption, and $\bar{M}_f \equiv \frac{1}{2}(M_f^{\rm I} + M_f^{\rm MR})$ and $\bar{\chi}_f \equiv \frac{1}{2}(\chi_f^{\rm I} + \chi_f^{\rm MR})$ are the averages between the two. The appendix of [104] describes how the probability distribution of ϵ and σ is derived from the following marginalizations:

$$P(\epsilon, \sigma) = \int_0^1 \int_0^\infty P_{\rm I}\left(\left[1 + \frac{\epsilon}{2}\right]\bar{M}_f, \left[1 + \frac{\sigma}{2}\right]\bar{\chi}_f\right) \times P_{\rm MR}\left(\left[1 - \frac{\epsilon}{2}\right]\bar{M}_f, \left[1 - \frac{\sigma}{2}\right]\bar{\chi}_f\right) \bar{M}_f \bar{\chi}_f d\bar{M}_f d\bar{\chi}_f. \tag{10}$$

Finally, the agreement of the resulting probability distribution in the $\epsilon - \sigma$ plane with the GR value of $(\epsilon, \sigma)|_{\rm GR} = (0, 0)$ determines how consistent the GW signal is with the predictions of GR.

3.2. Current Bounds

Let us now discuss the current status of the IMR consistency tests with the BBH mergers observed thus far. Using a full Bayesian analysis, Abbott et al. [17] performed the IMR consistency test on the LVC catalog of BBH merger events. All such events were found to be statistically consistent with the predictions of GR. While this does not point towards any modifications to GR, such deviations could still potentially be buried within the relatively large statistical noise found within the current generation (O1 and O2) of LIGO-Virgo interferometers. Additionally, Ghosh et al. [102] has discussed testing GR with the IMR consistency test using golden BBH events, as well as adding simulations of modified GR signals in a phenomenological manner [104].

The IMR consistency test has been performed yet again in [87,88] for GW150914-like events, using a simplified Fisher analysis. Figure 4 presents the 90% credible level contours in the $\epsilon - \sigma$ plane. Observe how the two contours for LIGO O1 show good agreement between the Bayesian and Fisher analyses. In order to reveal the resolving power one can gain upon future detections on upgraded interferometers, we focus on the area of these contours. Such resolving power is indicative of how well one can effectively discriminate between GR and non-GR effects entering the gravitational waveform. Once the area of such contours becomes small enough, potential deviations from GR may become highlighted. The top portion of Table 2 compares these areas for the LIGO O1 Bayesian and Fisher results. Observe that the resulting areas agree very well with each other, up to $\sim 10\%$. This indicates that the Fisher analysis IMR consistency tests presented in [87,88] can be trusted to agree somewhat well with their Bayesian counterpart.

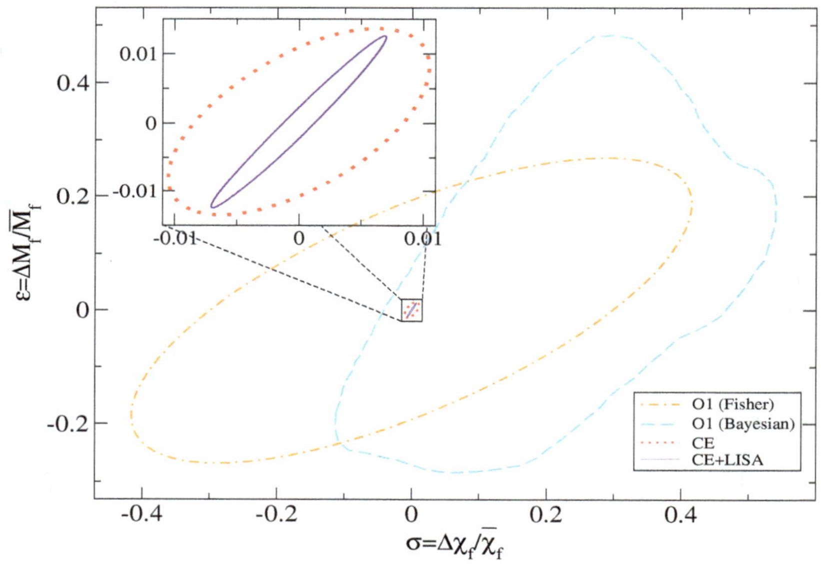

Figure 4. The 90% credible region contours of the transformed probability distributions in the $\epsilon - \sigma$ plane, describing the consistency of the remnant mass and spin general relativity (GR) predictions between the inspiral and merger-ringdown waveforms for GW150914-like events. Here we display the results for LIGO O1 (Fisher [87,88] and Bayesian [17] for comparison), CE, and the multi-band observation of CE and LISA. The areas of such confidence regions are displayed in Table 2, and show the following: (i) good agreement within $\sim$ 10% between the Fisher and Bayesian analyses, (ii) three orders-of-magnitude improvement from LIGO O1 to CE, and (iii) up to an additional order-of-magnitude improvement with multi-band observations. This figure is taken and edited from [88].

3.3. Future Bounds

In this section, we discuss the future prospects for the IMR consistency test, with upgraded third-generation ground-based GW detectors CE. We do not consider space-based interferometers as they fail to probe the merger-ringdown portion of GW150914-like events, which makes such interferometers incompatible with the IMR consistency test. However, we refer our readers to a work by Hughes and Menou [105], where they described the compatibility of this test with supermassive BBHs observed on space-based detectors.

We summarize our results in Figure 4 and Table 2. Observe that detections of future GW150914-like events by CE can increase the effective non-GR resolving power by up to three orders of magnitude. Such an increase in discriminating power could potentially shed light on any minuscule deviations from GR which could currently be hiding within the detector noise.

3.4. Multi-Band Bounds

Here, we discuss how one can further improve upon the IMR consistency test presented in the previous section by making use of multi-band observations between space- and ground-based detectors. While space-based detectors can not observe the merger-ringdown signal for GW150914-like events, they can indeed probe the early inspiral of such events. The multi-band IMR consistency test is performed by first combining the inspiral signal from both the ground-based detector CE, and space-based detectors such as LISA, TianQin, B-DECIGO, and DECIGO. The merger-ringdown portion of the IMR consistency test can then be obtained from the ground-based detector CE alone. The remainder of the test proceeds as before, allowing us to place contours in the $\epsilon - \sigma$ plane for multi-band observations.

Again, the results are summarized in Figure 4 and Table 2 for the multi-band observations between CE and LISA. The former shows the resulting probability distributions in the $\epsilon - \sigma$ plane for such combined multi-band signals, as obtained in [87,88]. Further, the bottom portion of Table 2 presents the resulting areas of the 90% confidence regions from the multi-band observations between the ground-based detector CE and space-based detector LISA. Observe how, in addition to the three-order-of-magnitude improvement made for the future detector CE alone, improvement of an additional factor of about seven can be made by further considering multi-band observations. Moreover, we also found that the multi-band observations with other space-based detectors TianQin, B-DECIGO, and DECIGO, show similar multiplicative improvements in the range of seven to ten [87,88]. Such large improvements may prove to be crucial for future GW observations in highlighting potential deviations from GR which may be small enough to not be visible through CE observations alone.

Table 2. Resulting areas of the 90% confidence ellipses from the $\epsilon - \sigma$ posterior distributions for GW150914-like events found in Figure 4, as obtained in [87,88].

Detector	90% Area
LIGO O1 (Fisher)	0.25
LIGO O1 (Bayesian) [17]	0.29
CE	3.6×10^{-4}
LISA+CE	5.0×10^{-5}

4. Conclusions and Open Questions

In the present communication, we have reviewed the present and future considerations for testing GR with gravitational waves. Non-GR effects may only become actively dominant in extreme-gravity regimes, such as the coalescences of orbiting BHs and/or NSs, which may be effectively probed through the gravitational wave observations of such events. To date, 11 confirmed events have been detected [13–15], and none have thus far been identified to deviate from Einstein's GR [16,17]. However, hope still exists in finding such deviations from GR — these effects, however small they may be, could very well be hidden within the relatively large statistical uncertainties dominant in the current LIGO/Virgo infrastructure. For this reason, many future space-based and ground-based GW interferometers have been proposed, planned, and even funded. With detector noises reaching up to ~ 100 times more sensitive than the current LIGO O2 generation of detectors, in both the low- and high-frequency bands, these detectors stand increasingly large chances of probing these elusive effects in the GW signal. With such detectors, the two fronts of GR tests discussed in this document can be pushed even further than ever done before. We showed that constraints found from several parameterized tests of GR can be improved upon by several orders-of-magnitude with future GW detectors, as well as multi-band observations. Further, we showed that the IMR consistency test can gain many orders-of-magnitude improvement in the resolving power between GR and non-GR effects with such considerations. Together, these improvements can push the bounds formed on many proposed modified theories of gravity.

We end this article by listing several issues that need to be improved further:

1. *Higher PN corrections*: In many cases, the mapping between the ppE parameters and theoretical constants are known only to the leading PN order. However, as compact binaries come close to coalescence, the PN approximation breaks down, and thus it is important to derive and implement higher PN corrections.
2. *Merger-ringdown corrections*: One also needs to include non-GR corrections in the merger-ringdown phases to have complete waveform templates in theories beyond GR. To do so, one needs to to carry out numerical relativity simulations of binary mergers in such theories. Several groups are making progress in this direction [106–114]. Another approach is to extend the effective-one-body waveforms to non-GR theories [115–117].
3. *Precessing/eccentric orbits*: We have focused on spin-aligned binaries with circular orbits. It would be important to extend the analyses described here to more exotic binaries with strong spin-precession [118,119] and largely eccentric orbits.
4. *Cosmological screening*: If one wants to test theories motivated from cosmology, one may need to consider how screening mechanisms affect the GW emission from compact binaries [120–122].
5. *Stacking*: In the future, we expect to have thousands of detections. Thus, we need to study how much improvement one gains in terms of tests of GR with GWs by appropriately stacking multiple events [61,89,123–126].
6. *Sensitivities*: Additional radiation in non-GR theories, such as scalar or dipolar radiation, is typically controlled by the sensitivities of compact bodies [27,127–131]. Currently, BH sensitivities have not been calculated yet in e.g., Einstein-Æther theory and NS sensitivities in this theory need to be revisited within the allowed parameter region after GW170817 [132].

Author Contributions: Conceptualization, K.Y.; Calculation, Z.C.; Writing, Z.C. & K.Y.; Visualization, Z.C.; Supervision, K.Y.; Funding Acquisition, K.Y.

References

1. Clifton, T.; Ferreira, P.G.; Padilla, A.; Skordis, C. Modified Gravity and Cosmology. *Phys. Rept.* **2012**, *513*, 1–189. doi:10.1016/j.physrep.2012.01.001.
2. Joyce, A.; Jain, B.; Khoury, J.; Trodden, M. Beyond the Cosmological Standard Model. *Phys. Rept.* **2015**, *568*, 1–98, doi:10.1016/j.physrep.2014.12.002.
3. Famaey, B.; McGaugh, S. Modified Newtonian Dynamics (MOND): Observational Phenomenology and Relativistic Extensions. *Living Rev. Rel.* **2012**, *15*, 10, doi:10.12942/lrr-2012-10.
4. Milgrom, M. The MOND paradigm. *arXiv* **2008**, arXiv:astro-ph/0801.3133.
5. Jain, B.; Khoury, J. Cosmological Tests of Gravity. *Ann. Phys.* **2010**, *325*, 1479–1516, doi:10.1016/j.aop.2010.04.002.
6. Koyama, K. Cosmological Tests of Modified Gravity. *Rept. Prog. Phys.* **2016**, *79*, 046902, doi:10.1088/0034-4885/79/4/046902.
7. Milgrom, M. A modification of the Newtonian dynamics as a possible alternative to the hidden mass hypothesis. *Astrophys. J.* **1983**, *270*, 365–370. doi:10.1086/161130.

8. Salvatelli, V.; Piazza, F.; Marinoni, C. Constraints on modified gravity from Planck 2015: When the health of your theory makes the difference. *JCAP* **2016**, *1609*, 027, doi:10.1088/1475-7516/2016/09/027.
9. Will, C.M. The Confrontation between General Relativity and Experiment. *Living Rev. Relativ.* **2014**, *17*, 4, doi:10.12942/lrr-2014-4.
10. Stairs, I.H. Testing general relativity with pulsar timing. *Living Rev. Relativ.* **2003**, *6*, 5, doi:10.12942/lrr-2003-5.
11. Wex, N. Testing Relativistic Gravity with Radio Pulsars. *arXiv* **2014**, arXiv:gr-qc/1402.5594.
12. Ferreira, P.G. Cosmological Tests of Gravity. *arXiv* **2019**, arXiv:astro-ph.CO/1902.10503.
13. Abbott, B.P.; Abbott R.; Abbott, T.D.; Abernathy, M.R.; Acernese, F.; Ackley, K.; Adams, C.; Adams, T.; Addesso, P.; Adhikari, R.X.; et al. Properties of the Binary Black Hole Merger GW150914. *Phys. Rev. Lett.* **2016**, *116*, 241102, doi:10.1103/PhysRevLett.116.241102.
14. Abbott, B.P.; Abbott, R.; Abbott, T.D.; Abraham, S.; Acernese, F.; Ackley, K.; Adams, C.; Adhikari, R.X.; Adya, V.B.; Affeldt, C.; et al. GWTC-1: A Gravitational-Wave Transient Catalog of Compact Binary Mergers Observed by LIGO and Virgo during the First and Second Observing Runs. *arXiv* **2018**, arXiv:astro-ph.HE/1811.12907.
15. Abbott, B.P.; LIGO Scientific Collaboration; Virgo Collaboration. GW170817: Observation of Gravitational Waves from a Binary Neutron Star Inspiral. *Phys. Rev. Lett.* **2017**, *119*, 161101, doi:10.1103/PhysRevLett.119.161101.
16. Abbott, B.P.; Abbott, R.; Abbott, T.D.; Abernathy, M.R.; Acernese, F.; Ackley, K.; Adams, C.; Adams, T.; Addesso, P.; Adhikari, R.X.; et al. Tests of general relativity with GW150914. *Phys. Rev. Lett.* **2016**, *116*, 221101, doi:10.1103/PhysRevLett.116.221101.
17. Abbott, B.P.; Abbott, R.; Abbott, T.D.; Abraham, S.; Acernese, F.; Ackley, K.; Adams, C.; Adhikari, R.X.; Adya, V.B.; Affeldt, C.; et al. Tests of General Relativity with the Binary Black Hole Signals from the LIGO-Virgo Catalog GWTC-1. *arXiv* **2019**, arXiv:gr-qc/1903.04467.
18. Advanced LIGO. Available online: https://www.advancedligo.mit.edu/ (accessed on 10 January 2019).
19. Ligo-t1400316-v4: Instrument Science White Paper. Available online: https://dcc.ligo.org/ligo-T1400316/public (accessed on 10 January 2019).
20. The ET Project Website. Available online: http://www.et-gw.eu/ (accessed on 10 January 2019).
21. Robson, T.; Cornish, N.; Liu, C. The construction and use of LISA sensitivity curves. *Class. Quant. Grav.* **2019**, *36*, 105011, doi:10.1088/1361-6382/ab1101.
22. Luo, J.; Chen, L.S.; Duan, H.Z.; Gong, Y.G.; Hu, S.; Ji, J.; Liu, Q.; Mei, J.; Milyukov, V.; Sazhin, M.; et al. TianQin: A space-borne gravitational wave detector. *Class. Quantum Gravity* **2016**, *33*, 035010, doi:10.1088/0264-9381/33/3/035010.
23. Shi, C.; Bao, J.; Wang, H.; Zhang, J.d.; Hu, Y.; Sesana, A.; Barausse, E.; Mei, J.; Luo, J. Science with TianQin: Preliminary Results on Testing the No-hair Theorem with Ringdown Signals. *arXiv* **2019**, arXiv:gr-qc/1902.08922.
24. Yagi, K.; Seto, N. Detector configuration of DECIGO/BBO and identification of cosmological neutron-star binaries. *Phys. Rev.* **2011**, *D83*, 044011, doi:10.1103/PhysRevD.95.109901.
25. Isoyama, S.; Nakano, H.; Nakamura, T. Multiband Gravitational-Wave Astronomy: Observing binary inspirals with a decihertz detector, B-DECIGO. *PTEP* **2018**, *2018*, 073E01, doi:10.1093/ptep/pty078.
26. Maeda, K.I.; Ohta, N.; Sasagawa, Y. Black Hole Solutions in String Theory with Gauss-Bonnet Curvature Correction. *Phys. Rev.* **2009**, *D80*, 104032, doi:10.1103/PhysRevD.80.104032.
27. Yagi, K.; Stein, L.C.; Yunes, N.; Tanaka, T. Post-Newtonian, Quasi-Circular Binary Inspirals in Quadratic Modified Gravity. *Phys. Rev.* **2012**, *D85*, 064022, doi:10.1103/PhysRevD.93.029902.
28. Jackiw, R.; Pi, S.Y. Chern-Simons modification of general relativity. *Phys. Rev.* **2003**, *D68*, 104012, doi:10.1103/PhysRevD.68.104012.
29. Alexander, S.; Yunes, N. Chern-Simons Modified General Relativity. *Phys. Rept.* **2009**, *480*, 1–55, [arXiv:hep-th/0907.2562]. doi:10.1016/j.physrep.2009.07.002.
30. Yagi, K.; Yunes, N.; Tanaka, T. Gravitational Waves from Quasi-Circular Black Hole Binaries in Dynamical Chern-Simons Gravity. *Phys. Rev. Lett.* **2012**, *109*, 251105, doi:10.1103/PhysRevLett.116.169902.
31. Nair, R.; Perkins, S.; Silva, H.O.; Yunes, N. Fundamental physics implications on higher-curvature theories from the binary black hole signals in the LIGO-Virgo Catalog GWTC-1. *arXiv* **2019**, arXiv:gr-qc/1905.00870.
32. Horbatsch, M.W.; Burgess, C.P. Cosmic Black-Hole Hair Growth and Quasar OJ287. *JCAP* **2012**, *1205*, 010, doi:10.1088/1475-7516/2012/05/010.

33. Jacobson, T. Primordial black hole evolution in tensor scalar cosmology. *Phys. Rev. Lett.* **1999**, *83*, 2699–2702, doi:10.1103/PhysRevLett.83.2699.
34. Harikumar, E.; Rivelles, V.O. Noncommutative Gravity. *Class. Quant. Grav.* **2006**, *23*, 7551–7560, doi:10.1088/0264-9381/23/24/024.
35. Kobakhidze, A.; Lagger, C.; Manning, A. Constraining noncommutative spacetime from GW150914. *Phys. Rev.* **2016**, *D94*, 064033, doi:10.1103/PhysRevD.94.064033.
36. Yunes, N.; Pretorius, F.; Spergel, D. Constraining the evolutionary history of Newton's constant with gravitational wave observations. *Phys. Rev.* **2010**, *D81*, 064018, doi:10.1103/PhysRevD.81.064018.
37. Tahura, S.; Yagi, K. Parameterized Post-Einsteinian Gravitational Waveforms in Various Modified Theories of Gravity. *Phys. Rev.* **2018**, *D98*, 084042, doi:10.1103/PhysRevD.98.084042.
38. Yagi, K.; Tanahashi, N.; Tanaka, T. Probing the size of extra dimension with gravitational wave astronomy. *Phys. Rev.* **2011**, *D83*, 084036, doi:10.1103/PhysRevD.83.084036.
39. Berti, E.; Yagi, K.; Yunes, N. Extreme Gravity Tests with Gravitational Waves from Compact Binary Coalescences: (I) Inspiral-Merger. *Gen. Rel. Grav.* **2018**, *50*, 46, doi:10.1007/s10714-018-2362-8.
40. Will, C.M. Bounding the mass of the graviton using gravitational wave observations of inspiralling compact binaries. *Phys. Rev.* **1998**, *D57*, 2061–2068, doi:10.1103/PhysRevD.57.2061.
41. Mirshekari, S.; Yunes, N.; Will, C.M. Constraining Generic Lorentz Violation and the Speed of the Graviton with Gravitational Waves. *Phys. Rev.* **2012**, *D85*, 024041, doi:10.1103/PhysRevD.85.024041.
42. De Rham, C. Massive Gravity. *Living Rev. Relativ.* **2014**, *17*, 7, doi:10.12942/lrr-2014-7.
43. Zhang, J.; Zhou, S.Y. Can the graviton have a large mass near black holes? *Phys. Rev.* **2018**, *D97*, 081501, doi:10.1103/PhysRevD.97.081501.
44. Arun, K.G.; Iyer, B.R.; Qusailah, M.S.S.; Sathyaprakash, B.S. Testing post-Newtonian theory with gravitational wave observations. *Class. Quant. Grav.* **2006**, *23*, L37–L43, doi:10.1088/0264-9381/23/9/L01.
45. Arun, K.G.; Iyer, B.R.; Qusailah, M.S.S.; Sathyaprakash, B.S. Probing the non-linear structure of general relativity with black hole binaries. *Phys. Rev.* **2006**, *D74*, 024006, doi:10.1103/PhysRevD.74.024006.
46. Mishra, C.K.; Arun, K.G.; Iyer, B.R.; Sathyaprakash, B.S. Parametrized tests of post-Newtonian theory using Advanced LIGO and Einstein Telescope. *Phys. Rev.* **2010**, *D82*, 064010, doi:10.1103/PhysRevD.82.064010.
47. Yunes, N.; Pretorius, F. Fundamental Theoretical Bias in Gravitational Wave Astrophysics and the Parameterized Post-Einsteinian Framework. *Phys. Rev.* **2009**, *D80*, 122003, doi:10.1103/PhysRevD.80.122003.
48. Agathos, M.; Del Pozzo, W.; Li, T.G.F.; Van Den Broeck, C.; Veitch, J.; Vitale, S. TIGER: A data analysis pipeline for testing the strong-field dynamics of general relativity with gravitational wave signals from coalescing compact binaries. *Phys. Rev.* **2014**, *D89*, 082001, doi:10.1103/PhysRevD.89.082001.
49. Meidam, J.; Agathos, M.; Van Den Broeck, C.; Veitch, J.; Sathyaprakash, B.S. Testing the no-hair theorem with black hole ringdowns using TIGER. *Phys. Rev.* **2014**, *D90*, 064009, doi:10.1103/PhysRevD.90.064009.
50. Husa, S.; Khan, S.; Hannam, M.; Pürrer, M.; Ohme, F.; Forteza, X.J.; Bohé, A. Frequency-domain gravitational waves from nonprecessing black-hole binaries. I. New numerical waveforms and anatomy of the signal. *Phys. Rev. D* **2016**, *93*, 044006, doi:10.1103/PhysRevD.93.044006.
51. Khan, S.; Husa, S.; Hannam, M.; Ohme, F.; Pürrer, M.; Forteza, X.J.; Bohé, A. Frequency-domain gravitational waves from nonprecessing black-hole binaries. II. A phenomenological model for the advanced detector era. *Phys. Rev. D* **2016**, *93*, 044007, doi:10.1103/PhysRevD.93.044007.
52. Cutler, C.; Flanagan, E.E. Gravitational waves from merging compact binaries: How accurately can one extract the binary's parameters from the inspiral waveform? *Phys. Rev. D* **1994**, *49*, 2658–2697, doi:10.1103/PhysRevD.49.2658.
53. Poisson, E.; Will, C.M. Gravitational waves from inspiraling compact binaries: Parameter estimation using second-post-Newtonian waveforms. *Phys. Rev. D* **1995**, *52*, 848–855, doi:10.1103/PhysRevD.52.848.
54. Berti, E.; Buonanno, A.; Will, C.M. Estimating spinning binary parameters and testing alternative theories of gravity with LISA. *Phys. Rev.* **2005**, *D71*, 084025, doi:10.1103/PhysRevD.71.084025.
55. Yagi, K.; Tanaka, T. Constraining alternative theories of gravity by gravitational waves from precessing eccentric compact binaries with LISA. *Phys. Rev.* **2010**, *D81*, 064008, doi:10.1103/PhysRevD.81.109902.
56. Sampson, L.; Yunes, N.; Cornish, N. Rosetta stone for parametrized tests of gravity. *Phys. Rev.* **2013**, *D88*, 064056, doi:10.1103/PhysRevD.88.064056.
57. Yunes, N.; Hughes, S.A. Binary Pulsar Constraints on the Parameterized post-Einsteinian Framework. *Phys. Rev.* **2010**, *D82*, 082002, doi:10.1103/PhysRevD.82.082002.

58. Yunes, N.; Yagi, K.; Pretorius, F. Theoretical physics implications of the binary black-hole mergers GW150914 and GW151226. *Phys. Rev. D* **2016**, *94*, 084002, doi:10.1103/PhysRevD.94.084002.
59. Tahura, S.; Yagi, K.; Carson, Z. Testing Gravity with Gravitational Waves from Binary Black Hole Mergers: Contributions from Amplitude Corrections. *arXiv* **2019**, arXiv:gr-qc/1907.10059.
60. Berti, E.; Barausse, E.; Cardoso, V.; Gualtieri, L.; Pani, P.; Sperhake, U.; Stein, L.C.; Wex, N.; Yagi, K.; Baker, T.; et al. Testing General Relativity with Present and Future Astrophysical Observations. *Class. Quant. Grav.* **2015**, *32*, 243001, doi:10.1088/0264-9381/32/24/243001.
61. Yagi, K.; Tanaka, T. DECIGO/BBO as a probe to constrain alternative theories of gravity. *Prog. Theor. Phys.* **2010**, *123*, 1069–1078, doi:10.1143/PTP.123.1069.
62. Abbott, B.P.; Abbott, R.; Abbott, T.D.; Acernese, F.; Ackley, K.; Adams, C.; Adams, T.; Addesso, P.; Adhikari, R.X.; Adya, V.B.; et al. Tests of General Relativity with GW170817. *arXiv* **2018**, arXiv:gr-qc/1811.00364.
63. Yamada, K.; Narikawa, T.; Tanaka, T. Testing massive-field modifications of gravity via gravitational waves. *arXiv* **2019**, arXiv:gr-qc/1905.11859.
64. Amendola, L.; Charmousis, C.; Davis, S. Solar System Constraints on Gauss-Bonnet Mediated Dark Energy. *JCAP* **2007**, *0710*, 004, doi:10.1088/1475-7516/2007/10/004.
65. Kanti, P.; Mavromatos, N.E.; Rizos, J.; Tamvakis, K.; Winstanley, E. Dilatonic black holes in higher curvature string gravity. *Phys. Rev.* **1996**, *D54*, 5049–5058, doi:10.1103/PhysRevD.54.5049.
66. Yagi, K. A New constraint on scalar Gauss-Bonnet gravity and a possible explanation for the excess of the orbital decay rate in a low-mass X-ray binary. *Phys. Rev.* **2012**, *D86*, 081504, doi:10.1103/PhysRevD.86.081504.
67. Pani, P.; Cardoso, V. Are black holes in alternative theories serious astrophysical candidates? The Case for Einstein-Dilaton-Gauss-Bonnet black holes. *Phys. Rev.* **2009**, *D79*, 084031, doi:10.1103/PhysRevD.79.084031.
68. Ali-Haimoud, Y.; Chen, Y. Slowly-rotating stars and black holes in dynamical Chern-Simons gravity. *Phys. Rev.* **2011**, *D84*, 124033, doi:10.1103/PhysRevD.84.124033.
69. Yagi, K.; Yunes, N.; Tanaka, T. Slowly Rotating Black Holes in Dynamical Chern-Simons Gravity: Deformation Quadratic in the Spin. *Phys. Rev.* **2012**, *D86*, 044037, doi:10.1103/PhysRevD.89.049902.
70. Manchester, R.N. Pulsars and Gravity. *Int. J. Mod. Phys.* **2015**, *D24*, 1530018, doi:10.1142/S0218271815300189.
71. Konopliv, A.S.; Asmar, S.; Folkner, W.M.; Karatekin, Z.; Nunes, D.; Smrekar, S.; Yoder, C.F.; Zuber, M.T. Mars High Resolution Gravity Fields from MRO, Mars Seasonal Gravity, and Other Dynamical Parameters. *Icarus* **2011**, *211*, 401–428. doi:10.1016/j.icarus.2010.10.004.
72. Copi, C.J.; Davis, A.N.; Krauss, L.M. A New nucleosynthesis constraint on the variation of G. *Phys. Rev. Lett.* **2004**, *92*, 171301, doi:10.1103/PhysRevLett.92.171301.
73. Bambi, C.; Giannotti, M.; Villante, F.L. The Response of primordial abundances to a general modification of G(N) and/or of the early Universe expansion rate. *Phys. Rev.* **2005**, *D71*, 123524, doi:10.1103/PhysRevD.71.123524.
74. Chung, K.W.; Li, T.G.F. A Phenomenological Inclusion of Alternative Dispersion Relations to the Teukolsky Equation and its Application to Bounding the Graviton Mass with Gravitational-wave Measurements. *arXiv* **2018**, arXiv:gr-qc/1808.04050.
75. Finn, L.S.; Sutton, P.J. Bounding the mass of the graviton using binary pulsar observations. *Phys. Rev.* **2002**, *D65*, 044022, doi:10.1103/PhysRevD.65.044022.
76. Miao, X.; Shao, L.; Ma, B.Q. Bounding the mass of graviton in a dynamic regime with binary pulsars. *arXiv* **2019**, arXiv:astro-ph.CO/1905.12836.
77. Abbott, B.P.; Abbott, R.; Abbott, T.D.; Abernathy, M.R.; Acernese, F.; Ackley, K.; Adams, C.; Adams, T.; Addesso, P.; Adhikari, R.X.; et al. Binary Black Hole Mergers in the first Advanced LIGO Observing Run. *Phys. Rev.* **2016**, *X6*, 041015, doi:10.1103/PhysRevX.6.041015.
78. Talmadge, C.; Berthias, J.P.; Hellings, R.W.; Standish, E.M. Model-Independent Constraints on Possible Modifications of Newtonian Gravity. *Phys. Rev. Lett.* **1988**, *61*, 1159–1162. doi:10.1103/PhysRevLett.61.1159.
79. Goldhaber, A.S.; Nieto, M.M. Mass of the graviton. *Phys. Rev. D* **1974**, *9*, 1119–1121. doi:10.1103/PhysRevD.9.1119.
80. Hare, M.G. Mass of the Graviton. *Can. J. Phys.* **1973**, *51*, 431–433, doi:10.1139/p73-056.
81. Brito, R.; Cardoso, V.; Pani, P. Massive spin-2 fields on black hole spacetimes: Instability of the Schwarzschild and Kerr solutions and bounds on the graviton mass. *Phys. Rev.* **2013**, *D88*, 023514, doi:10.1103/PhysRevD.88.023514.
82. Desai, S. Limit on graviton mass from galaxy cluster Abell 1689. *Phys. Lett.* **2018**, *B778*, 325–331, doi:10.1016/j.physletb.2018.01.052.

83. Gupta, S.; Desai, S. Limit on graviton mass using stacked galaxy cluster catalogs from SPT-SZ, Planck-SZ and SDSS-redMaPPer. *Annals Phys.* **2018**, *399*, 85–92, doi:10.1016/j.aop.2018.09.017.
84. Gair, J.R.; Vallisneri, M.; Larson, S.L.; Baker, J.G. Testing General Relativity with Low-Frequency, Space-Based Gravitational-Wave Detectors. *Living Rev. Rel.* **2013**, *16*, 7, doi:10.12942/lrr-2013-7.
85. Yagi, K. Scientific Potential of DECIGO Pathfinder and Testing GR with Space-Borne Gravitational Wave Interferometers. *Int. J. Mod. Phys.* **2013**, *D22*, 1341013, doi:10.1142/S0218271813410137.
86. Chamberlain, K.; Yunes, N. Theoretical Physics Implications of Gravitational Wave Observation with Future Detectors. *Phys. Rev.* **2017**, *D96*, 084039, doi:10.1103/PhysRevD.96.084039.
87. Carson, Z.; Yagi, K. Multi-band gravitational wave tests of general relativity. *arXiv* **2019**, arXiv:gr-qc/1905.13155.
88. Carson, Z.; Yagi, K. Future prospects for strong-field tests of gravity with multi-band gravitational wave observations. (In preparation)
89. Carson, Z.; Seymour, B.C.; Yagi, K. Future Prospects for Probing Scalar-Tensor Theories with Gravitational Waves from Mixed Binaries. *arXiv* **2019**, arXiv:gr-qc/1907.03897.
90. Nair, R.; Jhingan, S.; Tanaka, T. Synergy between ground and space based gravitational wave detectors for estimation of binary coalescence parameters. *PTEP* **2016**, *2016*, 053E01, doi:10.1093/ptep/ptw043.
91. Barausse, E.; Yunes, N.; Chamberlain, K. Theory-Agnostic Constraints on Black-Hole Dipole Radiation with Multiband Gravitational-Wave Astrophysics. *Phys. Rev. Lett.* **2016**, *116*, 241104, doi:10.1103/PhysRevLett.116.241104.
92. Vitale, S. Multiband Gravitational-Wave Astronomy: Parameter Estimation and Tests of General Relativity with Space- and Ground-Based Detectors. *Phys. Rev. Lett.* **2016**, *117*, 051102, doi:10.1103/PhysRevLett.117.051102.
93. Gnocchi, G.; Maselli, A.; Abdelsalhin, T.; Giacobbo, N.; Mapelli, M. Bounding Alternative Theories of Gravity with Multi-Band GW Observations. *arXiv* **2019**, arXiv:gr-qc/1905.13460.
94. Sesana, A. Prospects for Multiband Gravitational-Wave Astronomy after GW150914. *Phys. Rev. Lett.* **2016**, *116*, 231102, doi:10.1103/PhysRevLett.116.231102.
95. Gerosa, D.; Ma, S.; Wong, K.W.K.; Berti, E.; O'Shaughnessy, R.; Chen, Y.; Belczynski, K. Multiband gravitational-wave event rates and stellar physics. *Phys. Rev.* **2019**, *D99*, 103004, doi:10.1103/PhysRevD.99.103004.
96. Cutler, C.; Berti, E.; Jani, K.; Kovetz, E.D.; Randall, L.; Vitale, S.; Wong, K.W.K.; Holley-Bockelmann, K.; Larson, S.L.; Littenberg, T.; et al. What we can learn from multi-band observations of black hole binaries. *arXiv* **2019**, arXiv:astro-ph.HE/1903.04069.
97. Amaro-Seoane, P.; Santamaria, L. Detection of IMBHs with ground-based gravitational wave observatories: A biography of a binary of black holes, from birth to death. *Astrophys. J.* **2010**, *722*, 1197–1206, doi:10.1088/0004-637X/722/2/1197.
98. Tso, R.; Gerosa, D.; Chen, Y. Optimizing LIGO with LISA forewarnings to improve black-hole spectroscopy. *arXiv* **2018**, arXiv:gr-qc/1807.00075.
99. Wong, K.W.K.; Kovetz, E.D.; Cutler, C.; Berti, E. Expanding the LISA Horizon from the Ground. *Phys. Rev. Lett.* **2018**, *121*, 251102, doi:10.1103/PhysRevLett.121.251102.
100. Moore, C.J.; Gerosa, D.; Klein, A. Are stellar-mass black-hole binaries too quiet for LISA? *arXiv* **2019**, arXiv:astro-ph.HE/1905.11998.
101. Nair, R.; Tanaka, T. Synergy between ground and space based gravitational wave detectors. Part II: Localisation. *JCAP* **2018**, *1808*, 033, doi:10.1088/1475-7516/2018/08/033.
102. Ghosh, A.; Ghosh, A.; Johnson-McDaniel, N.K.; Mishra, C.K.; Ajith, P.; Del Pozzo, W.; Nichols, D.A.; Chen, Y.; Nielsen, A.B.; Berry, C.P.L.; et al. Testing general relativity using golden black-hole binaries. *Phys. Rev.* **2016**, *D94*, 021101, doi:10.1103/PhysRevD.94.021101.
103. Ghosh, A.; Johnson-Mcdaniel, N.K.; Ghosh, A.; Mishra, C.K.; Ajith, P.; Del Pozzo, W.; Berry, C.P.L.; Nielsen, A.B.; London, L. Testing general relativity using gravitational wave signals from the inspiral, merger and ringdown of binary black holes. *Class. Quantum Gravity* **2018**, *35*, 014002, doi:10.1088/1361-6382/aa972e.
104. Ghosh, A.; Johnson-McDaniel, N.K.; Ghosh, A.; Mishra, C.K.; Ajith, P.; Pozzo, W.D.; Berry, C.P.L.; Nielsen, A.B.; London, L. Testing general relativity using gravitational wave signals from the inspiral, merger and ringdown of binary black holes. *Class. Quantum Gravity* **2017**, *35*, 014002. doi:10.1088/1361-6382/aa972e.
105. Hughes, S.A.; Menou, K. Golden binaries for LISA: Robust probes of strong-field gravity. *Astrophys. J.* **2005**, *623*, 689–699, doi:10.1086/428826.

106. Healy, J.; Bode, T.; Haas, R.; Pazos, E.; Laguna, P.; Shoemaker, D.; Yunes, N. Late Inspiral and Merger of Binary Black Holes in Scalar-Tensor Theories of Gravity. *Class. Quantum Gravity* **2012**, *29*, 232002, doi:10.1088/0264-9381/29/23/232002.
107. Barausse, E.; Palenzuela, C.; Ponce, M.; Lehner, L. Neutron-star mergers in scalar-tensor theories of gravity. *Phys. Rev.* **2013**, *D87*, 081506, doi:10.1103/PhysRevD.87.081506.
108. Shibata, M.; Taniguchi, K.; Okawa, H.; Buonanno, A. Coalescence of binary neutron stars in a scalar-tensor theory of gravity. *Phys. Rev.* **2014**, *D89*, 084005, doi:10.1103/PhysRevD.89.084005.
109. Berti, E.; Cardoso, V.; Gualtieri, L.; Horbatsch, M.; Sperhake, U. Numerical simulations of single and binary black holes in scalar-tensor theories: Circumventing the no-hair theorem. *Phys. Rev.* **2013**, *D87*, 124020, doi:10.1103/PhysRevD.87.124020.
110. Hirschmann, E.W.; Lehner, L.; Liebling, S.L.; Palenzuela, C. Black Hole Dynamics in Einstein-Maxwell-Dilaton Theory. *Phys. Rev.* **2018**, *D97*, 064032, doi:10.1103/PhysRevD.97.064032.
111. Okounkova, M.; Stein, L.C.; Scheel, M.A.; Hemberger, D.A. Numerical binary black hole mergers in dynamical Chern-Simons gravity: Scalar field. *Phys. Rev.* **2017**, *D96*, 044020, doi:10.1103/PhysRevD.96.044020.
112. Okounkova, M.; Scheel, M.A.; Teukolsky, S.A. Evolving Metric Perturbations in dynamical Chern-Simons Gravity. *Phys. Rev.* **2019**, *D99*, 044019, doi:10.1103/PhysRevD.99.044019.
113. Okounkova, M.; Stein, L.C.; Scheel, M.A.; Teukolsky, S.A. Numerical binary black hole collisions in dynamical Chern-Simons gravity. *arXiv* **2019**, arXiv:gr-qc/1906.08789.
114. Witek, H.; Gualtieri, L.; Pani, P.; Sotiriou, T.P. Black holes and binary mergers in scalar Gauss-Bonnet gravity: Scalar field dynamics. *Phys. Rev.* **2019**, *D99*, 064035, doi:10.1103/PhysRevD.99.064035.
115. Julié, F.L.; Deruelle, N. Two-body problem in Scalar-Tensor theories as a deformation of General Relativity: An Effective-One-Body approach. *Phys. Rev.* **2017**, *D95*, 124054, doi:10.1103/PhysRevD.95.124054.
116. Julié, F.L. Reducing the two-body problem in scalar-tensor theories to the motion of a test particle: A scalar-tensor effective-one-body approach. *Phys. Rev.* **2018**, *D97*, 024047, doi:10.1103/PhysRevD.97.024047.
117. Khalil, M.; Sennett, N.; Steinhoff, J.; Vines, J.; Buonanno, A. Hairy binary black holes in Einstein-Maxwell-dilaton theory and their effective-one-body description. *Phys. Rev.* **2018**, *D98*, 104010, doi:10.1103/PhysRevD.98.104010.
118. Loutrel, N.; Tanaka, T.; Yunes, N. Scalar Tops and Perturbed Quadrupoles: Probing Fundamental Physics with Spin-Precessing Binaries. *Class. Quantum Gravity* **2019**, *36*, 10LT02, doi:10.1088/1361-6382/ab15fa.
119. Loutrel, N.; Tanaka, T.; Yunes, N. Spin-Precessing Black Hole Binaries in Dynamical Chern-Simons Gravity. *Phys. Rev.* **2018**, *D98*, 064020, doi:10.1103/PhysRevD.98.064020.
120. Perkins, S.; Yunes, N. Probing Screening and the Graviton Mass with Gravitational Waves. *Class. Quantum Gravity* **2019**, *36*, 055013, doi:10.1088/1361-6382/aafce6.
121. Zhang, X.; Liu, T.; Zhao, W. Gravitational radiation from compact binary systems in screened modified gravity. *Phys. Rev.* **2017**, *D95*, 104027, doi:10.1103/PhysRevD.95.104027.
122. Honardoost, M.; Khosravi, N.; Rahmanpour, N. Hairy Waves: Gravitational Waves in the presence of Screening Mechanism. *arXiv* **2019**, arXiv:gr-qc/1905.12946.
123. Berti, E.; Gair, J.; Sesana, A. Graviton mass bounds from space-based gravitational-wave observations of massive black hole populations. *Phys. Rev.* **2011**, *D84*, 101501, doi:10.1103/PhysRevD.84.101501.
124. Zhang, X.; Yu, J.; Liu, T.; Zhao, W.; Wang, A. Testing Brans-Dicke gravity using the Einstein telescope. *Phys. Rev.* **2017**, *D95*, 124008, doi:10.1103/PhysRevD.95.124008.
125. Zimmerman, A.; Haster, C.J.; Chatziioannou, K. On combining information from multiple gravitational wave sources. *Phys. Rev.* **2019**, *D99*, 124044, doi:10.1103/PhysRevD.99.124044.
126. Isi, M.; Chatziioannou, K.; Farr, W.M. A hierarchical test of general relativity with gravitational waves. *arXiv* **2019**, arXiv:gr-qc/1904.08011.
127. Eardley, D.M. Observable effects of a scalar gravitational field in a binary pulsar. *Astrophys. J. Lett.* **1975**, *196*, L59–L62. doi:10.1086/181744.
128. Yagi, K.; Blas, D.; Yunes, N.; Barausse, E. Strong Binary Pulsar Constraints on Lorentz Violation in Gravity. *Phys. Rev. Lett.* **2014**, *112*, 161101, doi:10.1103/PhysRevLett.112.161101.
129. Yagi, K.; Blas, D.; Barausse, E.; Yunes, N. Constraints on Einstein-AEther theory and Horava gravity from binary pulsar observations. *Phys. Rev.* **2014**, *D89*, 084067, doi:10.1103/PhysRevD.90.069902.
130. Yagi, K.; Stein, L.C.; Yunes, N. Challenging the Presence of Scalar Charge and Dipolar Radiation in Binary Pulsars. *Phys. Rev.* **2016**, *D93*, 024010, doi:10.1103/PhysRevD.93.024010.

131. Anderson, D.; Yunes, N. Scalar Charges and Scaling Relations in Massless Scalar-Tensor Theories. *arXiv* **2019**, arXiv:gr-qc/1901.00937.
132. Oost, J.; Mukohyama, S.; Wang, A. Constraints on Einstein-aether theory after GW170817. *Phys. Rev.* **2018**, *D97*, 124023, doi:10.1103/PhysRevD.97.124023.

4

A Solution of the Mitra Paradox

Øyvind Grøn

Oslo and Akershus University College of Applied Sciences, Faculty of Technology, Art and Sciences, PB 4 St. Olavs. Pl., NO-0130 Oslo, Norway; oyvind.gron@hioa.no

Academic Editors: Lorenzo Iorio and Elias C. Vagenas

Abstract: The "Mitra paradox" refers to the fact that while the de Sitter spacetime appears non-static in a freely falling reference frame, it looks static with reference to a fixed reference frame. The coordinate-independent nature of the paradox may be gauged from the fact that the relevant expansion scalar, $\theta = \sqrt{3\Lambda}$, is finite if $\Lambda > 0$. The trivial resolution of the paradox would obviously be to set $\Lambda = 0$. However, here it is assumed that $\Lambda > 0$, and the paradox is resolved by invoking the concept of "expansion of space". This is a reference-dependent concept, and it is pointed out that the solution of the Mitra paradox is obtained by taking into account the properties of the reference frame in which the coordinates are co-moving.

Keywords: general theory of relativity; exact solutions; spherical symmetry; physical interpretation

PACS: 04.20-q; 04.20.Cv; 04.20.Jb

---

1. Introduction

Abhas Mitra [1] has recently discussed an interesting problem concerning the physical interpretation of the de Sitter spacetime. He has pointed out that seemingly there is a contradiction between the static form of the de Sitter metric and the non-static, expanding representation of this spacetime as a Friedmann-Lemaitre-Robertson-Walker universe model of the steady state type. This will here be called the Mitra paradox.

Mitra writes that there is a physical or at least interpretational self-contradiction between the original static interpretation of the de Sitter metric and the present day non-static de Sitter view. Furthermore he writes that a metric represents a physical point of view, and due to the principle of covariance, the essential physical picture should not depend on the choice of coordinates. He also points out that that there has not been any attempt for physical resolution to reconcile the static and non-static versions of for example the de Sitter metric. In this paper I will provide such a reconciliation.

Since there is a similar conflict between the Minkowski spacetime and the Milne universe model [2,3], I will start the present discussion by considering the corresponding Mitra paradox for these metrics. Then the de Sitter spacetime will be considered and finally the Schwarzschild and the Schwarschild-de Sitter spacetime.

Write Schutz [4] writes that we define a *static* spacetime to be one in which we can find a time coordinate t with two properties: (i) all metric components are independent of t; and (ii) the geometry is unchanged by time reversal, $t \to -t$. A spacetime with the property (i) but not (ii) is said to be *stationary*. This definition can be formulated in terms of Killing vectors. A static spacetime is a spacetime, which has a time-like Killing vector field that is hypersurface orthogonal. This is a coordinate-independent characterization of a static spacetime. If any coordinate system exists in which none of the metric components depend upon time, there exists a time-like Killing vector in the spacetime. In this case the actual physical geometry of the spacetime is unchanging with time. Although the geometry of a stationary spacetime does not change in time, it can rotate. If the spacetime does not permit a time coordinate so that all the metric components are independent of t it is non-static and non-stationary. In this case the spacetime has no time-like Killing vector field.

The somewhat surprising fact is that even *a static spacetime can have a time-dependent metric.* The Mitra paradox is concerned with finding the proper physical meaning of this fact.

2. The Connection between the Global Minkowski Metric and the Milne Universe

It is well known that the Minkowski and Milne metrics are connected by a change of reference frame [5–8]. This will here be utilized to shed some light upon the Mitra paradox. We consider Minkowski spacetime with spherical coordinates $(R,\ \theta,\ \varphi)$ and a time coordinate T so that the line element takes the form

$$ds^2 = -\,dT^2 + dR^2 + R^2 d\Omega^2 \quad , \quad d\Omega^2 = d\theta^2 + \sin^2\theta\, d\varphi^2 \tag{1}$$

where we have used units so that $c = 1$. Then we introduce new coordinates $(t,\ r)$ by the transformation

$$t = \sqrt{T^2 - R^2} \quad , \quad r = \frac{t_0 R}{\sqrt{T^2 - R^2}} \tag{2}$$

The inverse transformation is

$$T = t\sqrt{1 + \frac{r^2}{t_0^2}} \quad , \quad R = \frac{r\,t}{t_0} \tag{3}$$

where t_0 is the present age of the universe. It follows from this transformation that

$$R = \frac{r}{\sqrt{1 + r^2/t_0^2}}\frac{T}{t_0} \quad , \quad T^2 - R^2 = t^2 \tag{4}$$

We see that the world-lines of the reference particles defining the reference frame in which $(t,\ r)$ are co-moving, i.e., $r = \text{constant}$, are straight lines, and the simultaneity curves $t = \text{constant}$ are hyperbolae.

It is seen that while the coordinates $(T,\ R)$ are co-moving in a static reference frame, the coordinates $(t,\ r)$ are co-moving in an expanding reference frame. A reference particle with $r = \text{constant}$ in the expanding frame has a coordinate velocity

$$v_R = \frac{r/t_0}{\sqrt{1 + r^2/t_0^2}} = \frac{R}{T} \tag{5}$$

in the rigid frame. In the expanding frame the line element of the Minkowski spacetime takes the form

$$ds^2 = -\,dt^2 + \left(\frac{t}{t_0}\right)^2 \left(\frac{dr^2}{1 + r^2/t_0^2} + r^2 d\Omega^2\right) \tag{6}$$

This is the line element of an empty, expanding universe model with negative spatial curvature—the Milne universe.

The Minkowski coordinates $(T,\ R)$ are the co-moving coordinates of a rigid inertial reference frame of an arbitrarily chosen reference particle P in the expanding cloud of particles defining the Milne universe model. The time T is the *private time* of P. The time t is measured on clocks following all of the reference particles. As seen from the first of Equation (4) the space T = constant has a finite extent, $R_{\max} = \lim_{r\to\infty} R = T$. This space is the *private space* of an observer following the particle P. The space t = constant is represented by a hyperbola given in the second of Equation (4) as shown in the Minkowski diagram of the P observer. It is defined by simultaneity of the clocks carried by all the reference particles, and is called the *public space* or simply the space of the universe model. It has infinite extension in spite of the fact that the Big Bang has the character of a point event in the Milne universe model.

In the inertial and rigid Minkowski coordinate system the velocity of a reference particle with co-moving coordinates is less than c for all values of. However, in the expanding cosmic frame it is different. Here the velocity of the reference particles as defined by an observer at the origin is given by Hubble's law. Hence the reference particles have superluminal velocity at sufficiently great distances from the observer. According to special relativistic kinematics, superluminal velocity is problematic because the particles cannot move through space with a velocity greater than c. However, according to the general relativistic interpretation, the reference particles define the public space of the universe model, and there is no limit to how fast space itself can expand.

The metric (1) is static and the metric (6) not. The Mitra paradox is concerned with a reconciliation of these properties of two metrics that are connected by a coordinate transformation, and hence that represent one and the same spacetime.

The Mitra paradox makes it clear that one cannot define a static spacetime as a spacetime where the metric is independent of time. The metric is coordinate-dependent, and may be independent of time in one coordinate system, but dependent on it in another, while the static property of a spacetime is invariant.

3. Proposal for a Solution of the Mitra Paradox

An important point when we try to solve the Mitra paradox is to distinguish between coordinate-dependent quantities and coordinate-independent physical quantities. The term *metric* is usually taken to mean the functions that appear in the line element multiplied by the coordinate differentials. Hence the metric is understood to mean the *components* of the metric tensor. This means that *the metric is a coordinate-dependent quantity.* It is natural, therefore, that at least in some cases, one and the same spacetime can be represented by both a static and a non-static metric.

Another important distinction is the difference between a coordinate system and a reference frame. In four-dimensional spacetime, a *coordinate system* provides a region of spacetime with a continuum of 4-tuples so that each event in spacetime is marked with a 4-tuple, different events with different 4-tuples. A coordinate system is a mathematical quantity. A *reference frame* is a continuum of world-lines representing reference particles with specified motions. This is a physical quantity. *Co-moving coordinates* in a reference frame are coordinates so that the reference particles of the frame have constant spatial coordinates.

The Mitra paradox is not only concerned with the metric, but also with 3-space. Hence it is important to distinguish between a coordinate 3-space and a coordinate-independent physical 3-space. Here we meet an important difficulty of the Mitra paradox. The 3-space has two very different qualities. On the one hand, it is a set of simultaneous events measured by clocks at rest in the chosen reference frame. Again this is coordinate-dependent or better, reference-dependent, due to the relativity of simultaneity.

It follows from Friedmann's 1. equation for a flat universe, $H = \sqrt{8\pi G\rho/3}$, that if the Minkowski spacetime is perceived as the limit of the Friedmann-Lemaitre-Robertson-Walker universe model with empty space, the Hubble parameter vanishes, and the 3-space is static. In a similar way, the De Sitter spacetime is then the limit of empty space with a cosmological constant, Λ, having a positive, constant Hubble parameter, $H = \sqrt{\Lambda/3}$, and the 3-space of the de Sitter spacetime is non-static and expanding. However, both of these spacetimes have maximal symmetry and have a time-like hyper surface orthogonal Killing vector, meaning that these spacetimes are static. Hence it is important to note the difference between a static 3-space and a static spacetime. The Mitra paradox is concerned with 3-space.

We considered the globally flat spacetime above. In the standard coordinates co-moving with a static reference frame with time-independent distances between the reference points, the 3-space is static. But as described in terms of coordinates co-moving in an expanding reference frame, the 3-space is not static, but expands. The flat, static spacetime then looks like an expanding universe—the Milne universe model.

This seems strange. So far we have defined 3-space as a set of simultaneous events. There is no motion involved in this definition. Hence the definition of a 3-space should be supplied by a second quality permitting space to expand. We can then supply the definition of a 3-space: A 3-space is made up of a set of reference particles at a given point of time. The 3-space of a reference frame is defined by identifying the reference particles of the 3-space with the reference particles of the frame. The most simple mathematical description of the 3-space is obtained by using coordinates co-moving with the reference frame of the 3-space. *The 3-space is said to be stationary if the physical distances between the reference points does not change.* In this case the reference frame can be said to be *rigid*. If these physical distances change, the 3-space is non-stationary. If the rotation of the velocity field of the reference particles of a stationary 3-space vanishes, the 3-space is said to be *static*.

It is then clear that whether a 3-space appears static or non-static depends upon the reference frame it is associated with. This means that *the static or non-static character of a 3-space is a coordinate-dependent quality of the spacetime*. It will be made clear below that this is an important ingredient in the solution of the Mitra paradox.

One may wonder whether this means that physical 3-space does not exist. It is then important to make one more distinction: between *physical* and *invariant*. A quantity is said to be invariant if it has the same value in every reference frame or coordinate system. A physical property need not be invariant. For example a 3-velocity of a particle is a physical property, but it is not invariant. It can even be transformed away by going into the rest frame of the particle.

We should not talk about *the* 3-space of Minkowski spacetime. We should talk about *a* 3-space. Minkowski spacetime can have a static 3-space and equally well a non-static 3-space. Although the property of a 3-space of being static is a physical property, its static character is not invariant. This is the proposed solution of the Mitra paradox, which will be further worked out below.

4. Static and Expanding 3-Space

Let us first consider a static spacetime as described in the co-moving coordinates of a rigid reference frame, RF, so that the metric is static and has the form

$$ds^2 = -f\left(R\right)dt^2 + \frac{dR^2}{f\left(R\right)} + R^2 d\Omega^2 \tag{7}$$

Here the radial coordinate is chosen so that the invariant area of a spherical surface with radius R is equal to $4\pi R^2$. This radial coordinate is sometimes called the curvature radius or alternatively the area coordinate. The 3-space of simultaneous events as measured by clocks carried by the reference particles of RF, is static. This is the preferred 3-space of spacetimes with a localized mass distribution, such as the Schwarzschild spacetime.

Assume that there exists a surface with coordinate radius $R = R_0$ so that a particle permanently at rest on this surface has vanishing 4-acceleration, i.e., a free particle instantaneously at rest at this surface will remain at rest on the surface. The radial component of the 4-acceleration of a particle at rest in the coordinate system is according to the geodesic equation given in terms of certain Christoffel symbols and the time component of the 4-velocity as

$$a^R = \Gamma^R_{T\,T}\left(u^T\right)^2 = (1/2)\,f\prime(R) \tag{8}$$

Hence the radius R_0 is given by $f\prime\left(R_0\right) = 0$. Let us now consider the 3-space of simultaneous events as shown by clocks carried by *free particles* starting their movements from a state of rest at $R = R_0$. These particles make up a locally inertial frame, IF. Hence this 3-space may be called an *inertial 3-space*. This is the preferred 3-space of the relativistic universe models. In [9] it was shown that the 3-velocity of the inertial 3-space as given with respect to the orthonormal basis of an observer at rest in RF is

$$v_{3-space} = \left(\frac{d\hat{x}^R}{d\tau}\right)_{3-space} = \pm\sqrt{1 - \frac{f(R)}{f(R_0)}} \tag{9}$$

In the case of the Minkowski metric (1) this gives $v_{3-space} = 0$, and the inertial 3-space is then at rest in an arbitrary rigid frame in flat spacetime.

Let us now describe the 3-space with reference to an expanding reference frame in which the metric is of the Friedmann-Lemaitre-Robertson-Walker type. Then the reference frame consists of a set of freely moving particles expanding together with the cosmic fluid. Let t be the proper time of clocks co-moving with the reference particles of this frame, and r a co-moving radial coordinate following the cosmic fluid. Then the line element has the form

$$ds^2 = -\,dt^2 + [a(t)]^2\left(\frac{dr^2}{1-kr^2} + r^2 d\Omega^2\right) \tag{10}$$

where k is a constant, which can have the values $k = \{-1/t_0^2\,,\,0\,,\,1/t_0^2\}$, and $a(t)$ is the scale factor. If it is normalized to $a(t_0) = 1$ at the present time t_0, it represents the ratio of the cosmic distances between the reference particles at an arbitrary point of time and their present distances. In this coordinate system the 3-space has an expansion, $\theta = 3\dot{a}/a$. Hence the 3-space is not static in this frame if $\dot{a} \neq 0$.

So the 3-space of the Minkowski spacetime may be pictured as either static or non-static depending upon the reference frame that is used. This freedom of point of view is due to the Lorentz invariance of this solution of Einstein's field equations. It is not typical of the solutions of the field equations in general. But there are two other solutions that share this property of Lorentz invariance with the Minkowski spacetime, and those are the de Sitter and anti-de Sitter spacetimes. Let us consider the de Sitter spacetime.

5. The de Sitter Spacetime

We consider the de Sitter spacetime with spherical coordinates $(R,\ \theta,\ \varphi)$ and a time coordinate T, so that the line element takes the form

$$ds^2 = -\left(1 - H^2R^2\right)dT^2 + \frac{dR^2}{1 - H^2R^2} + R^2 d\Omega^2 \tag{11}$$

for $R < R_H = 1/H$, where $H = \sqrt{\Lambda/3}$, and the cosmological constant $\Lambda = 8\pi G\rho_\Lambda$ represents the constant density ρ_Λ of a Lorentz Invariant Vacuum Energy, LIVE, with stress $p_\Lambda = -\rho_\Lambda$. It should be noted that the coordinate clocks showing T go with a position-independent rate equal to that of a standard clock at $R = 0$.

We introduce new coordinates $(t,\ r)$ by the transformation

$$t = T + \frac{1}{2H}\ln\sqrt{1 - H^2R^2}\quad,\quad r = \frac{R}{\sqrt{1 - H^2R^2}}e^{-HT} \tag{12}$$

The inverse transformation is

$$T = t - (1/2H)\,\ln\sqrt{1 - H^2r^2e^{2Ht}}\quad,\quad R = r\,e^{Ht} \tag{13}$$

Differentiating the second of Equation (12) with respect to T with constant r gives the coordinate velocity of a reference particle in the $(t,\ r)$-system with respect to the $(T,\ R)$-system

$$v_R = H\,e^{Ht} = \sqrt{1 - H^2R^2}\,e^{HT} \tag{14}$$

Hence the (t, r) coordinates are co-moving in an expanding reference frame relative to the rigid (T, R)-system. In the coordinates (t, r) the line element has the form

$$ds^2 = -dt^2 + e^{2Ht}\left(dr^2 + r^2 d\Omega^2\right) \tag{15}$$

The (t, r)-coordinates are co-moving with free particles, as is the case for all the FLRW-universe models. Equation (14) shows that the free particles in this spacetime move outwards with an accelerated motion.

It follows from Equation (8) that for the metric (11) the 4-acceleration of a particle fixed in the rigid reference frame, is

$$a^R = H^2 R \tag{16}$$

Hence a free particle at rest in the rigid reference frame must be at the position $R_0 = 0$. The 3-space made up of simultaneous events as measured by clocks carried by these particles is the inertial 3-space. In this case the velocity of inertial 3-space as given by Equation (9) is

$$v_{3-space} = HR \tag{17}$$

Even if the metric (11) is static, the velocity of the inertial 3-space obeys the Hubble lav. There is an expansion equal to $3H$. Hence the 3-space expands in accordance with the Hubble law. This is often called the *Hubble flow*. The inertial 3-space flows with the velocity of light at $R = R_H = 1/H$ and with superluminal velocity for $R > R_H$. There is a horizon at $R = R_H$.

The coordinate time t is shown on standard clocks following the freely falling reference particles of the 3-space, and the coordinate r is co-moving with those particles. That is the reason for the time dependence of the metric (15) in this coordinate system. Hence there is no contradiction between the static form of the line element (11) and the non-static form (15). The first form reflects the rigidity of the reference frame in which the coordinates T, R are co-moving, and the second reflects the expansion of the reference frame in which the coordinates t, r are co-moving. This solves the Mitra paradox.

However, there is something strange about the metric (11). There is a coordinate singularity at $R_H = 1/H$. Note that $\lim_{R \to 1/H} r = \infty$, and that $v_{3-space}(R_H) = 1$. Hence the 3-space is flowing with the velocity of light at this surface.

Consider radially moving light in the metric (11). The coordinate velocity of light is the same in the inwards and outwards direction, and is equal to

$$(dR/dT)_L = \pm\left(1 - H^2R^2\right) \tag{18}$$

which vanishes at $R = R_H$. Hence in this coordinate system the light cone collapses at $R = R_H$.

In order to have open light cones at $R = R_H$ one may introduce a new time coordinate. There are several related such coordinates, and it may be useful to compare the description of the light cones in three of them. All of them are given by an *internal coordinate transformation* in the sense that the coordinate clocks are at rest in the same reference frame as those showing T.

We first consider a light cone coordinate, $\overline{T}$, used by Spradlin, Strominger and Volovich [10], given by the coordinate transformation

$$\overline{T} = T - \frac{1}{2H}\ln\frac{1 + HR}{1 - HR} \tag{19}$$

By using L'Hopital's rule we get $\lim_{H \to 0} \overline{T} = T - R$ showing that $\overline{T}$ reduces to an ordinary light cone coordinate in Minkowski spacetime. Differentiating we have

$$d\overline{T} = dT - \frac{dR}{1 - H^2R^2} \tag{20}$$

Hence the coordinate clocks showing $\overline{T}$ have another simultaneity than those showing T. With the new time coordinate the line element of the de Sitter spacetime takes the form

$$ds^2 = -\left(1 - H^2R^2\right)d\overline{T}^2 - 2d\overline{T}\,dR + R^2 d\Omega^2 \tag{21}$$

For light moving radially we have $ds^2 = d\Omega^2 = 0$ and hence,

$$2d\overline{T}\,dR = -\left(1 - H^2R^2\right)d\overline{T}^2 \tag{22}$$

For light moving outwards $dR > 0$, which is not permitted by Equation (22). However, for light moving outwards, Equations (16) and (18) give $d\overline{T} = 0$, which *is* permitted. Hence $\overline{T}$ is a light cone coordinate for outgoing light. For light moving inwards, Equation (22) gives the coordinate velocity

$$\left(\frac{dR}{d\overline{T}}\right)_{L-} = -\frac{1}{2}\left(1 - H^2R^2\right) \tag{23}$$

which vanishes at $R = R_H$. The "inwards directed" velocity of light changes sign at $R = R_H$ and becomes outwards directed for $R > R_H$.

Another time coordinate T_P called the Painlevé-de Sitter coordinate, was used by Parikh [11] and is given by the transformation

$$T_P = T + (1/2\,H)\ln\left(1 - H^2R^2\right) \tag{24}$$

Comparing with equation the first of the Equation (12) we see that $T_P = t$. Hence the Paainlevé-de Sitter time is the same as the cosmic time, which is measured by standard clocks following freely moving particles. Differentiating gives

$$dT_P = dT - \frac{HR}{1 - H^2R^2}dR \tag{25}$$

Inserting this into Equation (11) we find that the line element takes the form

$$ds^2 = -\left(1 - H^2R^2\right)dT_P^2 - 2HR\,dT_P\,dR + dR^2 + R^2 d\Omega^2 \tag{26}$$

The coordinate velocity of outgoing and ingoing light is

$$(dR/dT_P)_+ = 1 + HR \quad , \quad (dR/dT_P)_- = -\,(1 - HR) \tag{27}$$

At the horizon $(dR/dT_P)_+ = 2, (dR/dT_P)_- = 0$. The velocity of the ingoing light changes sign at the horizon, and moves outwards outside the horizon. This is an effect of the repulsive gravity due to the LIVE, which fills this spacetime and causes an accelerated expansion of the inertial 3-space. Note that in this context "accelerated" means non-vanishing 3-acceleration. The 4-acceleration of the reference particles of the inertial 3-space vanishes, since the particles are freely falling.

Finally a time coordinate corresponding to the ingoing Eddington-Finkelstein coordinate in the Schwarzshild spacetime is defined by the condition that the coordinate velocity of outgoing light is equal to 1. This was used by Braeck and Grøn [9] and is given by

$$\widetilde{T} = T + \frac{1}{2H}\ln\frac{1 - HR}{1 + HR} + R \tag{28}$$

Differentiation gives

$$d\widetilde{T} = dT - \frac{H^2R^2}{1 - H^2R^2}dR \tag{29}$$

With this time coordinate the line element takes the form

$$ds^2 = -\left(1 - H^2R^2\right) d\widetilde{T}^2 - 2H^2R^2\, d\widetilde{T}\, dR + \left(1 + H^2R^2\right) dR^2 + R^2 d\Omega^2 \tag{30}$$

The coordinate velocity of outgoing and ingoing light is

$$\left(dR/d\widetilde{T}\right)_+ = 1, \quad \left(dR/d\widetilde{T}\right)_- = -\frac{1 - H^2R^2}{1 + H^2R^2} \tag{31}$$

At the horizon $\left(dR/d\widetilde{T}\right)_+ = 1$, $\left(dR/d\widetilde{T}\right)_- = 0$. Again the velocity of the ingoing light changes sign at the horizon, i.e., the light cones turn outwards, implying that nothing can enter the horizon from the outside region.

All of the line elements (21), (26) and (30) are stationary, although they are not static. The stationary character shows that the coordinate R is co-moving in a rigid reference frame. The reason that they are not static is that the coordinate clocks are not Einstein synchronized. Their simultaneity is not that of Einstein synchronized clocks at rest in the rigid reference frame.

The de Sitter spacetime is static since there exists a coordinate system where the metric is static and the time-like basis vector is a Killing vector.

Nevertheless this spacetime is filled with vacuum energy that expands. This energy causes repulsive gravity, which acts back upon the energy itself and makes the expansion accelerate. It should be noted that in a homogeneous universe there is no pressure gradient, so the accelerated expansion is not a pressure effect, but a gravitational effect. The negative pressure, $p = -\rho c^2$, contributes to the effective gravitational mass density, $\rho_{grav} = \rho + 3p/c^2$, making it negative, which means that gravity is repulsive [12]. Hence there is energy with accelerated expansion in this spacetime. Is it then reasonable to say that it is static?

Compare 3-space with a river, and consider the river now and an hour later, assuming that there is the same amount of water in the river at these points of time. In this situation, the river has not changed. The river is static. But the water is not static. It flows. Similarly spacetime is static, but 3-space is expanding.

In spacetime the river corresponds to the geometry of space at a certain position, and the flowing water corresponds to the flowing reference particles constituting the 3-space. In the de Sitter spacetime the geometry of space is unchanged at a fixed position in a rigid reference frame. Hence it is a static spacetime; but the 3-space is flowing. It is not static. The metric in a coordinate system co-moving with the reference particles of the 3-space is not static, but depends upon time as in the metric of Equation (15).

6. The Schwarzschild Spacetime

Outside the Schwarzschild radius the Schwarzschild spacetime has a time-like Killing vector field that is hypersurface orthogonal. Hence it is static and there exists a coordinate system in which the metric is independent of time and the line element has no product terms where a spatial differential is multiplied by a time differential. One such coordinate system is the standard so-called curvature coordinates where the invariant area of a surface with coordinate radius R around the origin is $4\pi R^2$. In this coordinate system the line element takes the form

$$ds^2 = -\left(1 - R_S/R\right) dT^2 + \frac{dR^2}{1 - R_S/R} + R^2 d\Omega^2 \tag{32}$$

where $R_S = 2GM$ is the Schwarzschild radius of the central mass. Inside the Schwarzschild radius the Killing vector field is spacelike, and in this region the Schwarzschild spacetime is not static.

Consider now an observer moving with the inertial 3-space in this spacetime, i.e., he is falling freely from a state of rest infinitely far away from the central mass. The co-moving radius of this observer is r and he carries with him a standard clock showing t. The new coordinates are given by transformation [13–15]

$$T = t + R_S \ln \frac{\sqrt{R/R_S} + 1}{\sqrt{R/R_S} - 1} - 2\sqrt{R_S R} \quad , \quad R^{3/2} = -\ (3/2)\ \sqrt{R_S}\ (t + r) \tag{33}$$

In terms of the co-moving coordinates of the inertial 3-space the line element of the Schwarzschild spacetime takes the form

$$ds^2 = -\ dt^2 + \left(\frac{2}{3} \frac{R_S}{t + r} \right)^{2/3} dr^2 + \left(\frac{3}{2} \sqrt{R_S}\ (t + r) \right)^{4/3} d\Omega^2 \tag{34}$$

An observer with $r = r_0$ has initially a large negative value of t, which increases towards $-\ r_0 - (2/3)\ R_S$ as the observer passes the Schwarzschild horizon. Hence the line element (34) corresponds to that of an inhomogeneous universe with an anisotropic and position-dependent scale factor, and the inertial 3-space expands in the radial direction and contracts in the tangential direction. In these coordinates the metric of the Schwarzschild spacetime is not static. These geometrical changes with time of the inertial 3-space are due to tidal forces becoming stronger at the position of the reference particles co-moving with the inertial 3-space, as they approach the central mass.

The coordinate transformation is well defined only for $R > R_S$. This is due to the rigid character of the reference frame in which the coordinates $(T,\ R)$ are co-moving, which is physically possible only outside the Schwarzschild horizon. However the line element (34) has no coordinate singularity at the Schwarzschild horizon. The coordinates t and r are well defined in all of spacetime outside the central singularity, also inside the horizon, and the line element gives a singularity-free description of the Schwarzschild spacetime in the whole of this region. This illustration shows that a static spacetime, which is usually expressed so that 3-space is static, may also be expressed so that the 3-space is non-static.

7. The Schwarzschild-de Sitter Spacetime

This is a static spacetime in which the line element may be written

$$ds^2 = -\ \left(1 - \frac{R_S}{R} - H^2 R^2 \right) dT^2 + \frac{dR^2}{1 - \frac{R_S}{R} - H^2 R^2} + R^2 d\Omega^2 \tag{35}$$

In this spacetime the inertial 3-space has a rather interesting behavior. At the surface with

$$R_0 = \left(R_S R_H^2 / 2 \right)^{1/3} \tag{36}$$

the 4-acceleration of a particle permanently at rest vanishes [9]. Hence the reference particles of the inertial 3-space are at rest at this surface. But the inertial 3-space diverges at this surface. It expands outside this surface and contracts inside it.

8. Static Form of the FLRW Metric

Mitra has recently deduced an interesting form of the Friedmann-Lemaitre-Robertson-Walker metric in curvature coordinates [16] and used this to investigate when an expanding universe can look static [17]. The FLRW-metric is first written in the usual form (10). Mitra then found that the metric can be written in curvature coordinates as follows

$$ds^2 = -\ \left(\frac{\partial t}{\partial T} \right)^2 \frac{1 - k\,(R/a)^2 - (\dot{a}/a)^2 R^2}{1 - k\,(R/a)^2} dT^2 + \frac{dR^2}{1 - k\,(R/a)^2 - (\dot{a}/a)^2 R^2} + R^2 d\Omega^2 \tag{37}$$

where

$$R = r\,a\,(t) \tag{38}$$

He then showed that essentially only the Milne universe and the de Sitter and anti-de Sitter universe models can be written in static form using curvature coordinates.

As an illustrating example we will here consider only the first model studied by Mitra. It is the de Sitter universe model with negative spatial curvature, $k = -1$ and $\Lambda > 0$. For this model the solution of the Friedmann equations gives the scale factor

$$a\,(t) = (1/H)\sinh\,(H\,t) \tag{39}$$

where $H = \sqrt{\Lambda/3}$. This universe model is filled by vacuum energy with constant density and stress given by

$$p_\Lambda = -\,\rho_\Lambda = -\,\Lambda/8\pi G \tag{40}$$

Inserting Equations (38) and (39) into Equation (37), the line element takes the form

$$ds^2 = -\left(\frac{\partial t}{\partial T}\right)^2 \frac{1 - H^2R^2}{1 + r^2}dT^2 + \frac{dR^2}{1 - H^2R^2} + R^2 d\Omega^2 \tag{41}$$

Comparing with Equation (11) we obtain

$$\frac{\partial t}{\partial T} = \sqrt{1 + r^2} \tag{42}$$

Mitra has shown that Equations (38), (39) and (42) lead to the transformation

$$HR = r\sinh\,(Ht)\,, \quad \tanh\,(HT) = \sqrt{1 + r^2}\tanh\,(Ht) \tag{43}$$

The inverse transformation may be written

$$\sinh\,(H\,t) = \sqrt{\sinh^2\,(HT) - H^2R^2\cosh^2\,(HT)}, \quad r = \frac{HR}{\sqrt{\sinh^2\,(HT) - H^2R^2\cosh^2\,(HT)}} \tag{44}$$

Differentiating the first of these equations partially with respect to T and using the second equation, we get

$$\frac{\partial t}{\partial T} = \sqrt{\frac{1 - H^2R^2}{\tanh^2\,(HT) - H^2R^2}}\tanh\,(HT) = \sqrt{1 + r^2} \tag{45}$$

in agreement with Equation (42). Furthermore, by taking the differentials of the transformation (43) and inserting the expressions into Equation (11) one finds that (43) transforms the static metric (11) to the line element of the expanding de Sitter universe model with a negatively curved 3-space,

$$ds^2 = -\,dt^2 + \sinh^2\,(Ht)\left(\frac{dr^2}{1 + r^2} + r^2 d\Omega^2\right) \tag{46}$$

Equation (38) shows that the value of R increases with time for a fixed value of r. Again we see that the reconciliation of the static and non-static forms of the line element for one and the same spacetime is in recognizing the motion of the reference frames in which the radial coordinates are co-moving. The radial coordinate of the time-dependent metric is co-moving with an expanding reference frame, and the radial coordinate of the static metric is co-moving with a rigid reference frame. This is the solution of the Mitra paradox as applied to the present spacetime.

9. Energy Conservation

Writing the line element of the de Sitter spacetime in terms of coordinates co-moving with free particles, i.e., in the form (15), Mitra deduced that the vacuum energy inside a radius r is

$$U = (\Lambda/6\,G)\; r^3 e^{3Ht} \tag{47}$$

Mitra concluded: "Thus the total energy of the de Sitter model increases in an exponential manner. Such a bad violation of the "Principle of Conservation of Energy" in the co-moving frame is in sharp contrast with the corresponding nice behavior in the Schwarzschild frame". (He uses units in which the gravitational constant is $G = 1$, but I have kept G in the formulae.)

The solution of this seeming paradox is as follows. The Friedmann equations lead to

$$dU + p_\Lambda dV = 0 \tag{48}$$

where $V = (4\pi/3)\,a^3$, $U = \rho_\Lambda\,V$, and a is the scale factor of Equation (37). Equation (49) is the 1. Law of thermodynamics for adiabatic expansion as applied to a co-moving region with radius $r = 1$ around an observer. It expresses the law of energy conservation. Heat is defined as transport of energy due to temperature difference. In a homogeneous universe there are no large scale temperature differences, and this is the reason that the universe expands adiabatically.

Using $\rho_\Lambda = \Lambda/8\pi G$ and $a = e^{Ht}$, we get $dU = \rho_\Lambda 4\pi a^2 da = (\Lambda H/2G)\,e^{3Ht}dt$, which is the same as we get by taking the differential of U in Equation (47). The volume work performed at the boundary of the region is $dW = p_\Lambda dV = -\rho_\Lambda 4\pi a^2 da$. Hence the energy conservation equation is obeyed in spite of the fact that the amount of vacuum energy is increasing inside the co-moving surface. The reason is that there is a negative work at the boundary, which transfers energy from the outside region to the inside region. Imagining that the region is extended so that the boundary is infinitely far from the observer, one may say that the density of the vacuum energy is kept constant in spite of the expansion by extracting energy from an infinitely far region. This shows that global energy conservation is indeed a problematic concept at least for a universe with infinitely great spatial extension.

10. Results and Discussion

Mitra has pointed out that there seems to be an interpretational self-contradiction between the static interpretation of the de Sitter metric and the non-static de Sitter universe model. This has here been called the Mitra paradox. He also writes that there has not been any attempt for a physical resolution to reconcile the static and non-static versions of for example the de Sitter metric.

Both the problem and the resolution are of a conceptual nature of great significance for a proper way of teaching the general theory of relativity. A theory is much more than some rules for calculating physical effects, making it possible to falsify the theory. The theory also provides us with concepts representing the foundations of our world picture. As said by Einstein: It is the theory that tells what we observe.

Hence it is extremely important to obtain a proper physical interpretation of the general theory of relativity, free of contradictions. This also means that interpretational problems such as that formulated by Mitra, should not be neglected. The present article has been an effort to give a constructive discussion of this problem—and to solve it. I have here provided a resolution by focusing upon the difference between 3-space and spacetime and pointing out the significance of the motion of the reference frames in which different coordinate systems are co-moving.

11. Conclusions

The Mitra paradox is concerned with the physical reconciliation of two metrics, where one is static and the other time-dependent, that are connected by a coordinate transformation, and hence that represent one and the same spacetime.

The Mitra paradox makes it clear that one cannot define a static spacetime as a spacetime where the metric is independent of time. The metric is coordinate-dependent, and may be independent of time in one coordinate system, but dependent on it in another, while the static property of a spacetime is invariant and characterized by the existence of a hypersurface orthogonal Killing vector.

The solution of the Mitra paradox lies in recognizing that the metric is not determined by the geometric properties of the spacetime. In general there are ten independent components of the metric tensor and only six independent field equations, leaving the freedom of choosing the coordinate system. By choosing coordinates in a given spacetime as co-moving with a rigid reference frame one obtains a time-independent metric—otherwise a time-dependent one.

These general properties of the solution to the Mitra paradox have been illustrated in the present paper by considering several cases, the first of which being flat spacetime. With coordinates co-moving in a rigid inertial frame one obtains the usual Minkowski metric, and with coordinates co-moving in an expanding reference frame one obtains the time-dependent Milne metric. Secondly, the de Sitter spacetime has been considered. Again, by using coordinates co-moving in a rigid frame one obtains a static metric and using coordinates co-moving with freely moving particles that make up a system that expands due to repulsive gravity in this spacetime, one obtains a time-dependent metric. Thirdly, we have discussed the Schwarzschild spacetime. Again the metric is static in a rigid frame. But using coordinates co-moving with a system of freely falling particles one obtains a time-dependent metric, still representing the Schwarzschild spacetime.

Finally, as shown by Mitra, and interpreted physically here, a similar result is obtained for just three different Friedmann-Lemaitre Robertson-Walker universe models, the Milne universe and the de Sitter and anti-de Sitter universe models. All these models are solutions of Einstein's equations for empty space, the first one without a cosmological constant and the two latter ones with a positive and negative cosmological constant, respectively.

The universe models with matter or radiation energy are solutions of the field equations with a time-dependent energy-momentum tensor. Hence Mitra's result implies that universe models with a time-dependent energy-momentum tensor cannot be represented globally by a line element with a time-independent metric. This should be formulated in a coordinate-independent way.

For a perfect fluid with energy-momentum tensor

$$T^{\mu\nu} = (\rho + p)\, u^{\mu} u^{\nu} + p g^{\mu\nu} \tag{49}$$

we may define the energy-momentum scalar

$$T^{\mu\nu} T_{\mu\nu} = \rho^2 + 3p^2 \tag{50}$$

Hence we may conclude by formulating Mitra's result for the FLRW-universe models in the following way: The line element cannot be written in a globally time-independent way for a universe model with a time-dependent energy-momentum scalar.

Acknowledgments: I would like to thank the referees for useful suggestions that contributed to improvements of the article.

References

1. Mitra, A. Interpretational conflicts between the static and non-static forms of the de Sitter metric. *Sci. Rep.* **2012**, *2*, 923. [CrossRef] [PubMed]
2. Rindler, V. Finite foliations of open FRW universes and the point-like big bang. *Phys. Lett. A* **2000**, *276*, 52–58. [CrossRef]
3. Grøn, Ø.; Elgarøy, Ø. Is space expanding in the Friedmann universe models. *Am. J. Phys.* **2007**, *75*, 151–157. [CrossRef]

4. Schutz, B.F. *A First Course in General Rkkelativity*; Cambridge University Press: Cambridge, UK, 1985.
5. Grøn, Ø.; Hervik, S. *Einstein's General Theory of Relativity*; Springer: New York, NY, USA, 2007.
6. Carroll, S.M. *Spacetime and Geometry*, 1st ed.; Addison-Wesley: Boston, MA, USA, 2004.
7. Mukhanov, V. *Physical Foundations of Cosmology*; Cambridge University Press: Cambridge, UK, 2005.
8. Misner, C.; Thorne, K.; Wheeler, J.A. *Gravitation*; W. H. Freeman and Company: New York, NY, USA, 1971.
9. Bræck, S.; Grøn, Ø. A river model of space. *Eur. Phys. J. Plus* **2013**, *128*, 24. [CrossRef]
10. Spradlin, M.; Strominger, A.; Volovich, A. Unity from Duality: Gravity, Gauge Theory and Strings. In *Les Houches—Ecole d'Ete de Physique Theorique*; Springer: Berlin/Heidelberg, Germany, 2002.
11. Parikh, M.K. New coordinates for de Sitter space and de Sitter radiation. *Phys. Lett. B* **2002**, *546*, 189–195. [CrossRef]
12. Grøn, Ø. Repulsive gravitation and electron models. *Phys. Rev. D* **1985**, *31*, 2129–2131. [CrossRef]
13. Landau, L.; Lifshitz, E.M. *The Classical Theory of Fields*; Reed Educational and Professional Publishing Ltd.: Oxford, UK, 2002.
14. Lightman, A.P.; Press, W.H.; Price, R.H.; Teukolsky, S.A. *Problem Book in Relativity and Gravitation*; Princeton University Press: Princeton, NJ, USA, 1975.
15. Dai, S.; Guan, C.B. Maximally symmetric subspace decomposition of the Schwarzschild black hole. *arXiv* **2004**, arXiv:gr-qc/0406109.
16. Mitra, A. Friedmann-Robertson-Walker metric in curvature coordinates and its applications. *Gravit. Cosmol.* **2013**, *19*, 134–137. [CrossRef]
17. Mitra, A. When can an "Expanding Universe" look "Static" and vice versa: A comprehensive study. *Int. J. Mod. Phys. D* **2015**, *24*, 1550032. [CrossRef]

5

The Scales of Gravitational Lensing

Francesco De Paolis [1,2,*,†], Mosè Giordano [1,2,†], Gabriele Ingrosso [1,2,†], Luigi Manni [1,2,†], Achille Nucita [1,2,†] and Francesco Strafella [1,2,†]

1 Department of Mathematics and Physics "Ennio De Giorgi", University of Salento, CP 193, I-73100 Lecce, Italy; giordano@le.infn.it (M.G.); ingrosso@le.infn.it (G.I.); luigi.manni@le.infn.it (L.M.); nucita@le.infn.it (A.N.); strafella@le.infn.it (F.S.)

2 INFN (Istituto Nazionale di Fisica Nucleare), Sezione di Lecce CP-193, Italy

* Correspondence: francesco.depaolis@le.infn.it.

† These authors contributed equally to this work.

Academic Editor: Lorenzo Iorio

Abstract: After exactly a century since the formulation of the general theory of relativity, the phenomenon of gravitational lensing is still an extremely powerful method for investigating in astrophysics and cosmology. Indeed, it is adopted to study the distribution of the stellar component in the Milky Way, to study dark matter and dark energy on very large scales and even to discover exoplanets. Moreover, thanks to technological developments, it will allow the measure of the physical parameters (mass, angular momentum and electric charge) of supermassive black holes in the center of ours and nearby galaxies.

Keywords: gravitational lensing

---

1. Introduction

In 1911, while he was still involved in the development of the general theory of relativity (subsequently published in 1916), Einstein made the first calculation of light deflection by the Sun [1]. He correctly understood that a massive body may act as a gravitational lens deflecting light rays passing close to the body surface. However, his calculation, based on Newtonian mechanics, gave a deflection angle wrong by a factor of two. On 14 October 1913, Einstein wrote to Hale, the renowned astronomer, inquiring whether it was possible to measure a deflection angle of about 0.84″ toward the Sun. The answer was negative, but Einstein did not give up, and when, in 1915, he made the calculation again using the general theory of relativity, he found the right value $\phi = 2r_s/b$ (where $r_s = 2GM/c^2$ is the Schwarzschild radius and b is the light rays' impact parameter) that corresponds to an angle of about 1.75″ in the case of the Sun. That result was resoundingly confirmed during the Solar eclipse of 1919 [2].

In 1924, Chwolson [3] considered the particular case when the source, the lens and the observer are aligned and noticed the possibility of observing a luminous ring when a far source undergoes the lensing effect by a massive star. In 1936, after the insistence of Rudi Mandl, Einstein published a paper on science [4] describing the gravitational lensing effect of one star on another, the formation of the luminous ring, today called the Einstein ring, and giving the expression for the source amplification. However, Einstein considered this effect exceedingly curious and useless, since in his opinion, there was no hope to actually observe it.

On this issue, however, Einstein was wrong: he underestimated technological progress and did not foresee the motivations that today induce one to widely use the gravitational lensing phenomenon. Indeed, Zwicky promptly understood that galaxies were gravitational lenses more powerful than stars and might give rise to images with a detectable angular separation. In two letters published in 1937 [5,6], Zwicky noticed that the observation of galaxy lensing, in addition to giving a further proof of the general theory of relativity, might allow observing sources otherwise invisible, thanks to the light gravitational amplification, thereby obtaining a more direct and accurate estimate of the lens galaxy dynamical mass. He also found that the probability to observe lensed galaxies was much larger than that of star on star. This shows the foresight of this eclectic scientist, since the first strong lensing event was discovered only in 1979: the double quasar QSO 0957+561 a/b [7], shortly followed by the observation of tens of other gravitational lenses, Einstein rings and gravitational arcs. All of that plays today an extremely relevant role for the comprehension of the evolution of the structures and the measure of the parameters of the so-called cosmological standard model.

Actually, there are different scales in gravitational lensing, on which we shall briefly concentrate in the next sections, after a short introduction to the basics of the theory of gravitational lensing (Section 2). Generally speaking, gravitational lens images separated by more than a few tenths of arcsecs are clearly seen as distinct images by the observer. This was the case considered by Zwicky, and the gravitational lensing in this regime is called *strong (or macro) lensing* (see Section 3), which also includes distorted galaxy images, like Einstein rings or arcs. If instead, the distortions induced by the gravitational fields on background objects are much smaller, we have the *weak lensing* effect (see Section 4). On the other side, if one considers the phenomenology of the star-on-star lensing (as Einstein did), the resulting angular distance between the images is of the order of a few μas, generally not separable by telescopes. Gravitational lensing in this regime is called, following Paczyński [8], *microlensing*, and the observable is an achromatic change in the brightness of the source star over time, due to the relative motion of the lens and the source with respect to the line of sight of the source (see Section 5). In all of these regimes, the gravitational field can be treated in the weak field approximation. Another scale on which gravitational lensing applies is that involving black holes. In particular, when light rays come very close to the event horizon, they are subject to strong gravitational field effects, and thereby, the deflection angles are large. This effect is called *retro-lensing*, and we shall discuss it in Section 6. The observation of retro-lensing events is of great importance also because the general theory of relativity still stands practically untested in the strong gravitational field regime (see [9] for a very recent review), apart from gravitational waves [10]. A short final discussion is then offered in Section 7.

2. Basics of Gravitational Lensing

In the general theory of relativity, light rays follow null geodesics, *i.e.*, the minimum distance paths in a curved space-time. Therefore, when a light ray from a far source interacts with the gravitational field due to a massive body, it is bent by an angle approximately equal to $\alpha_S(b) = 2r_s/b$. By looking at Figure 1, assuming the ideal case of a thin lens and noting that $\alpha_S D_{LS} = (\theta - \theta_S)D_S$, one can easily derive the so-called lens equation:

$$\theta_S = \theta - \frac{\theta_E^2}{\theta}, \tag{1}$$

where θ_S indicates the source position and:

$$\theta_E = \left(\frac{4GM}{c^2}\frac{D_{LS}}{D_S D_L}\right)^{1/2} \tag{2}$$

is the Einstein ring radius, which is the angular radius of the image when lens and source are perfectly aligned ($\theta_S = 0$). Therefore, one can see that two images appear in the source plane, whose positions can be obtained by solving Equation (1).

More generally (see for details [11]), the light deflection between the two-dimensional position of the source $\boldsymbol{\theta}_S$ and the position of the image $\boldsymbol{\theta}$ is given by the lens mapping equation:

$$\boldsymbol{\theta}_S = \boldsymbol{\theta} - \boldsymbol{\nabla}\phi(\boldsymbol{\theta}), \tag{3}$$

where $\phi = 2D_{LS}\Phi_N^{2D}/(D_S c^2)$ is the so-called lensing potential and Φ_N^{2D} is the two-dimensional Newtonian projected gravitational potential of the lens. We also note, in turn, that the ratio D_{LS}/D_S depends on the redshift of the source and the lens, as well as on the cosmological parameters $\Omega_M = \rho_M/\rho_c$ and $\Omega_\Lambda = \rho_\Lambda/\rho_c$, being $\rho_c = 3H_0^2/(8\pi G)$, ρ_M and ρ_Λ the critical, the matter and the dark energy densities, respectively. The transformation above is thus a mapping from the source plane to the image plane, and the Jacobian $\mathcal{J}$ of the transformation is given by:

$$\mathcal{J} = \frac{d\boldsymbol{\theta}_S}{d\boldsymbol{\theta}} = \mathcal{A}^{-1} = \begin{pmatrix} 1-\phi_{,11} & -\phi_{,12} \\ -\phi_{,12} & 1-\phi_{,22} \end{pmatrix} = \begin{pmatrix} 1-\kappa-\gamma_1 & -\gamma_2 \\ -\gamma_2 & 1-\kappa+\gamma_1 \end{pmatrix}, \tag{4}$$

where the commas are the partial derivatives with respect to the two components of $\boldsymbol{\theta}$. Here, κ is the convergence, which turns out to be equal to $\Sigma/2\Sigma_{\rm cr}$, where:

$$\Sigma_{\rm cr} = c^2 D_S/(4\pi G D_L D_{LS}) \tag{5}$$

is the critical surface density, $\gamma = (\gamma_1, \gamma_2)$ is the shear and $\mathcal{A}$ is the magnification matrix. Thus, the previous equations define the convergence and shear as second derivatives of the potential, *i.e.*,

$$\kappa = \frac{1}{2}(\partial_1\partial_1 + \partial_2\partial_2)\phi = \frac{1}{2}\nabla^2\phi, \quad \gamma_1 = (\partial_1\partial_1 - \partial_2\partial_2)\phi, \quad \gamma_2 = \partial_1\partial_2\phi. \tag{6}$$

From the above discussion, it is clear that gravitational lensing may allow one to probe the total mass distribution within the lens system, which reproduces the observed image configurations and distortions. This, in turn, may allow one to constrain the cosmological parameters, although this is a second order effect.

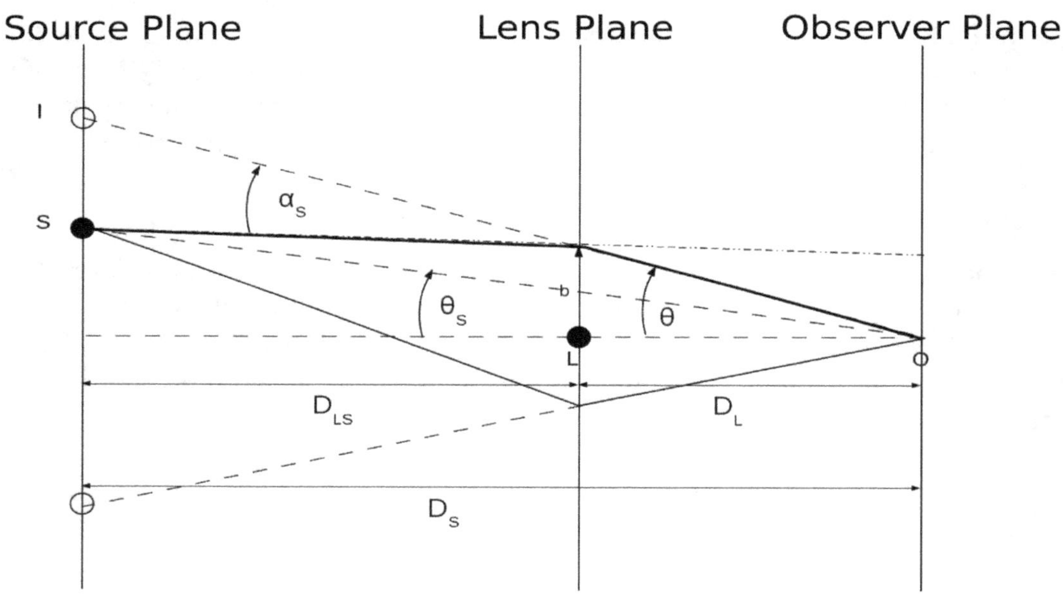

Figure 1. Schematics of the lensing phenomenon.

3. Strong Lensing

Quasars are the brightest astronomical objects, visible even at a distance of billions of parsecs. After the identification of the first quasar in 1963 [12], these objects remained a mystery for quite a long time, but today, we know that they are powered by mass accretion on a supermassive black hole, with a mass billions of times that of the Sun. The first strong gravitational lens, discovered in 1979, was indeed linked to a quasar (QSO 0957+561 [7]), and although the phenomenon was expected on theoretical grounds, it left the astronomers astonished. The existence of two objects separated by about 6″ and characterized by an identical spectrum led to the conclusion that they were the doubled image of the same quasar, clearly showing that Zwicky was perfectly right and that galaxies may act as gravitational lenses. Afterwards, also the lens galaxy was identified, and it was established that its dynamical mass, responsible for the light deflection, was at least ten-times larger than the visible mass. This double quasar was also the first object for which the time delay (about 420 days) between the two images [13], due to the different paths of the photons forming the two images, has been measured. This has also allowed obtaining an independent estimate of the lens galaxy dynamical mass. Observations can also show four images of the same quasar, as in the case of the so-called Einstein Cross, or when the lens and the source are closely aligned, one can observe the Einstein ring, e.g., in the case of MG1654-1346 [14]. The macroscopic effect of multiple images' formation is generally called strong lensing, which also consists of the formation of arcs, as those clearly visible in the deep sky field images by the Sloan Digital Sky Survey (SDSS; see, e.g., [15]). The sources of strong lensing events are often quasars, galaxies, galaxy clusters and supernovae, whereas the lenses are usually galaxies or galaxy clusters. The image separation is generally larger than a few tenths of an arcsec, often up to a few arcsecs.

Over the years, many strong lensing events have been found in deep surveys of the sky, such as the CLASS [16], the Sloan ACS [17], the SDSS, one of the most successful surveys in the history of astronomy (see, e.g., [18] and references therein), the SQLS (the Sloan Digital Sky Survey for Quasar Lens Search) [19], and so on.

Strong gravitational lensing is nowadays a powerful tool for investigation in astrophysics and cosmology (see, e.g. [20,21]). As already mentioned in the previous section, strong lensing gives a unique opportunity to measure the dynamical mass of the lens object using, for example, the mass estimator $M(< R_E) = \pi\Sigma_{\rm cr}\theta_E^2$, which directly gives the mass within R_E, using Equation (5) in this regime. The result is that masses obtained in this way are almost always larger than the visible mass of the lensing object, showing that galaxy and galaxy cluster masses are dominated by dark matter. In any case, accurately constraining the mass distribution of the lens system (e.g., a galaxy cluster) is a generally degenerate problem, in the sense that there are several mass distributions that can fit the observables; thus, the best way to solve it is to use multiple images (see, e.g., [22]).

Another important application of strong lensing is the study of dark matter halo substructures. Indeed, sometimes flux ratio anomalies in the lensed quasar images are detected (see, e.g., [23,24]), and while smooth mass models of the lensing galaxy may generally explain the observed image positions, the prediction of such models of the corresponding fluxes is frequently violated. Especially in the radio band observations, since the quasar radio emitting region is quite large, the observed radio flux anomalies are explained as being due to the presence of substructures of about $10^6 - 10^8\ M_\odot$ along the line of sight. After some controversy regarding whether ΛCDM (cold dark matter plus Cosmological Constant) simulations predict enough dark matter substructures to account for the observations (for example, in [25], some indication is found of an excess of massive galaxy satellites), more recent analysis, taking also into account the uncertainty in the lens system ellipticity, finds results consistent with those predicted by the standard cosmological model [26,27]. However, at present, the list of multiply-imaged quasars observed in the radio and mid-IR bands is quite short, and further observational and theoretical work would be very helpful in this respect. Another indication of dark matter halo substructures comes from detailed analysis of galaxy-galaxy lensing. Although the results obtained are generally consistent with ΛCDM simulations, more data should be analyzed in order to get strong constraints [28,29]. Strongly lensed quasars have been observed to show a certain variability of one image with respect to the others. This can be often attributed to microlensing (see Section 5) by the stars throughout the lens galaxy. This effect, and in particular its variation with respect to the wavelength, has provided an opportunity to study in detail the central engine of the source quasar, and the magnitude of the microlensing variability has allowed astrophysicists to constrain the stellar density in the lens galaxy [30–32].

Strong gravitational lensing may be used as a natural telescope that magnifies dim galaxies, making them easier to be studied in detail. For this reason, mass concentrations, like galaxies and clusters of galaxies, can be effectively used as cosmic telescopes to study faint sources that would not be possible to detect in the absence of gravitational lensing (see, e.g., [33,34]). At present, there is also an event of a high magnified supernova multiply imaged and also seen exploding again, being lensed by a galaxy in the cluster MACS J1149.6+2223 [35,36].

The ultimate goal of strong lensing is not only to get information on the large-scale structure of the Universe, but also to constrain the cosmological parameters. For instance, analyzing the time delay among the lensed source images, it is possible to estimate also the value of the Hubble constant H_0. Indeed, the time delay is given by the difference of the light paths from the images and is inversely proportional to H_0, as first understood by Refsdal [37] (see also the review in [38]). At present, one of the most accurate measurement of the Hubble constant using a gravitational lens is provided in [39]. There is also a project (COSMOGRAIL) particularly devoted to the time delay measurements of doubly- or multiply-lensed quasars (see [40] and the references therein). Moreover, the measure of both the frequency of occurrence and the redshift of multiple images in deep sky surveys may allow one to constrain the values of Ω_M and Ω_Λ in an independent way with respect to other methods, such as those coming from SN Ia or the CMB (Cosmic Microwave Background) power spectrum.

4. Weak Lensing

In addition to the macroscopic deformations discussed in the previous section, in the deep field surveys of the sky, also arclets (*i.e.*, single distorted images with an elliptical shape) and weakly distorted images of galaxies, with an almost invisible individual elongation, have been detected. This effect is known as weak lensing and is playing an increasingly important role in cosmology.

The weak lensing's main feature is the shape deformation of background galaxies, whose light crosses a mass distribution (e.g., a galaxy or a galaxy cluster) that acts as a gravitational lens. Actually, as discussed in Section 2, gravitational lensing gives rise to two distinct effects on a source image: convergence, which is isotropic, and shear, which is anisotropic. In the weak lensing regime, the observer makes use of the shear, that is the image deformation (sometimes related to the galaxy orientation), while the convergence effect is not used, since the intrinsic luminosity and the size of the lensed objects are unknown. For a complete and in-depth review on the basics of weak gravitational lensing, with full mathematical details of all the most important concepts, we refer the reader to [41].

The first weak lensing event was detected in 1990 as statistical tangential alignment of galaxies behind massive clusters [42], but only in 2000, coherent galaxy distortions were measured in blind fields, showing the existence of the cosmic shear (see, e.g., [43,44]). Here, we remark that the weak lensing cannot be measured by a single galaxy, but its observation relies on the statistical analysis of the shape and alignment of a large number of galaxies in a certain direction.

Therefore, the game is to measure the galaxy ellipticities and orientations and to relate them to the surface mass density distribution of the lens system (generally a galaxy cluster placed in between). There are at least two major issues in weak lensing studies, one mainly relying on the theory, the other one on observations: the former concerns finding the best way to reconstruct the intervening mass distribution from the shear field $\gamma = (\gamma_1, \gamma_2)$, the latter with looking for the best way to determine the *true* ellipticity of a faint galaxy, which is smeared out by the instrumental point spread function (PSF). To solve these issues, several approaches have been proposed, which can be distinguished into two broad families: direct and inverse methods. On the theoretical side, the direct approaches are: the integral method, which consists of expressing the projected mass density distribution as the convolution of γ by a kernel (see, e.g., [45]), and the local inversion method, which instead starts from the gradient of γ (see, e.g., [46] and the references therein). The inverse approaches work on the lensing potential ϕ (see Equation 3), and they include the use of the maximum likelihood [47,48] or the maximum entropy methods [49] to determine the most likely projected mass distribution that reproduces the shear field. The inverse methods are particularly useful since they make it possible to quantify the errors in the resultant lensing mass estimates, as, for instance, errors deriving from the assumption of a spherical mass model when fitting a non-spherical system [50,51].

The inverse methods allow one also to derive constraints from external observations, such as X-ray data on galaxy clusters' strong lensing or CMB lensing. In particular, one can compare mass measurements from weak lensing and X-ray observations for large samples of galaxy clusters [52]. In this respect, [53] used a large sample of nearby clusters with good weak lensing and X-ray measurements to investigate the agreement between mass estimates based on weak lensing and X-ray data, as well as studied the potential sources of errors in both methods. Moreover, a combination of weak lensing and CMB data may provide powerful constraints on the cosmological parameters, especially on the Hubble constant H_0, the amplitude of fluctuations σ_8 and the matter cosmic density Ω_m [54,55]. We also mention, in this respect, that one way to determine the fluid-mechanical properties of dark energy, characterized by its sound speed and its viscosity apart from its equation of state, is to combine Planck data with galaxy clustering and weak lensing observations by Euclid, yielding one percent sensitivity on the dark energy sound speed and viscosity [56] (see the end of this section).

On the observational side, the first priority is to use a telescope with a wide field of view, appropriate to probe the large-scale structure distribution at least of a galaxy cluster. On the other hand, it is also necessary to minimize the source of noise in the determination of the ellipticity of very faint galaxies, so that the best-seeing conditions for a ground-based telescope or, better, a space-based instrument, are extremely useful.

Very promising results have been obtained with the weak lensing technique so far, as, for example, the best measure, until today, of the existence and distribution of dark matter within the famous Bullet cluster [57] (actually constituted by a pair of galaxy clusters observed in the act of colliding). Astronomers found that the shocked plasma was almost entirely in the region between the two clusters, separated from the galaxies. However, weak lensing observations showed that the mass was largely concentrated around the galaxies themselves, and this enabled a clear, independent measurement of the amount of dark matter.

With the major aim to map, through the weak lensing effect, the mass distribution in the Universe and the dark energy contribution by measuring the shape and redshift of billions of very far away galaxies (for a review, see [58]), the European Space Agency (ESA) is planning to launch the Euclid satellite in the near future. Also ground-based telescopes will allow one to detect an enormous number of weak and strong lensing events. An example is given by the LSST (Large Synoptic Survey Telescope) project, located on the Cerro Pachón ridge in north-central Chile, which will become operative in 2022. Its 8.4-meter telescope uses a special three-mirror design, creating an exceptionally wide field of view, and has the ability to survey the entire sky in only three nights. The effective number density n_{eff} of weak lensing galaxies (which is a measure of the statistical power of a weak lensing survey) that will be discovered by LSST is, conservatively, in the range of 18–24 arcmin^{-2} (see Table 4 in [59]). The very large (about 1.5×10^4 square degrees) and deep survey of the sky that will be performed by Euclid will allow astrophysicists to address fundamental questions in physics and cosmology about the nature and the properties of dark matter and dark energy, as well as in the physics of the early Universe and the initial conditions that provided the seeds for the formation of cosmic structure. Before closing this section, we also mention that strong systematics may be present in weak lensing surveys. For example, the intrinsic alignment of background sources may mimic to an extent the effects of shear and may contaminate the weak lensing signal. However, these systematics may be controlled if also the galaxy redshifts are acquired, and this fully removes the unknown intrinsic alignment errors from weak lensing detections (for further details, see [60,61]).

5. Microlensing

Let us consider now the microlensing scale of the lensing phenomenon that occurs when θ_E is smaller than the typical telescope angular resolution, as in the case of stars lensing the light from background stars (for a review on gravitational microlensing and its astrophysical applications, we refer to, e.g., [62]). As is clear from the discussion in Section 2, by solving the lens Equation (1), one can determine the angular positions of the primary (I_1) and secondary (I_2) images. In Figure 2, these positions are shown for four different values of the impact parameter θ_S in the case of a point-like source. If the source and the lens are aligned (first panel on the left), the circular symmetry of the problem leads to the formation of a luminous annulus having radius θ_E around the lens position. Otherwise, increasing the θ_S value, the secondary image gets closer to the lens position, while the primary image drifts apart from it, and in the limit of $\theta_S \gg \theta_E$, the microlensing phenomenon tends to disappear. However, observing multiple images during a microlensing event is practically impossible with the present technology. For instance, in the case in which the phenomenon is maximized, corresponding to the perfect alignment, for a star in the galactic bulge (about 8 kpc away), one has $\Delta\theta = 2\theta_E \simeq 1$ μas, which is well below the angular resolving power, even of the Hubble Space Telescope (about 43 mas at 500 nm); see, e.g., http://www.coseti.org/9008-065.htm.

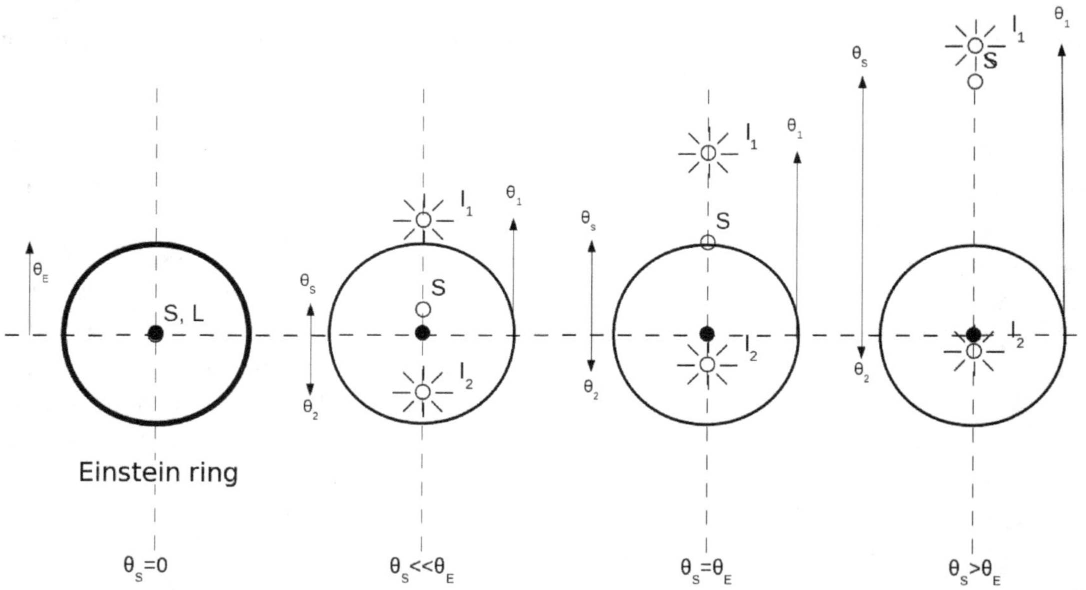

Figure 2. Angular positions of the primary (I_1) and secondary (I_2) images for four different values of the source impact parameter θ_S in the case of a point-like source.

When a source is microlensed, its images do not have the same luminosity; therefore, the observer receives a total flux (or magnitude) different from that of the unlensed source. The flux difference can be described very simply in terms of the light magnification and the law of the conservation of the specific intensity I, which represents the energy, with frequency in the range $d\nu$ crossing the surface dA during the time interval dt in the solid angle $d\Omega$ around the direction orthogonal to the surface. Indeed, the light specific intensity turns out to be conserved in the absence of absorption phenomena, interstellar scattering or Doppler shifts. This is also a consequence of Liouville's theorem, which claims that the density of states in the phase space remains constant if the interacting forces are non-collisional (and gravitation fulfills this condition due to its weak coupling constant), and the propagating medium is approximately transparent (as is the case for interstellar space). This effect can produce a magnification or a de-magnification of the images of an extended light source (see Figure 3). If the image is magnified, it means that it certainly subtends a wider angle with respect to that subtended by the source in the absence of the lens. In microlensing, the source disk size should not be neglected in general. Within the framework of the finite source approximation for a source with flux F_S and assuming $\theta_E \leq \theta_S$, one can show that the magnification A of an image at angular position θ is given by $(1 - \theta_E^4/\theta^4)^{-1}$. As a consequence, the observed flux corresponds to $F = AF_S$. Of course, when the source star disk gradually moves away from the line of sight, the magnification decreases, and the unlensed F_S flux is then recovered.

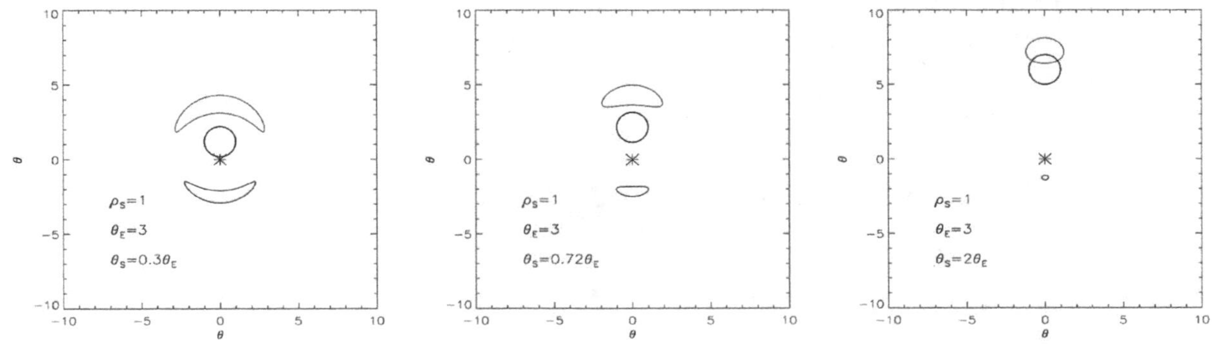

Figure 3. As in Figure 2, but with a conformal transformation of the source boundary, considered extended and with radius ρ_S, by a point-like lens. Each point of the source disk behaves as a point-like source. The black circle represents the source disk, while the red and yellow arcs are the deformed primary and secondary images.

As already anticipated, the observer cannot see, in the microlensing case, well-separated images, but, instead, detects a single image made by the overlapping of the primary and the secondary images. In this case, one can easily obtain the classical magnification factor A by summing up the individual magnifications, *i.e.*, $A = (u^2+2)/\sqrt{u^2(u^2+4)}$, where $u = \theta_S/\theta_E$ is the impact factor. If there is a relative movement between the lens and the source, u changes with time, and a standard Paczyňski curve [8] does emerge.

An important role in gravitational microlensing is played by the *caustics*, the geometric loci of the points belonging to the lens plane where the light magnification of a point-like source becomes infinite, and by the corresponding *critical* curves in the source plane. In the case of a single lens, the caustic is a point coinciding with the lens position; therefore, the magnification diverges when the impact parameter approaches zero. However, real sources are not point-like, so we always have finite magnifications that can be calculated by an average procedure:

$$\langle A \rangle = \frac{\int A(\mathbf{y}) I(\mathbf{y}) d^2 y}{\int I(\mathbf{y}) d^2 y}, \tag{7}$$

where $A(\mathbf{y})$ is the point-like source magnification, $I(\mathbf{y})$ is the brightness profile of the stellar disk (the limb darkening profile) and the integral is extended over the source star disk.

Observations show that about half of all stars are in binary systems, and moreover, thousands of exoplanets are being discovered around their host stars by different techniques and instruments. Therefore, it is worth considering binary and multiple systems as lenses in microlensing observations. In this case, the lens equation, obviously, becomes more complicated, but it can still be solved by numerical methods in order to obtain the magnification map where caustics take on distinctive shapes depending on the specific geometry of the system. In Figure 4, we show the magnification map and the resulting light curve for a simulated microlensing event due to a binary lens with mass ratio $q = M_1/M_2 \simeq 0.01$ (e.g., a solar mass as the primary component and a Jupiter-like planet as the secondary one). In these cases, the resulting light curve may be rather different with respect to the typical Paczyňski one, depending on the system parameters. The study of these anomalies in the microlensing light curves behavior is becoming more and more important nowadays, since it allows one to estimate some of the parameters of the lensing system (see, e.g., [63]). The main advantage of this technique, compared to the other methods adopted by the exoplanets hunters (e.g., radial velocity, direct imaging, transits), is the possibility to detect even very small planets orbiting their own star at enormous distances from Earth. It also allows one to discover the so-called free-floating planets (FFPs), otherwise hardly detectable [64]. By studying the PA99-N2 microlensing event, detected in 1999 by the French-British collaboration POINT-AGAPE [66], Ingrosso *et al.* [65] revealed in 2009 that the anomaly observed was compatible with the presence of a super-Jupiter with a mass of $\simeq 5M_J$ around a star lying in the Andromeda galaxy (see Figure 5), thus finding the first putative exoplanet in another galaxy.

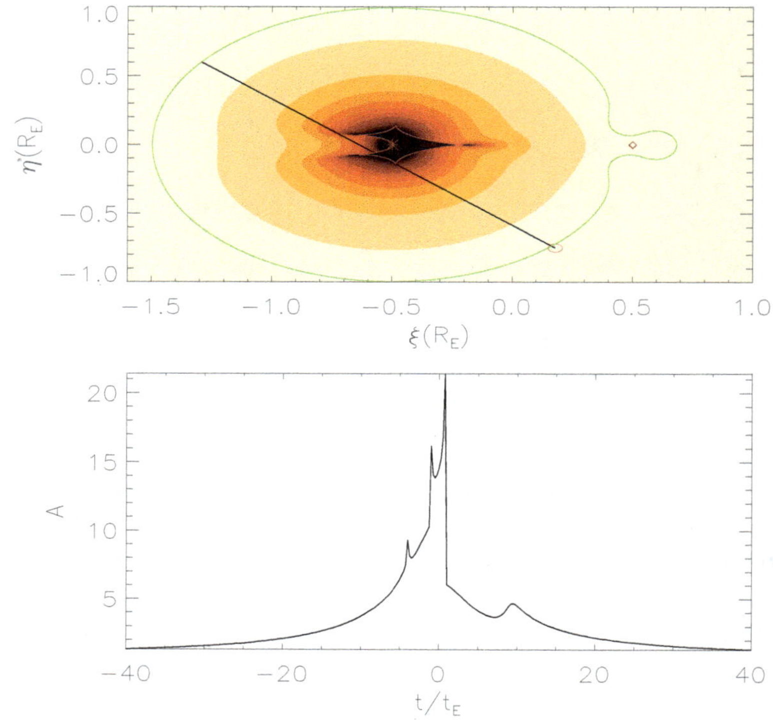

Figure 4. Magnification map for a binary lens system characterized by two objects separated by a projected distance of $1R_E$ and mass ratio $q = 0.01$. The green and red closed lines indicate the critical and caustic curves obtained by solving the lens equation in Equation (1). The black line indicates the trajectory of the source star, which has a radius of $0.03R_E$. The simulated light curve is shown in the lower panel.

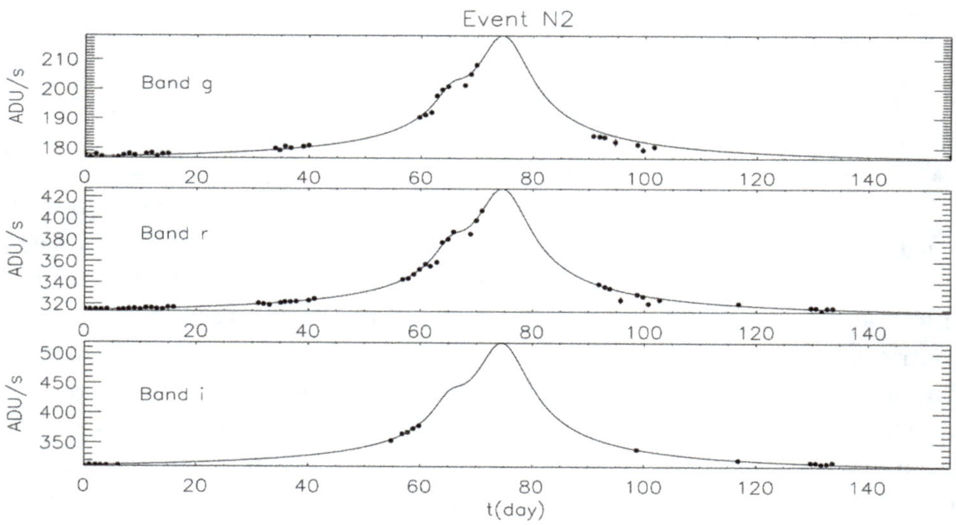

Figure 5. Light curve in three different bands (g, r and i) of the PA99-N2 event detected in 1999 toward the Andromeda galaxy.

5.1. *Astrometric Microlensing*

During an ongoing microlensing event, the centroid of the multiple images and the source star positions move in the lens plane giving rise to a phenomenon known as astrometric microlensing (see, e.g., [67,68] and the references therein). In the simplest case of a point-like lens, for a source at angular distance θ_S, the position θ of the images with respect to the lens can be obtained by solving the lens Equation (1). Since the Einstein radius $R_E = D_L\theta_E$ defines the scale length on the lens plane, the lens equation reads:

$$d^2 - dd_S - R_E^2 = 0, \tag{8}$$

where d_S and d are the linear distances, in the lens plane, of the source and images from the gravitational lens, respectively. Moreover, using the dimensionless source-lens distances $u = \theta_S/\theta_E$ and $\tilde{u} = \theta/\theta_E$, the previous relation can be further simplified as:

$$\tilde{u}^2 - u\tilde{u} - 1 = 0. \tag{9}$$

Denoting with u_+ and u_- the solutions of this equation, one notes that, in the lens plane, the $+$ image resides always outside the circular ring centered on the lens position with radius equal to the Einstein angle, while the $-$ image is always within the ring. As the source-lens distance increases, the $+$ image approaches the source position, while the $-$ one (becoming fainter) moves towards the lens location. For a source moving in the lens plane with transverse velocity $v_\perp$ directed along the ξ axis (η is perpendicular to it), the projected coordinates of the source result in being:

$$\xi(t) = \frac{t - t_0}{t_E}, \quad \eta(t) = u_0. \tag{10}$$

where $t_E = R_E/v_\perp$ and u_0 is the impact parameter (in this case, lying on the η axis). Since $u^2 = \xi^2 + \eta^2$ is time dependent, the two images move in the lens plane during the gravitational lensing event.

By weighting the $+$ and $-$ image position with the associated magnification [69], one gets:

$$\bar{u} \equiv \frac{\tilde{u}_+\mu_+ + \tilde{u}_-\mu_-}{\mu_+ + \mu_-} = \frac{u(u^2+3)}{u^2+2}. \tag{11}$$

Finally, the observable is defined as the displacement of the centroid with respect to the source,

$$\Delta \equiv \bar{u} - u = \frac{u}{2+u^2}. \tag{12}$$

Note that the centroid shift may be viewed as a vector:

$$\boldsymbol{\Delta} = \frac{\mathbf{u}}{2+u^2} \tag{13}$$

with components along the axes:

$$\Delta_\xi = \frac{\xi(t)}{2+u^2}, \quad \Delta_\eta = \frac{u_0}{2+u^2}. \tag{14}$$

Here, we remind that all of the angular quantities are given in units of the Einstein angle θ_E, which, for a source at distance $D_S \gg D_L$, results in being:

$$\theta_E \simeq 2\left(\frac{M}{0.5\,\mathrm{M}_\odot}\right)^{1/2}\left(\frac{D_L}{\mathrm{kpc}}\right)^{-1/2} \mathrm{mas}, \tag{15}$$

which fixes the scale of the phenomenon.

It is straightforward to show (see [69]) that during a microlensing event, the centroid shift Δ traces (in the Δ_ξ, Δ_η plane) an ellipse centered in the point $(0, b)$. The ellipse semi-major axis a (along Δ_η) and semi-minor axis b (along Δ_ξ) are:

$$a = \frac{1}{2}\frac{1}{\sqrt{u_0^2+2}}, \quad b = \frac{1}{2}\frac{u_0}{u_0^2+2}. \tag{16}$$

Then, for $u_0 \to \infty$, the ellipse becomes a circle with radius $1/(2u_0)$, while it degenerates into a straight line of length $1/\sqrt{2}$ for u_0 approaching zero. Note also that Equation (16) implies:

$$u_0^2 = 2(b/a)^2\left[1-(b/a)^2\right]^{-1}, \tag{17}$$

so that by measuring a and b, one can determine the event impact parameter u_0.

As observed in [67], $\mathbf{\Delta}$ falls more slowly than the magnification, implying that the centroid shift may be an interesting observable also for large source-lens distances, *i.e.*, far from the light curve peak. In fact, in astrometric microlensing, the threshold impact parameter $u_{\rm th}$ (*i.e.*, the value of the impact parameter that gives an astrometric centroid signal larger than a certain quantity $\delta_{\rm th}$) is given by $u_{\rm th} = \sqrt{T_{\rm obs}v_\perp/(\delta_{\rm th}D_L)}$, where $T_{\rm obs}$ is the observing time and $v_\perp$ the relative velocity of the source with respect to the lens. For example, the Gaia satellite should reach an astrometric precision $\sigma_G \simeq 300$ μas (for objects with visual magnitude $\simeq 20$) in five years of observation [70]. Then, assuming a threshold centroid shift $\delta_{\rm th} \simeq \sigma_G$, one has $u_{\rm th} \simeq 60$ for a lens at a distance of 0.1 kpc and transverse velocity $v_\perp \simeq 100$ km s^{-1}. For comparison, the threshold impact parameter for a ground-based photometric observation is $\simeq 1$. Consequently, the cross-section for an astrometric microlensing measurement is much larger than the photometric one, since it scales as $u_{\rm th}^2$. Hence, in the absence of finite-source and blending effects, by measuring a and b, one can directly estimate the impact parameter u_0.

A further advantage of the astrometric microlensing is that some events can be predicted in advance [71]. In fact, by studying in detail the characteristics of stars with large proper motions, Proft *et al.* [72] identified tens of candidates to measure astrometric microlensing by the Gaia satellite, an European Space Agency (ESA) mission that will perform photometry, spectroscopy and high precision astrometry (see [70]).

5.2. Polarization and Orbital Motion Effects in Microlensing Events

Gravitational microlensing observations may also offer a unique tool to study the atmospheres of far away stars by detecting a characteristic polarization signal [73]. In fact, it is well known that the light received from stars is linearly polarized by the photon scattering occurring in the stellar atmospheres. The mechanism is particularly effective for the hot stars (of A or B type) that have a free electron atmosphere giving rise to a polarization degree increasing from the center to the stellar limb [74]. By a minor extent, polarization may be also induced in main sequence F or G stars by the scattering of star light off atoms/molecules and in evolved, cool giant stars by photon scattering on dust grains contained in their extended envelopes.

Following the approach in [74], the polarization P in the direction making an angle $\chi = \arccos(\mu)$ with the normal to the star surface is $P(\mu) = [I_r(\mu) - I_l(\mu)]/[I_r(\mu) + I_l(\mu)]$, where $I_l(\mu)$ is the intensity in the plane containing the line of sight and the normal, and $I_r(\mu)$ is the intensity in the direction perpendicular to this plane. Here, $\mu = \sqrt{1-(r/R)^2}$, where r is the distance of a star disk element from the center and R the star radius, and we are assuming that light propagates in the direction $\mathbf{r} \times \mathbf{l}$.

For isolated stars, a polarization signal has been measured only for the Sun for which, due to the distance, the projected disk is spatially resolved. Instead, when a star is significantly far away and can be considered as point-like, only the polarization $\langle P\rangle$ averaged over the stellar disk can be measured, and usually $\langle P\rangle = 0$, since the flux from each stellar disk element is the same. A net polarization of the light appears if a suitable asymmetry in the stellar disk is present (caused by, e.g., eclipses, tidal distortions, stellar spots, fast rotation, magnetic fields). In the microlensing context, the polarization arises since different regions of the source star disk are magnified differently during the event. Indeed, during an ongoing microlensing event, the gravitational lens scans the disk of the background star, giving rise not only to a time-dependent light magnification, but also to a time-dependent polarization.

This effect (see also [75]) is particularly relevant in the microlensing events where: (1) the magnification turns out to be significant; (2) the source star radius and the lens impact parameter are comparable; (3) the source star is a red giant, characterized by a rather low surface temperature ($T \leq 3000\,K$), around which the formation of dust grains is possible. This occurs beyond the distance R_h from the star center at which the gas temperature in the stellar wind becomes lower than the grain sublimation temperature ($\simeq$ 1400 K). The intensity of the expected polarization signal relies on the dust grain optical depth τ and can reach values of 0.1%–1%, which could be reasonably observed using, for example, the ESO VLTtelescope (see [76]). In Figure 6, we show some typical polarization curves, expected in *bypass* (continuous curves) and *transit* events (dashed curves), in which the lens trajectory approaches or passes through the source regions where the dust grains are present. In Figure 7, the distribution of the peak polarization values (given in percent) as a function of the intrinsic source star color index $(V-I)_{\text{int}}$ (*i.e.*, the de-reddened color of the unlensed source star) is shown for a sample of OGLE-type microlensing events generated by a synthetic stellar catalog simulating the bulge stellar population. As one can see, red giants with $(V-I)_{\text{int}} \leq 3$, which corresponds to the events inside the regions delimited by dashed lines, have $P_{\text{max}} \leq 1$ percent values. These are the typical events observed by the OGLE-III microlensing campaign. There are, however, a few events with $1 \leq P_{\text{max}} \leq 10$ percent, characterized by $(V-I)_{\text{int}} \geq 3$, corresponding to source stars in the AGB phase. These stars, which are rather rare in the galactic bulge, have not been sources of microlensing events observed in the OGLE-III campaign, but they are expected to exist in the galactic bulge. In this respect, the significant increase in the event rate by the forthcoming generation of microlensing surveys towards the galactic bulge, both ground-based, like KMTNet [77], and space-based, like EUCLID [78] and WFIRST [79], opens the possibility to develop an alert system able to trigger polarization measurements in ongoing microlensing events.

Another way to study the atmosphere of the source star is to analyze the amplification curve and look for dips and peaks, typically due to the presence of stellar spots on the photosphere of the star [80,81]. These features may be easily confused, however, with the signatures of a binary lensing system. When the source star has a relevant rotation motion during the lensing event, there is the possibility to really detect the stellar spots on the source's surface and to estimate the rotation period of the star [82]. A new generation of networks of telescopes dedicated to microlensing surveys, like KMTNet [83], will provide high-precision and high-cadence photometry that will enable us to observe spots on the source's surface. We remark that also multicolor observations of the event would help to disentangle the aforementioned degeneracy, as the ratio between the brightness of the spot and the surrounding photosphere strongly depends on the frequency of the observation. It has been shown that stellar spots can be detected also through polarimetric observations of microlensing caustic-crossing events [84].

Under certain circumstances, binary lens systems are characterized by the *close-wide degeneracy*: if the two objects are separated by a projected distance s or $1/s$, the resulting caustics have the same structure, and also, the observed light curves will appear the same [85,86]. This happens, for example, in systems with small mass ratio q, like planetary systems [87]. It is possible to resolve this degeneracy in the case of short-period binary lenses, the so-called *rapidly rotating lenses*, as the orbital motion induces repeating features in the amplification curve that can be exploited to estimate important physical parameters of the lensing systems, including the orbital period, the projected separation and the mass [88,89].

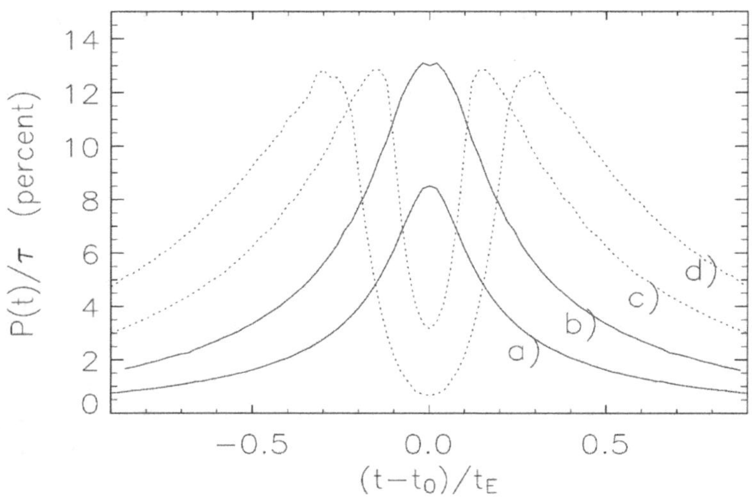

Figure 6. Assuming parameter values $u_0 = 0.09$, $t_E = 60$ day and $R_h/R_S = 5$, the $P(t)/\tau$ polarization curves are shown as a function of $(t - t_0)/t_E$, for increasing values of $R_h/u_0 = 0.35$, 0.75, 1.5, 2.5, corresponding to the curves labeled (a), (b) (c) and (d), respectively. Continuous curves, (a) and (b), are *bypass* events; dotted lines, (c) and (d), are *transit* events. Here, R_h is the minimum distance for the formation of dust grains, and R_S is the source star radius.

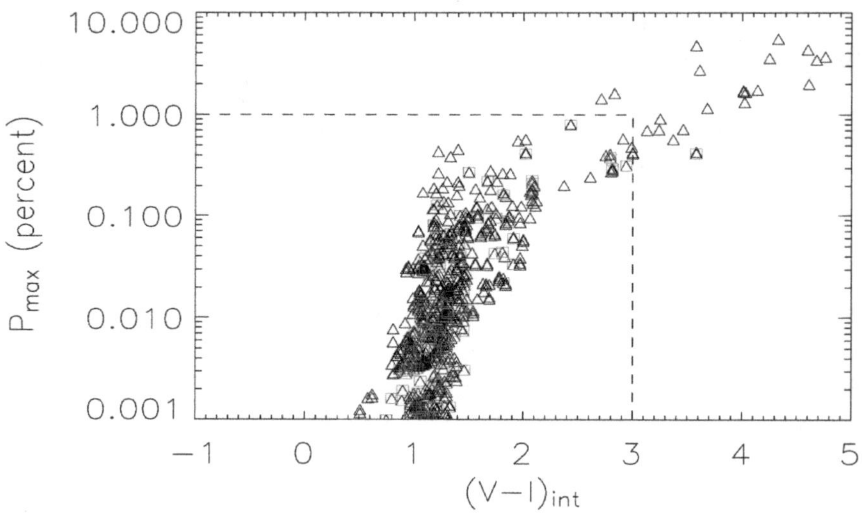

Figure 7. Distribution of the peak value of the polarization signal as a function of the intrinsic color index $(V - I)_{\text{int}}$ of the source star for simulated *transit* (triangles) and *bypass* (purple squares) events. The dashed lines indicate the region in which the events observed by the OGLECollaboration are expected to lie.

6. Retro-Lensing: Measuring the Black Hole Features

Gravitational lensing at the scales considered in the previous sections can be treated in the weak gravitational field approximation of the general theory of relativity, since in those cases, photons are deflected by very small angles. This is not the case when one considers black holes, for which it may happen that photons get very close to the event horizon of these compact objects.

Black holes are relatively simple objects. The *no-hair theorem* postulates that they are completely described by only three parameters: mass, angular momentum (generally indicated by the spin parameter *a*) and electric charge; any other information (for which *hair* is a metaphor) disappears behind the event horizon, and it is therefore inaccessible to external observers. Depending on the values of these parameters, black holes can be classified into Schwarzschild black holes (non-rotating and non-charged), Kerr black holes (rotating and non-charged), Reissner–Nordström black holes (non-rotating and charged) and Kerr–Newman black holes (rotating and charged).

Even though they appear so simple, black holes are mathematically complicated to describe (see, e.g., [90]). Nowadays, we know that black holes are placed at the center of the majority of galaxies, active or not, and in many binary systems emitting X-rays. Moreover, they are the engine of gamma-ray bursts (GRBs) and play an essential role in better understanding stellar evolution, galaxy formation and evolution, jets and, in the end, the nature of space and time. One goal astrophysicists have been pursuing for a long time is to probe the immediate vicinity of a black hole with an angular resolution as close as possible to the size of the event horizon. This kind of observations would give a new opportunity to study strong gravitational fields, and as we will see at the end of this section, we think we are very close to reaching this goal.

How do we measure the mass, angular momentum and electric charge of a black hole? One possibility, rich with interesting consequences, was suggested by Holz and Wheeler [91], who considered a phenomenon that was already known to be possible around black holes. They used the Sun as the source of light rays and a black hole far from the solar system. As shown in Figure 8, some photons would have the right impact parameter to turn around the black hole and come back to Earth. Other photons, with a slightly smaller impact parameter, can even rotate twice around the black hole, and so on. A series of concentric rings should then appear if the observer, the Sun and the black hole are perfectly aligned. The two authors also suggested to do a survey and look for concentric rings in the sky in order to discover black holes. Unfortunately, there are two problems with this idea. First, it is unlikely that the Sun, Earth and a black hole are perfectly aligned, and in any case, Earth moves around the Sun, so that the alignment can occur only for a short time interval. The second and most important problem is that the retro-image of the Sun is so dim, that even using the Hubble Space Telescope (HST), only a black hole with a mass larger than 10 $M_\odot$ within 0.01 pc from the Earth could be observed with the proposed technique. Moreover, we already know that such an object cannot be so close to the solar system without causing observable perturbations in the planet orbits.

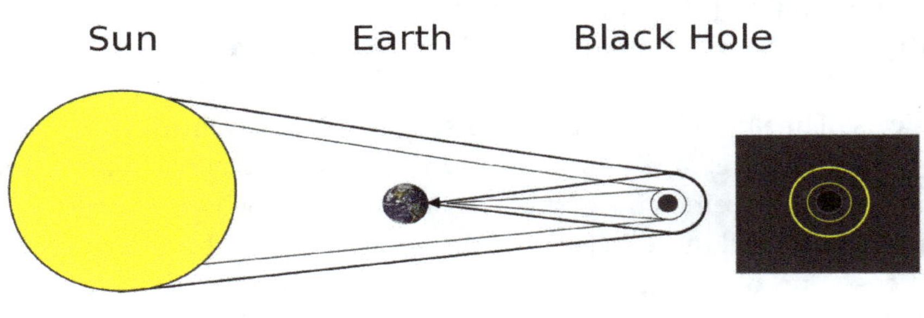

Figure 8. Retro-lensing of the Sun light by a black hole as seen from Earth. On the right-hand side, the series of rings around the black hole's event horizon an observer would observe in the case of perfect alignment. For clarity, only two rings are shown.

A better approach to test the idea proposed by Holz and Wheeler is to consider a well-known supermassive black hole and a bright star around it. Of course, the brighter the source star, the brighter will be the retro-image. Some of us [92] soon proposed to consider retro-lensing around the black hole at the galactic center, and in particular, the retro-lensing image of the closest star orbiting around it. Indeed, it is known that at the center of our galaxy, there is a supermassive black hole, with mass about $(4.2 \pm 0.2) \times 10^6\ M_\odot$, identified by studying the orbits of several bright stars orbiting around it (see [93,94] and the references therein). A method to determine the mass and the angular momentum of this black hole could then be to measure the periastron or apoastron shifts of some of the stars orbiting around it. Another method to estimate the black hole spin a is based on the analysis of the quasi-periodical oscillations towards Sgr A*. Recently, the analysis of the data in the X-ray and IR bands have allowed some astrophysicists to find that $a = 0.65 \pm 0.05$ [94,95]. However, there is a drawback in this approach: periastron and apoastron shift of orbits depend not only on the black hole parameters, but also on how stars are distributed around the black hole and on the mass density profile of the dark matter possibly present in the region surrounding the black hole. It is possible to understand the difficulty of the measure by noting that the difference of the periastron shift of the S2 star (the closest one to the black hole at the center of our galaxy) induced by a Schwarzschild black hole or a Kerr black hole with spin parameter $a = 1$ (and the same mass of the Schwarzschild one) is only of $\simeq 10$ μas (for the dependance of the periastron shift on the black hole spin orientation see [96]). Then, even if one had succeeded in measuring the periastron shift of the closest star to the central black hole, it would be unlikely to derive the amount of the black hole angular momentum. Our goal could be achieved anyway by measuring the periastron shift of many stars orbiting around the center of the Galaxy. The measure of the periastron shift could give, in turn, also an estimate of the parameters of the dark matter concentration expected to lie towards the Sgr A^* region [97], as well as to test different modifications of the general theory of relativity [98–100] (see also [101] for the constraints on R^n theories by Solar System data)). However, this is anything but easy [102]. An important step forward in this direction has been provided recently by near-infrared astrometric observations of many stars around Sgr A^* with a precision of about 170 μas in position and $\simeq 0.07$ mas$\cdot$ yr^{-1} in velocity [103]. A further improvement, hopefully in the near future, would make possible the direct detection of relativistic effects in the orbits of stars orbiting the central black hole.

Retro-lensing images of bright stars retro-lensed by the black hole at the galactic center might give an alternative method to estimate the Sgr A* black hole parameters. Even though in general it is difficult to calculate the retro-lensing images, since this requires integrating with high precision the trajectories followed by the light, it is possible to numerically do these calculations not only for a Schwarzschild black hole, but also for Kerr and Reissner–Nordström black holes. As discussed in several papers (see, e.g., [104] and the references therein), one finds that the shape of the retro-lensing image depends on the black hole spin (see Figure 9), and then, in principle, a single precise enough observation of the retro-lensing image of a star could allow one to unambiguously estimate the parameters of the black hole in Sgr A*. It is possible to show that also the electric charge of a Reissner–Nordström black hole can be obtained [105]. In fact, although the formation of a Reissner–Nordström black hole may be problematic, charged black holes are objects of intensive investigations, and the black hole charge can be estimated by using the size of the retro-lensing images that can be revealed by future astrometrical missions. The shape of the retro-lensing (or shadow) image depends in fact also on the electric charge of the black hole, and it becomes smaller as the electric charge increases. The mirage size difference between the extreme charged black hole and the Schwarzschild black hole case is about 30%, and in the case of the black hole in Sgr A*, the shadow typical angular sizes are about 52 μas for the Schwarzschild case and about 40 μas for a maximally charged Reissner–Nordström black hole. Therefore, a charged black hole could be, in principle, distinguished by a Schwarzschild black hole with RADIOASTRON, at least if its charge is close to the maximal value. We also mention that the black hole spin gives rise also to chromatic effects (while for non-rotating lenses, the gravitational lensing effect is always achromatic), making one side of the image bluer than the other side [104].

Can we really hope to observe these retro-lensing images towards Sgr A*? Despite what one could think, we are not so far from this goal. The successor of the Hubble Space Telescope, the James Webb Space Telescope (JWST), scheduled for launch in October 2018, has the sensitivity to observe the retro-lensing image of the S2 star produced by the black hole at the galactic center with an exposure time of about thirty hours. In Figure 10, we show the magnification (upper panel) and the magnitude (bottom panel) light curves (in K band) of the retro-lensing image of the S2 star produced by the black hole at the galactic center (see also [106]). Unfortunately, JWST has not the angular resolution necessary to provide information about the shape of the retro-lensing image. The right angular resolution could be gained with the next generation of radio interferometers. In fact, the diameter of the retro-lensing image around the central black hole should be of about 30 μas, and already in 2008, Doeleman and his collaborators [107] managed to achieve an angular resolution of about 37 μas, very close to the required one, by using interferometrically different radio telescopes with a baseline of about 4500 km. Progress in this field is so fast, that it is not hard to think we can eventually reach this aim in the near future by, e.g., the EHT (Event Horizon Telescope) project, or by the planned Russian space observatory, Millimetron (the spectrum-M project), or by combined observations with different interferometers, such as the Very Large Array (VLA) and ALMA (Atacama Large Millimeter Array).

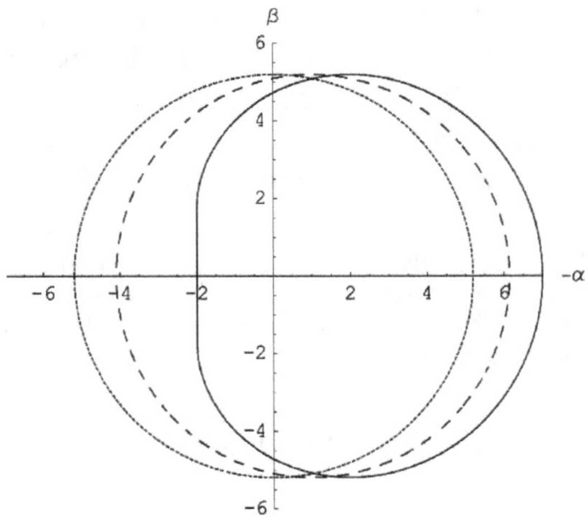

Figure 9. Retro-lensing images of a source by a Schwarzschild black hole (dotted circle), a Kerr black hole with spin parameter $a = 0.5$ (dashed line) and a maximally spinning black hole with $a = 1$ (continuous line). The line of sight of the observer is perpendicular to the spin axis of the black hole.

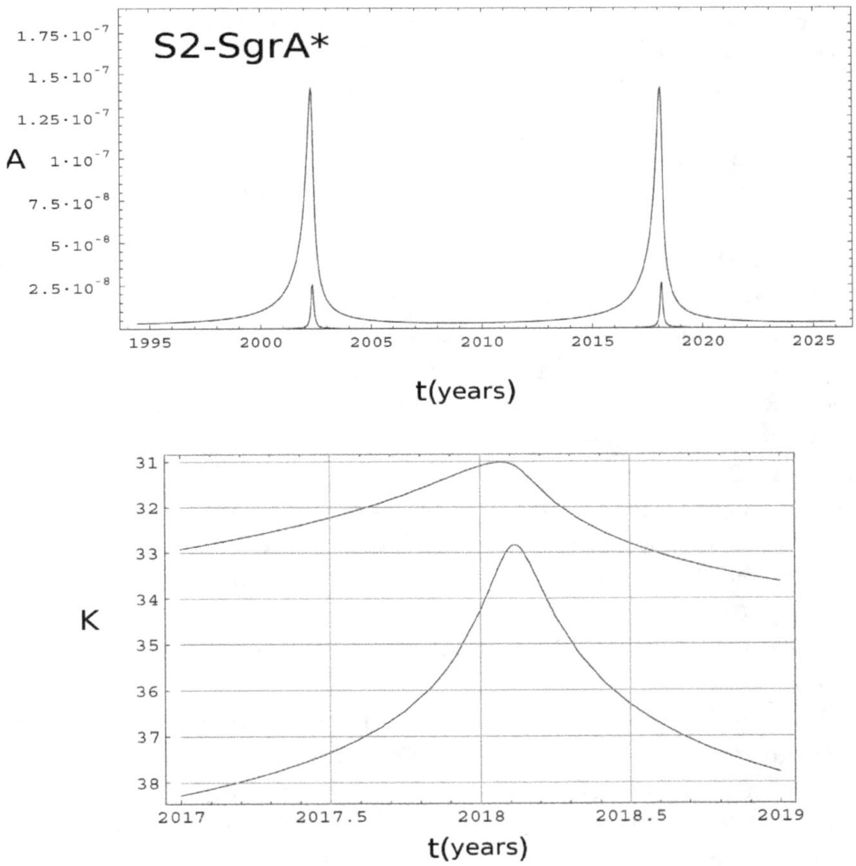

Figure 10. Upper panel: amplification as a function of time for the primary (upper curve) and secondary (lower curve) retro-lensing images of the S2 star by the black hole in the Galaxy center. Lower panel: light curve in K-band magnitude of the two retro-lensing images (adapted from [106]). The standard interstellar absorption coefficient towards the Galaxy center has been assumed.

7. Conclusions

In the paper, we have discussed the various scales in which gravitational lensing manifests itself and that may lead us to obtain valuable information about a great variety of astronomical issues ranging from the star distribution in the Milky Way, the study of stellar atmospheres, the discovery of exoplanets in the Milky Way and also in nearby galaxies, the study of far away galaxies, galaxy clusters and black holes. Gravitational lensing, in particular in the strong and weak lensing regime, may also allow scientists to answer, in the near future, fundamental questions in cosmology related to the nature of dark matter, why the Universe is accelerating and what is the nature of the source responsible for the acceleration, which physicists refer to as dark energy.

Acknowledgments: The authors acknowledge the support of the TAsP (Theoretical Astroparticle Physics) Project funded by INFN. Mosè Giordano acknowledges the support of the Max Planck Institute for Astronomy, Heidelberg, where part of this work has been done. We thank Asghar Qadir and Alexander Zakharov for the valuable discussions on the paper subject during many years.

Author Contributions: All authors have equally contributed to this paper and have read and approved the final version.

References

1. Einstein, A. On the Influence of Gravitation on the Propagation of Light. *Ann. Phys.* **1911**, *340*, 898–908.
2. Dyson, F.W.; Eddington, A.; Davidson, C. A determination of the deflection of light by the sun's gravitational field from observations made at the total eclipse of May 29, 1919. *Phil. Trans. Roy. Soc. A* **1920**, *220*, 291–333.
3. Chwolson, O. Über eine mögliche Form fiktiver Doppelsterne. *Astron. Nachr.* **1924**, *221*, 329–330.
4. Einstein, A. Lens-Like Action of a Star by the Deviation of Light in the Gravitational Field. *Science* **1936**, *84*, 506–507.
5. Zwicky, F. Nebulae as Gravitational Lenses. *Phys. Rev.* **1937**, *51*, 290.
6. Zwicky, F. On the Probability of Detecting Nebulae which Act as Gravitational lenses. *Phys. Rev.* **1937**, *51*, 679.
7. Walsh, D.; Carswell, R.F.; Weymann, R.J. 0957 + 561 A,B: Twin quasistellar objects or gravitational lens? *Nature* **1979**, *279*, 381–384.
8. Paczyǹski, B. Gravitational microlensing by the galactic halo. *Astrophys. J.* **1986**, *304*, 1–5.
9. Johannsen, T. Sgr A* and general relativity. **2015**, arXiv:1512.03818.
10. Abbott, B.P.; Abbott, R.; Abbott, T.D.; Abernathy, M.R.; Acernese, F.; Ackley, K.; Adams, C.; Adams, T.; Addesso, P.; Adhikari, R.X.; *et al.* Observation of Gravitational Waves from a Binary Black Hole Merger. *Phys. Rev. Lett.* **2016**, *116*, 061102.
11. Schneider, P.; Ehlers, G.; Falco, E.E. Gravitational Lenses; Springer Verlag: Berlin, Germany, 1992.
12. Schmidt, M. 3C 273: A star-like object with large red-shift. *Nature* **1963**, *197*, 1040.
13. Pelt, J.; Kayser, R.; Refsdal, S.; Schramm, T. The light curve and the time delay of QSO 0957+561. *Astron. Astrophys.* **1996**, *306*, 97–106.
14. Chen, G.H.; Kochanek, C.S.; Hewitt, J.N. The Mass Distribution of the Lens Galaxy in MG 1131+0456. *Astrophys. J.* **1995**, *447*, 62–81.
15. Hennawi, J.F.; Gladders, M.D.; Oguri, M.; Dalal, N.; Koester, B.; Natarajan, P.; Strauss, M.A.; Inada, N.; Kayo, I.; Lin, H. A New Survey for Giant Arcs. *Astron. J.* **2008**, *135*, 664–681.
16. Browne, I.W.A.; Wilkinson, P.N.; Jackson, N.J.F.; Myers, S.T.; Fassnacht, C.D.; Koopmans, L.V.E.; Marlow, D.R.; Norbury, M.; Rusin, D.; Sykes, C.M.; *et al.* The Cosmic Lens All-Sky Survey - II. Gravitational lens candidate selection and follow-up. *Mon. Not. R. Astron. Soc.* **2003**, *341*, 13–32.
17. Bolton, A.S.; Burles, S.; Koopmans, L.V.E.; Treu, T.; Moustakas, L.A. The Sloan Lens ACS Survey. I. A Large Spectroscopically Selected Sample of Massive Early-Type Lens Galaxies. *Astrophys. J.* **2006**, *638*, 703–724.
18. Alam, S.; Albareti, F.D.; Allende, P.; Anders, F.; Anderson, S.F.; Anderton, T.; Andrews, B.H.; Armengaud, E.; Aubourg, E.; Bailey, S.; *et al.* The Eleventh and Twelfth Data Releases of the Sloan Digital Sky Survey: Final Data from SDSS-III. *Astrophys. J. Suppl. Ser.* **2015**, *219*, 12.

19. Oguri, M.; Inada, N.; Pindor, B.; Strauss, M.A.; Richards, G.T.; Hennawi, J.F.; Turner, E.L.; Lupton, R.H.; Schneider, D.P.; Fukugita, M. The Sloan Digital Sky Survey Quasar Lens Search. I. Candidate Selection Algorithm. *Astrophys. J.* **2006**, *132*, 999–1013.
20. Blandford, R.D.; Narajan, R. Cosmological applications of gravitational lensing. *Annu. Rev. Astron. Astrophys.* **1992**, *30*, 311–358.
21. Treu, T. Strong lensing by galaxies. *Annu. Rev. Astron. Astrophys.* **2010**, *48*, 87–125.
22. Kneib, J.-P.; Ellis, R.S.; Smail, I.; Couch, W. J.; Sharples, R.M. Hubble Space Telescope Observations of the Lensing Cluster Abell 2218. *Astrophys. J.* **1996**, *471*, 643–656.
23. Mao, S.; Schneider, P. Evidence for substructure in lens galaxies? *Mon. Not. R. Astron. Soc.* **1998**, *95*, 587–594.
24. Metcalf, R.B.; Madau, P. Compound Gravitational Lensing as a Probe of Dark Matter Substructure within Galaxy Halos. *Astrophys. J.* **2001**, *563*, 9–20.
25. Xu, D.D.; Mao, S.; Wang, J.; Springel, V.; Gao, L.; White, S.D.M.; Frenk, C.S.; Jenkins, A.; Li, G.; Navarro, J.F. Effects of dark matter substructures on gravitational lensing: results from the Aquarius simulations. *Mon. Not. R. Astron. Soc.* **2009**, *408*, 1235–1253.
26. Metcalf, R.B.; Amara, A. Small-scale structures of dark matter and flux anomalies in quasar gravitational lenses. *Mon. Not. R. Astron. Soc.* **2012**, *419*, 3414–3425.
27. Xu, D.D.; Sluse, D.; Gao, L.; Wang, J.; Frenk, C.; Mao, S.; Schneider, P.; Springel, V. How well can cold dark matter substructures account for the observed radio flux-ratio anomalies. *Mon. Not. R. Astron. Soc.* **2015**, *447*, 3189–3206.
28. Vegetti, S.; Lagattuta, D.J.; McKean, J.P.; Auger, M.W.; Fassnacht, C.D.; Koopmans, L.V.E. Gravitational detection of a low-mass dark satellite galaxy at cosmological distance. *Nature* **2012**, *481*, 341–343.
29. Vegetti, S.; Koopmans, L.V.E.; Auger, M.W.; Treu, T.; Bolton, A.S. Inference of the cold dark matter substructure mass function at z = 0.2 using strong gravitational lenses. *Mon. Not. R. Astron. Soc.* **2014**, *442*, 2017–2035.
30. Kochanek, C.S. Quantitative Interpretation of Quasar Microlensing Light Curves. *Astrophys. J.* **2004**, *605*, 58–77.
31. Mosquera, A.M.; Kochanek, C.S. The Microlensing Properties of a Sample of 87 Lensed Quasars. *Astrophys. J.* **2011**, *738*, 96.
32. Pooley, D.; Rappaport, S.; Blackburne, J.A.; Schechter, P.L.; Wambsganss, J. X-Ray and Optical Flux Ratio Anomalies in Quadruply Lensed Quasars. II. Mapping the Dark Matter Content in Elliptical Galaxies. *Astrophys. J.* **2012**, *744*, 111.
33. Soucail, G.; Fort, B.; Mellier, Y.; Picat, J.P. A blue ring-like structure, in the center of the A 370 cluster of galaxies. *Astron. Astrophys.* **1987**, *172*, L14–L16.
34. Stark, D.P.; Swinbank, A.M.; Ellis, R.S. Dye, S.; Smail, I.R; Richard, J. The formation and assembly of a typical star-forming galaxy at redshift z$\sim$3. *Nature* **2008**, *455*, *775–777*.
35. Kelly, P.L.; Rodney, S.A.; Treu, T.; Foley, R.J.; Brammer, G.; Schmidt, K.B.; Zitrin, A.; Sonnenfeld, A.; Strolger, L.-G.; Graur, O.; *et al.* Multiple images of a highly magnified supernova formed by an early-type cluster galaxy lens. *Science* **2015**, *347*, 1123–1126.
36. Treu, T.; Brammer, G.; Diego, J.M.; Grillo, C.; Kelly, P.L.; Oguri, M.; Rodney, A.; Rosati, P.; Sharon, K.; Zitrin, A."Refsdal" Meets Popper: Comparing Predictions of the Re-appearance of the Multiply Imaged Supernova Behind MACSJ1149.5+2223. *Astrophys. J.* **2016**, *817*, 60.
37. Refsdal, S. On the possibility of determining Hubble's parameter and the masses of galaxies from the gravitational lens effect. *Mon. Not. R. Astron. Soc.* **1964**, *128*, 307–310.
38. Jackson, N. The Hubble Constant. *Living Rev. Relativ.* **2007**, *10*, 4.
39. Suyu, S.H.; Marshall, P.J.; Auger, M.W.; Hilbert, S.; Blandford, R.D.; Koopmans, L.V.E.; Fassnacht, C.D.; Treu, T.; *et al.* Dissecting the Gravitational lens B1608+656. II. Precision Measurements of the Hubble Constant, Spatial Curvature, and the Dark Energy Equation of State. *Astrophys. J.* **2010**, *711*, 201–221.
40. Bonvin, V.; Tewes, M.; Courbin, F.; Kuntzer, T.; Sluse, D.; Meylan, G. COSMOGRAIL: the COSmological MOnitoring of GRAvItational Lenses. XV. Assessing the achievability and precision of time-delay measurements. *Astron. Astrophys.* **2016**, *585*, A88.
41. Bartelmann, M.; Schneider, P. Weak gravitational lensing. *Phys. Rep.* **2001**, *340*, 291–472.
42. Tyson, J.A.; Wenk, R.A.; Valdes, F. Detection of systematic gravitational lens galaxy image alignments—Mapping dark matter in galaxy clusters. *Astrophys. J.* **1990**, *349*, L1–L4.

43. Bacon, D.J.; Refregier, A.R.; Ellis, R.S. Detection of weak gravitational lensing by large-scale structure. *Mon. Not. R. Astron. Soc.* **2000**, *318*, 625–640.
44. Kaiser, N. A New Shear Estimator for Weak-Lensing Observations. *Astrophys. J.* **2000**, *537*, 555–577.
45. Seitz, S.; Schneider, P. Cluster lens reconstruction using only observed local data: An improved finite-field inversion technique. *Astron. Astrophys.* **1996**, *305*, 383–401.
46. Lombardi, M.; Bertin, G. A fast direct method of mass reconstruction for gravitational lenses. *Astron. Astrophys.* **1999**, *348*, 38–42.
47. Schneider, P.; King, L.; Erben, T. Cluster mass profiles from weak lensing: Constraints from shear and magnification information. *Astron. Astrophys.* **2000**, *353*, 41–56.
48. Han, J.; Eke, V.R.; Frenk, C.S.; Mandelbaum, R.; Norberg, P.; Schneider, M.D.; Peacock, J.A.; Jing, Y.; Baldry, I.; Bland-Hawthorn, J.; *et al.* Galaxy And Mass Assembly (GAMA): the halo mass of galaxy groups from maximum-likelihood weak lensing. *Mon. Not. R. Astron. Soc.* **2015**, *446*, 1356–1379.
49. Marshall, P.J.; Hobson, M.P.; Gull, S.F.; Bridle, S.L. Maximum-entropy weak lens reconstruction: improved methods and application to data. *Mon. Not. R. Astron. Soc.* **2002**, *335*, 1037–1048.
50. Clowe, D.; De Lucia, G.; King, L. Effects of asphericity and substructure on the determination of cluster mass with weak gravitational lensing. *Mon. Not. R. Astron. Soc.* **2004**, *350*, 1038–1048.
51. Corless, V.L.; King, L.J. A statistical study of weak lensing by triaxial dark matter haloes: consequences for parameter estimation. *Mon. Not. R. Astron. Soc.* **2007**, *380*, 149–161.
52. Hoekstra, H. A comparison of weak-lensing masses and X-ray properties of galaxy clusters. *Mon. Not. R. Astron. Soc.* **2007**, *379*, 317–330.
53. Zhang, Y.-Y.; Finoguenov, A.; Böhringer, H.; Kneib, J.-P.; Smith, G.P.; Kneissl, R.; Okabe, N.; Dahle, H. LoCuSS: Comparison of observed X-ray and lensing galaxy cluster scaling relations with simulations. *Astron. Astrophys.* **2008**, *482*, 451–472.
54. Contaldi, C.R.; Hoekstra, H.; Lewis, A. Joint Cosmic Microwave Background and Weak Lensing Analysis: Constraints on Cosmological Parameters. *Phys. Rev. Lett.* **2003**, *22*, 221303.
55. Hollenstein, L.; Sapone, D.; Crittenden, R.; Schäfer, B.M. Constraints on early dark energy from CMB lensing and weak lensing tomography. *J. Cosmol. Astropart. Physucs* **2009**, *04*, id. 012.
56. Majerotto, E.; Sapone, D.; Schäfer, B,M. Combined constraints on deviations of dark energy from an ideal fluid from Euclid and Planck. *Mon. Not. R. Astron. Soc.* **2016**, *456*, 109–118.
57. Clowe, D.; Bradaĉ, M.; Gonzalez, A.H.; Markevitch, M.; Randall, S.W.; Jones, C.; Zaritsky, D. A Direct Empirical Proof of the Existence of Dark Matte. *Astrophys. J.* **2006**, *648*, L109–L113.
58. Amendola, L.; Appleby, S.; Bacon, D. Baker, T. Baldi, M. Bartolo, N. Blanchard, A. Bonvin, C. Borgani, S. Branchini, E.; *et al.* Cosmology and Fundamental Physics with the Euclid Satellite. *Living Rev. Relat.* **2013**, *16*, 6.
59. Chang, C.; Jarvis, M.; Jain, B. Kahn, S.M.; Kirkby, D.; Connolly, A.; Krughoff, S.; Peng, E.-H.; Peterson, J.R. The effective number density of galaxies for weak lensing measurements in the LSST project. *Mon. Not. R. Astron. Soc.* **2013**, *434*, 2121–2135.
60. Heymans, C.; Heavens, A. Weak gravitational lensing: reducing the contamination by intrinsic alignments. *Mon. Not. R. Astron. Soc.* **2003**, *339*, 711–720.
61. Valageas, P. Source-lens clustering and intrinsic-alignment bias of weak-lensing estimators. *Astron. Astrophys.* **2014**, *561*, A53.
62. Mao, S. Astrophysical applications of gravitational microlensing. *Res. Astron. Astrophys.* **2012**, *12*, 947–972.
63. Perryman, M. *The Exoplanet Handbook*; University Press: Cambridge, UK, 2014.
64. Sumi, T.; Kamiya, K.; Udalski, A.; Bennett, D.P.; Bond, I.A.; Abe, F.; Botzler, C.S.; Fukui, A.; Furusawa, K.; Hearnshaw, J.B.; Itow, Y.; *et al.* Unbound or distant planetary mass population detected by gravitational microlensing. *Nature* **2011**, *473*, 349–352.
65. Ingrosso, G.; Calchi Novati, S.; De Paolis, F.; Jetzer, P.; Nucita, A.A.; Strafella, F. Pixel lensing as a way to detect extrasolar planets in M31. *Mon. Not. R. Astron. Soc.* **2009**, *399*, 219–228.
66. An, J.H.; Evans, N.W.; Kerins, E.; Baillon, P.; Calchi-Novati, S.; Carr, B.J.; Creze, M.; Giraud-Heraud, Y.; Gould, A.; Hewett, P.; *et al.* The Anomaly in the Candidate Microlensing Event PA-99-N2. *Astrophys. J.* **2004**, *601*, 845–857.
67. Dominik, M.; Sahu, K.C. Astrometric Microlensing of Stars. *Astrophys. J.* **2000**' *534*, 213–226.

68. Lee, C.-H.; Seitz, S.; Riffeser, A.; Bender, R. Finite-source and finite-lens effects in astrometric microlensing. *Mon. Not. R. Astron. Soc.* **2010**, *407*, 1597–1608.
69. Walker, M.A. Microlensed Image Motions. *Astrophys. J.* **1995**, *453*, 37–39.
70. Eyer, L.; Holl, B.; Pourbaix, D.; Mowlavi, N.; Siopis, C.; Barblan, F.; Evans, D.W.; North, P. The Gaia Mission. *Cent. Eur. Astrophys. Bull. (CEAB)* **2013**, *37*, 115–126.
71. Paczyński, B. The Masses of Nearby Dwarfs and Brown Dwarfs with the HST. *Acta Astron.* **1996**, *46*, 291–296.
72. Proft, S.; Demleitner, M.; Wambsganss, J. Prediction of astrometric microlensing events during the Gaia mission. *Astron. Astrophys.* **2011**, *536*, A50.
73. Ingrosso, G.; Calchi Novati, S.; De Paolis, F.; Jetzer, P.; Nucita, A.A.; Strafella, F.; Zakharov, A.F. Polarization in microlensing events towards the Galactic bulge. *Mon. Not. R. Astron. Soc.* **2012**, *426*, 1496–1506.
74. Chandrasekhar, S. *Radiative Transfer*; Clarendon Press: Oxford, UK, 1950.
75. Simmons, J.F.L.; Bjorkman, J.E.; Ignace, R.; Coleman, I.J. Polarization from microlensing of spherical circumstellar envelopes by a point lens. *Mon. Not. R. Astron. Soc.* **2002**, *336*, 501–510.
76. Ingrosso, G.; Calchi Novati, S.; De Paolis, F.; Jetzer, P.; Nucita, A.A.; Strafella, F. Measuring polarization in microlensing events. *Mon. Not. R. Astron. Soc.* **2015**, *446*, 1090–1097.
77. Henderson, C.B.; Gaudi, B.S.; Han, C.; Skowron, J.; Penny, M.T.; Nataf, D.; Gould, A.P. Optimal Survey Strategies and Predicted Planet Yields for the Korean Microlensing Telescope Network. *Astrophys. J.* **2014**, *794*, 52.
78. Penny, M.T.; Kerins, E.; Rattenbury, N.; Beaulieu, J.-P.; Robin, A.C.; Mao, S.; Batista, V.; Calchi Novati, S.; Cassan, A.; Fouque, P.; *et al.* Zapatero Osorio ExELS: An exoplanet legacy science proposal for the ESA Euclid mission - I. Cold exoplanets. *Mon. Not. R. Astron. Soc.* **2013**, *434*, 2–22.
79. Yee, J.C.; Albrow, M.; Barry, R.K.; Bennett, D.; Bryden, G.; Chung, S.-J.; Gaudi, B.S.; Gehrels, N.; Gould, A.; Penny, M.T.; *et al.* Takahiro SumiNASA ExoPAG Study Analysis Group 11: Preparing for the WFIRST Microlensing Survey. **2014**, arXiv:1409.2759.
80. Heyrovský, D.; Sasselov, D. Detecting Stellar Spots by Gravitational Microlensing. *Astrophys. J.* **2000**, *529*, 69–76.
81. Hendry, M.A.; Bryce, H.M.; Valls-Gabaud, D. The microlensing signatures of photospheric starspots. *Mon. Not. R. Astron. Soc.* **2002**, *335*, 539–549.
82. Giordano, M.; Nucita, A.A.; De Paolis, F.; Ingrosso, G. Starspot induced effects in microlensing events with rotating source star. *Mon. Not. R. Astron. Soc.* **2015**, *453*, 2017–2021.
83. Kim, S.-L.; Park, B.-G.; Lee, C.-U.; Yuk, I.-S.; Han, C.; O'Brien, T.; Gould, A.; Lee, J.W.; Kimet, D.-J. Technical specifications of the KMTNet observation system. *Proc. SPIE* **2010**, doi:10.1117/12.856833.
84. Sajadian, S. Detecting stellar spots through polarimetric observations of microlensing events in caustic-crossing. *Mon. Not. R. Astron. Soc.* **2015**, *452*, 2587–2596.
85. Dominik, M. The binary gravitational lens and its extreme cases. *Astron. Astrophys.* **1999**, *349*, 108–125.
86. An, J.H. Gravitational lens under perturbations: symmetry of perturbing potentials with invariant caustics. *Mon. Not. R. Astron. Soc.* **2005**, *356*, 1409–1428.
87. Griest, K.; Safizadeh, N. The Use of High-Magnification Microlensing Events in Discovering Extrasolar Planets. *Astrophys. J.* **1998**, *500*, 37–50.
88. Penny, M.T.; Kerins, E.; Mao, S. Rapidly rotating lenses: Repeating features in the light curves of short-period binary microlenses. *Mon. Not. R. Astron. Soc.* **2011**, *417*, 2216–2229.
89. Nucita, A.A.; Giordano, M.; De Paolis, F.; Ingrosso, G. Signatures of rotating binaries in microlensing experiments. *Mon. Not. R. Astron. Soc.* **2014**, *438*, 2466–2473.
90. Chandrasekhar, S. *The Mathematical Theory of Black Holes*; Oxford University Press: Oxford, UK, 1983.
91. Holz, D.E.; Wheeler, J.A. Retro-MACHOs: pi in the Sky. *Astrophys. J.* **2002**, *57*, 330–334.
92. De Paolis, F.; Geralico, A.; Ingrosso, G.; Nucita, A.A. The black hole at the galactic center as a possible retro-lens for the S2 orbiting star. *Astron. Astrophys.* **2003**, *409*, 809–812.
93. Gillessen, S.; Eisenhauer, F.; Trippe, S.; Alexander, T.; Genzel, R.; Martins, F.; Ott, T. Monitoring Stellar Orbits Around the Massive Black Hole in the Galactic Center. *Astrophys. J.* **2009**, *692*, 1075–1109.
94. Dokuchaev, V.I. Spin and mass of the nearest supermassive black hole. *Gen. Rel. Gravit.* **2014**, *46*, 1832.
95. Dokuchaev, V.I.; Eroshenko, Y.N. Physical Laboratory at the center of the Galaxy. *Phys. Uspekhi* **2015**, *58*, 772–784.

96. Iorio, L. Perturbed stellar motions around the rotating black hole in Sgr A* for a generic orientation of its spin axis. *Phys. Rev. D* **2012**, *84*, 124001.
97. Zakharov, A.F.; Nucita, A.A.; De Paolis, F.; Ingrosso, G. Apoastron shift constraints on dark matter distribution at the Galactic Center. *Phys. Rev. D* **2007**, *76*, 062001.
98. Zakharov, A.F.; de Paolis, F.; Ingrosso, G.; Nucita, A.A. Shadows as a tool to evaluate the black hole parameters and a dimension of spacetime. *New Astron. Rev.* **2012**, *56*, 64–73.
99. Falcke, H.; Markoff, S.B. Toward the event horizon—The supermassive black hole in the galactic center. *Classicala Quantum Gravit.* **2013**, *30*, 244003.
100. Zakharov, A.F.; Borka, D.; Borka Jovanović, V.; Jovanović, P. Constraints on R^n gravity from precession of orbits of S2-like stars: A case of a bulk distribution of mass. *Adv. Space Res.* **2014**, *54*, 1108–1112.
101. Zakharov, A.F.; Nucita, A.A.; De Paolis, F.; Ingrosso, G. Solar system constraints on R^n gravity. *Phys. Rev. D* **2006**, *74*, 107101.
102. Nucita, A.A.; De Paolis, F.; Ingrosso, G.; Qadir, A.; Zakharov, A.F. Sgr A*: A Laboratory to Measure the Central Black Hole and Stellar Cluster Parameters. *Publ. Astron. Soc. Pac.* **2007**, *119*, 349–359.
103. Plewa, P.M. Gillessen, S.; Eisenhauer, F.; Ott, T.; Pfuhl, O.; George, E.; Dexter, J.; Habibi, M.; Genzel, R.; Reid, M.J.; *et al.* Pinpointing the near-infrared location of Sgr A* by correcting optical distortion in the NACO imager. *Mon. Not. R. Astron. Soc.* **2015**, *453*, 3234–3244.
104. De Paolis, F.; Ingrosso, G.; Nucita, A.A.; Qadir, A.; Zakharov, A.F. Estimating the parameters of the Sgr A* black hole. *Gen. Rel. Gravit.* **2011**, *43*, 977–988.
105. Zakharov, A.F.; De Paolis, F.; Ingrosso, G.; Nucita, A. Direct measurements of black hole charge with future astrometrical missions. *Astron. Astrophys.* **2005**, *442*, 795–799.
106. Bozza, V.; Mancini, L. Gravitational Lensing by Black Holes: A Comprehensive Treatment and the Case of the Star S2. *Astrophys. J.* **2005**, *611*, 1045–1053.
107. Doeleman, S.S.; Weintroub, J.; Rogers, A.E.E.; Plambeck, R.; Freund, R.; Tilanus, R.P.J.; Friberg, P.; Ziurys, L.M.; Moran, J.M.; Corey, B.; *et al.* Event-horizon-scale structure in the supermassive black hole candidate at the Galactic Centre. *Nature* **2008**, *455*, 78–80.

6

The Teleparallel Equivalent of General Relativity and the Gravitational Centre of Mass

José Wadih Maluf

Instituto de Física, Universidade de Brasília, C. P. 04385, 70.919-970 Brasília DF, Brazil; wadih@unb.br or jwmaluf@gmail.com

Academic Editor: Lorenzo Iorio

Abstract: We present a brief review of the teleparallel equivalent of general relativity and analyse the expression for the centre of mass density of the gravitational field. This expression has not been sufficiently discussed in the literature. One motivation for the present analysis is the investigation of the localization of dark energy in the three-dimensional space, induced by a cosmological constant in a simple Schwarzschild-de Sitter space-time. We also investigate the gravitational centre of mass density in a particular model of dark matter, in the space-time of a point massive particle and in an arbitrary space-time with axial symmetry. The results are plausible, and lead to the notion of gravitational centre of mass (COM) distribution function.

Keywords: teleparallel gravity; gravitational centre of mass moment; dark energy; teleparallel equivalent of general relativity

PACS: 04.20.Cv; 04.20.-q; 04.70.Bw

1. Introduction

The most popular and acceptable approach to the relativistic theory of gravitation is given by Einstein's theory of general relativity. However, nowadays there are several alternative formulations of theories for the gravitational field that attempt to explain the dark energy and dark matter problems, which do not find satisfactory explanations within the framework of Einstein's general relativity. Moreover, concepts such as energy, momentum, angular momentum and centre of mass of the gravitational field are usually defined only for asymptotically flat space-times, in the context of a 3+1 type formulation. The latter are definitions for the total quantities, and suffer from at least two restrictions: the definitions are valid only for asymptotically flat space-times, and there do not exist localized expressions for the densities of the energy-momentum and 4-angular momentum of the gravitational field. The ADM definition for the gravitational energy-momentum [1] is constructed out of the metric tensor, and by means of the metric tensor it is not possible to construct suitable scalar densities in the form of total divergences. The approach via pseudo-tensors is certainly not satisfactory. The notions of energy-momentum and angular momentum of the gravitational field have been extensively discussed in the literature, but not the concept of gravitational centre of mass.

The notion of centre of mass can be made clear in flat space-time. Any relativistic field theory in flat space-time is expected to be covariant under the inhomogeneous Lorentz transformations, or Poincaré transformations: the 4-rotations and space-time translations. The generators of these transformations satisfy an algebra, the algebra of the Poincaré group. The generators of the 4-rotations are composed by the generators of the ordinary 3 dimensional rotations, and by the generators of the boosts. The latter are related to the centre of mass moment of the field. Energy, momentum and angular momentum of the field constitute seven conserved integral quantities associated to the symmetries of the theory. The integrals are carried out over the whole three-dimensional space. The three other integral quantities are associated to the centre of mass of the field, which sometimes is also called the centre of energy [2].

In the notation of Ref. [2], the centre of mass integrals read

$$J^{0i} = tP^i - \int d^3x\, x^i T^{00} \,, \tag{1}$$

where 0 and i are time and space indices, P^i is the i-th component of the momentum of the field, and T^{00} is the energy component of the energy-momentum tensor of field. It is argued [2] that the J^{0i} components have no clear physical significance since J^{0i} can be made to vanish if the coordinate system is chosen to coincide with the "centre of energy" at $t = 0$. However, in the context of general relativity, Dixon [3–5] developed a procedure for describing the dynamics of extended bodies in an arbitrary gravitational field, and for this purpose a definition of the centre of mass of such bodies (considered as quasi-rigid bodies) was proposed. A general relation between the centre of mass 4-velocity and the energy-momentum of the body was obtained [6]. One is led to the concept of centre of mass world line, whose uniqueness depends on the strength of the gravitational field [6].

In the standard metric formulation of general relativity, the centre of mass moment for the gravitational field has been first considered by Regge and Teitelboim [7,8], and reconsidered by several other authors (see References [9–12] and references therein). The centre of mass integral was obtained in the context of the Hamiltonian formulation of general relativity. The idea was to require the variation of the total Hamiltonian to be well defined in an asymptotically flat space-time, where the standard asymptotic space-time translations and 4-rotations are considered as coordinate transformations at spacelike infinity. This requirement leads to the addition of boundary (surface) terms to the primary Hamiltonian, so that the latter has well defined functional derivatives, and therefore one may obtain the field equations in the Hamiltonian framework (Hamilton's equations) by means of a consistent procedure. In this way, one arrives at the total energy, momentum, angular momentum and centre of mass moment of the gravitational field, given by surface terms of the total Hamiltonian.

In this article we address the centre of mass moment of the gravitational field in the realm of the teleparallel equivalent of general relativity (TEGR), which is an alternative and mathematically consistent formulation of general relativity [13] (see also Reference [14], and chapters 5 and 6 of Reference [15] and references therein). The geometrical structure of the TEGR was already considered by Einstein [16,17] in his attempt to unify gravity and electromagnetism, and later on by Cho [18,19], Hayashi and Shirafuji [20,21], Hehl et al. [22], Nitsch [23], Schweizer et al. [24], Nester [25] and Wiesendanger [26]). In recent years the teleparallel geometrical structure has been used in modified theories of gravity, with the purpose of constructing cosmological models that provide a consistent explanation to the dark energy problem (see the review article [27] and references therein).

The TEGR is constructed out of the tetrad fields $e^a{}_\mu$, where $a = \{(0),(i)\}$ and $\mu = \{0,i\}$ are SO(3,1) and space-time indices, respectively. The extra six components of the tetrads (compared to the 10 components of the metric tensor) yield additional geometric structure, that allows to define field quantities that cannot be constructed in the ordinary metric formulation of the theory (such as non-trivial total divergences, for instance). The tetrad fields allow to use concepts and definitions of both Riemannian and Weitzenböck geometries.

The definitions of the gravitational energy, momentum, angular momentum and centre of mass moment in the TEGR are not obtained according to the procedure described above, based on surface integrals of the total Hamiltonian. In the TEGR we first consider the Hamiltonian formulation of the theory [28,29]. The constraint equations of the theory (typically as $C = 0$) are equations that define the energy-momentum and the 4-angular momentum of the gravitational field [13] (i.e., $C = H - E = 0$). Moreover, the definitions of the energy-momentum and 4-angular momentum satisfy the algebra of the Poincaré group in the phase space of theory [13,30]. However, the energy-momentum definition, together with the gravitational energy-momentum tensor (but not the 4-angular momentum) may also be obtained directly from the Lagrangian field equations [13].

The gravitational centre of mass moment to be considered here yields the concept of gravitational centre of mass (COM) distribution function. One purpose of the present article is to show that a cosmological constant, which might be responsible for the dark energy, induces a very intense (divergent) gravitational COM distribution function in the vicinity of the cosmological horizon $r = R \simeq \sqrt{3/\Lambda}$ in a simple Schwarzschild-de Sitter space-time, in agreement with the hypothetical existence of dark energy. It seems that this result, obtained by means of tetrad fields, cannot be obtained in the context of the metric formulation of general relativity.

In Section 2 we present a brief review of the TEGR, emphasizing a recent simplified definition of the 4-angular momentum of the gravitational field, given by a total divergence. In Section 3 we investigate the gravitational COM distribution function of (i) the space-time of a massive particle in isotropic coordinates; (ii) the Schwarzschild-de Sitter space-time; (iii) a particular model of dark energy that arises from the non-local formulation of general relativity; and (iv) of an arbitrary space-time with axial symmetry. In the analysis of the first three cases above, which are spherically symmetric, we arrive at interesting results, that share similarities with the standard expressions in classical mechanics. For such space-times, the total centre of mass moment vanishes, as expected.

Notation: space-time indices $\mu, \nu, ...$ and SO(3,1) (Lorentz) indices $a, b, ...$ run from 0 to 3. The torsion tensor is given by $T_{a\mu\nu} = \partial_\mu e_{a\nu} - \partial_\nu e_{a\mu}$. The flat space-time metric tensor raises and lowers tetrad indices, and is fixed by $\eta_{ab} = e_{a\mu} e_{b\nu} g^{\mu\nu} = (-1, +1, +1, +1)$. The frame components are given by the inverse tetrads $\{e_a{}^\mu\}$. The determinant of the tetrad fields is written as $e = \det(e^a{}_\mu)$.

It is important to note that we assume that the space-time geometry is determined by the tetrad fields only, and thus the only possible non-trivial definition for the torsion tensor is given by $T^a{}_{\mu\nu}$. This tensor is related to the antisymmetric part of the Weitzenböck connection $\Gamma^\lambda_{\mu\nu} = e^{a\lambda}\partial_\mu e_{a\nu}$, which determines the Weitzenböck space-time and the distant parallelism of vector fields.

2. A Review of the Lagrangian and Hamiltonian Formulations of the TEGR

The TEGR is constructed out of the tetrad fields only. The first relevant consideration is an identity between the scalar curvature and an invariant combination of quadratic terms in the torsion tensor,

$$eR(e) \equiv -e\left(\frac{1}{4}T^{abc}T_{abc} + \frac{1}{2}T^{abc}T_{bac} - T^aT_a\right) + 2\partial_\mu(eT^\mu)\,, \tag{2}$$

where $T_a = T^b{}_{ba}$ and $T_{abc} = e_b{}^\mu e_c{}^\nu T_{a\mu\nu}$. The Lagrangian density for the gravitational field in the TEGR is given by [31]

$$\begin{aligned} L(e) &= -ke\left(\frac{1}{4}T^{abc}T_{abc} + \frac{1}{2}T^{abc}T_{bac} - T^aT_a\right) - \frac{1}{c}L_M \\ &\equiv -ke\Sigma^{abc}T_{abc} - \frac{1}{c}L_M\,, \end{aligned} \tag{3}$$

where $k = c^3/(16\pi G)$, L_M represents the Lagrangian density for the matter fields, and Σ^{abc} is defined by

$$\Sigma^{abc} = \frac{1}{4}\left(T^{abc} + T^{bac} - T^{cab}\right) + \frac{1}{2}\left(\eta^{ac}T^b - \eta^{ab}T^c\right)\,. \tag{4}$$

Thus, the Lagrangian density is geometrically equivalent to the scalar curvature density. The variation of $L(e)$ with respect to $e^{a\mu}$ yields the fields equations

$$e_{a\lambda}e_{b\mu}\partial_\nu(e\Sigma^{b\lambda\nu}) - e(\Sigma^{b\nu}{}_aT_{b\nu\mu} - \frac{1}{4}e_{a\mu}T_{bcd}\Sigma^{bcd}) = \frac{1}{4kc}eT_{a\mu}\,, \tag{5}$$

where $T_{a\mu}$ is defined by $\delta L_M/\delta e^{a\mu} = eT_{a\mu}$.

The field equations are equivalent to Einstein's equations. It is possible to verify by explicit calculations that the equations above can be rewritten as

$$\frac{1}{2}[R_{a\mu}(e) - \frac{1}{2}e_{a\mu}R(e)] = \frac{1}{4kc}\,T_{a\mu}\,, \tag{6}$$

Since the Lagrangian density (3) does not contain the total divergence that arises on the right hand side of Equation (2), it is not invariant under arbitrary local SO(3,1) transformations, but the field Equation (5) are covariant under such transformations.

The equivalence between the TEGR and the standard metric formulation of general relativity is based on the equivalence of Equations (5) and (6). However, in the TEGR there are additional field quantities (like third order tensors) constructed by means of the tetrad fields, such as total divergences, for instance, that cannot be obtained in the standard metric formulation. These additional field quantities are covariant under the global Lorentz transformations, but not under local transformations. In the ordinary formulation of arbitrary field theories, energy, momentum, angular momentum and COM moment are frame dependent field quantities, that transform under the global SO(3,1) transformations. In particular, energy transforms as the zero component of the energy-momentum four-vector. This feature must hold also in the presence of the gravitational field. As an example, consider the total energy of a black hole, represented by the mass parameter m. As seen by a distant observer, the total energy of a static Schwarzschild black hole is given by $E = mc^2$. However, at great distances the black hole may be considered as a particle of mass m, and if it moves with constant velocity v, then its total energy as seen by the same distant observer is $E = \gamma mc^2$, where $\gamma = (1 - v^2/c^2)^{-1/2}$. Likewise, the gravitational momentum, angular momentum and the COM moment are also frame dependent field quantities in general, whose values are different for different frames and different observers. On physical grounds, energy, momentum, angular momentum and COM moment cannot be local Lorentz *invariant* field quantities, since these quantities depend on the frame, as we know from special relativity, which is the limit of the general theory of relativity when the gravitational field is weak or negligible.

After some rearrangements, Equation (5) may be written in the form [13]

$$\partial_\nu(e\Sigma^{a\mu\nu}) = \frac{1}{4k}ee^a{}_\nu(t^{\mu\nu} + \frac{1}{c}T^{\mu\nu})\,, \tag{7}$$

where

$$t^{\mu\nu} = k(4\Sigma^{bc\mu}T_{bc}{}^\nu - g^{\mu\nu}\Sigma^{bcd}T_{bcd})\,, \tag{8}$$

is interpreted as the gravitational energy-momentum tensor [13,32] and $T^{\mu\nu} = e_a{}^\mu T^{a\nu}$.

The Hamiltonian density of the TEGR is constructed as usual in the phase space of the theory. We first note that the Lagrangian density (3) does not depend on the time derivatives of e_{a0}. Therefore, the latter arise as Lagrange multipliers in the Hamiltonian density H. The momenta canonically conjugated to e_{a0} are denoted by Π^{a0}. The latter are primary constraints of the theory: $\Pi^{a0} \approx 0$. The momenta canonically conjugated to e_{ai} are given by $\Pi^{ai} = \delta L/\delta \dot{e}_{ai} = -4k\Sigma^{a0i}$. The Hamiltonian density is obtained by rewriting the Lagrangian density in the form $L = \Pi^{ai}\dot{e}_{ai} - H$, in terms of e_{ai}, Π^{ai} and Lagrange multipliers. After the Legendre transform is performed, we obtain the final form of the Hamiltonian density. It reads [29,30]

$$H(e,\Pi) = e_{a0}C^a + \lambda_{ab}\Gamma^{ab}. \tag{9}$$

where λ_{ab} are Lagrange multipliers. In the above equation we have omitted a surface term. $C^a = \delta H/\delta e_{a0}$ is a long expression of the field variables, and $\Gamma^{ab} = -\Gamma^{ba}$ are defined by

$$\Gamma^{ab} = 2\Pi^{[ab]} + 4ke(\Sigma^{a0b} - \Sigma^{b0a})\,. \tag{10}$$

After solving the field equations, the Lagrange multipliers are identified as $\lambda_{ab} = (1/4)$ $(T_{a0b} - T_{b0a} + e_a{}^0T_{00b} - e_b{}^0T_{00a})$. The constraints C^a may be written as

$$C^a = -\partial_i \Pi^{ai} - p^a = 0\,, \tag{11}$$

where p^a is an intricate expression of the field quantities.

The quantities C^a and Γ^{ab} are first class constraints. They satisfy an algebra similar to the algebra of the Poincaré group [29]. The integral form of the constraint equations $C^a = 0$ yields the gravitational energy-momentum P^a [13],

$$P^a = -\int_V d^3x\, \partial_i \Pi^{ai}\,, \tag{12}$$

where V is an arbitrary volume of the three-dimensional space and $\Pi^{ai} = -4k\Sigma^{a0i}$. In similarity to the definition above, the definition of the gravitational 4-angular momentum follows from the constraint equations $\Gamma^{ab} = 0$ [30]. However, it has been noted [33] that the second term on the right hand side of Equation (10) can be rewritten as a total divergence, so that the constraints Γ^{ab} become

$$\Gamma^{ab} = 2\Pi^{[ab]} - 2k\partial_i[e(e^{ai}e^{b0} - e^{bi}e^{a0})] = 0\,. \tag{13}$$

Therefore, the definition of the total 4-angular momentum of the gravitational field L^{ab} may be given by an integral of a total divergence, in similarity to Equation (12). We have

$$L^{ab} = -\int_V d^3x\, 2\Pi^{[ab]}\,, \tag{14}$$

where

$$2\Pi^{[ab]} = (\Pi^{ab} - \Pi^{ba}) = 2k\partial_i[e(e^{ai}e^{b0} - e^{bi}e^{a0})]\,. \tag{15}$$

It is easy to show [30] that expressions (12) and (14) satisfy the algebra of the Poincaré group in the phase space of the theory,

$$\begin{aligned}
\{P^a, P^b\} &= 0\,,\\
\{P^a, L^{bc}\} &= \eta^{ab}P^c - \eta^{ac}P^b\,,\\
\{L^{ab}, L^{cd}\} &= \eta^{ad}L^{cb} + \eta^{bd}L^{ac} - \eta^{ac}L^{db} - \eta^{bc}L^{ad}\,.
\end{aligned} \tag{16}$$

Therefore, from a physical point of view, the interpretation of the quantities P^a and L^{ab} is consistent.

Definitions (12) and (14) are invariant under coordinate transformations of the three-dimensional, under time reparametrizations, and under global SO(3,1) transformations. The gravitational energy is the zero component of the energy-momentum four vector P^a.

3. The Centre of Mass Moment

The gravitational centre of mass (COM) moment is given by the components

$$L^{(0)(i)} = -\int d^3x\, M^{(0)(i)}\,, \tag{17}$$

where

$$M^{(0)(i)} = 2\Pi^{[(0)(i)]} = 2k\partial_j[e(e^{(0)j}e^{(i)0} - e^{(i)j}e^{(0)0})]\,, \tag{18}$$

according to definition (15). The quantity $-M^{(0)(i)}$ is identified as the gravitational COM density. The evaluation of the expression above is very simple. One needs just to establish the suitable set of tetrad fields that define a frame in space-time.

The inverse tetrads $e_a{}^\mu$ are interpreted as a frame adapted to a particular class of observers in space-time. Let the curve $x^\mu(\tau)$ represent the timelike worldline C of an observer in space-time, where τ is the proper time of the observer. The velocity of the observer along C is given by $u^\mu = dx^\mu/d\tau$. A frame adapted to this observer is constructed by identifying the timelike component of the frame

$e_{(0)}{}^{\mu}$ with the velocity u^μ of the observer: $e_{(0)}{}^{\mu} = u^\mu(\tau)$. The three other components of the frame, $e_{(i)}{}^{\mu}$, are orthogonal to $e_{(0)}{}^{\mu}$, and may be oriented in the three-dimensional space according to the symmetry of the physical system. If the space-time has axial symmetry, for instance, then the $e_{(3)}{}^{\mu}$ components of the tetrad fields are chosen to be oriented, asymptotically, along the z axis of the coordinate system, i.e., $e_{(3)}{}^{\mu}(t,x,y,z) \simeq (0,0,0,1)$ in the limit $r \to \infty$. A static observer in space-time is defined by the condition $u^\mu = (u^0, 0, 0, 0)$. Thus, a frame adapted to a static observer in space-time must satisfy the conditions $e_{(0)}{}^{i}(t, x^k) = (0,0,0)$.

An alternative way to characterise a frame in space-time is by means of the acceleration tensor ϕ_{ab} [34–38],

$$\phi_{ab} = \frac{1}{2}[T_{(0)ab} + T_{a(0)b} - T_{b(0)a}] \, . \tag{19}$$

This tensor is invariant under coordinate transformations and covariant under global SO(3,1) transformations, but not under local SO(3,1) transformations. It yields the inertial (i.e., the non-gravitational) accelerations that are necessary to impart to a frame in space-time in order to maintain the frame in a given inertial state. Three components of ϕ_{ab} yield the translational accelerations, and three other components yield the frequency of rotation of the frame. Altogether, these six components cancel the gravitational acceleration, so that the frame is kept in a particular inertial state.

In the following, we will evaluate the density of the centre of mass moment of four space-time configurations that exhibit spherical symmetry. In the four cases we will establish the frame of a static observer in space-time.

3.1. The Space-Time of a Massive Point Particle

The Schwarzschild solution in isotropic coordinates represents the space-time of a point massive particle [39,40]. It is obtained as an exact solution of Einstein's equations by writing the energy-momentum tensor in terms of a δ function of a point particle of mass M, with support at the origin of the coordinate system. The solution is described by the line element

$$ds^2 = -\alpha^2 c^2 dt^2 + \beta^2[dr^2 + r^2(d\theta^2 + \sin^2\theta \, d\phi^2)] \, , \tag{20}$$

where

$$\alpha^2 = \left(\frac{1 - \frac{m}{2r}}{1 + \frac{m}{2r}}\right)^2 , \quad \beta^2 = \left(1 + \frac{m}{2r}\right)^4 . \tag{21}$$

The parameter $m = GM/c^2$ represents the mass of the point particle that appears in the energy-momentum tensor. The line element above is clearly a solution of Equation (6), with the appropriate energy-momentum tensor $T_{a\mu}$ described in Reference [40].

By performing a coordinate transformation to (x, y, z) coordinates where

$$\begin{aligned} x &= r \sin\theta \cos\phi \, , \\ y &= r \sin\theta \sin\phi \, , \\ z &= r \cos\theta \, , \end{aligned} \tag{22}$$

the line element becomes

$$ds^2 = -\alpha^2 \, c^2 dt^2 + \beta^2 (dx^2 + dy^2 + dz^2) \, . \tag{23}$$

The tetrad fields adapted to static observers is given by

$$e_{a\mu}(t,x,y,z) = \begin{pmatrix} -\alpha & 0 & 0 & 0 \\ 0 & \beta & 0 & 0 \\ 0 & 0 & \beta & 0 \\ 0 & 0 & 0 & \beta \end{pmatrix} . \tag{24}$$

Taking into account Equation (18), straightforward calculations yield $M^{(0)(1)} = 2k\,\partial_1\beta^2$, $M^{(0)(2)} = 2k\,\partial_2\beta^2$ and $M^{(0)(3)} = 2k\,\partial_3\beta^2$. It is easy to obtain

$$\begin{aligned} -M^{(0)(1)} &= d_g\,x\,, \\ -M^{(0)(2)} &= d_g\,y\,, \\ -M^{(0)(3)} &= d_g\,z\,. \end{aligned} \tag{25}$$

The quantity d_g is defined by

$$d_g = \frac{4\,k\,m}{r^3}\left(1+\frac{m}{2r}\right)^3 . \tag{26}$$

Therefore,

$$\begin{aligned} L^{(0)(1)} &= \int d^3x\,d_g x\,, \\ L^{(0)(2)} &= \int d^3x\,d_g y\,, \\ L^{(0)(3)} &= \int d^3x\,d_g z\,, \end{aligned} \tag{27}$$

where $d^3x = dx\,dy\,dz$ and $r^2 = x^2 + y^2 + z^2$. The expressions above remind the definition of centre of mass in classical mechanics. Given that $M^{(0)(i)} = 2k\,\partial_i\beta^2$, it is easy to see that all integrals given by Equation (17) vanish, namely, all components of the total centre of mass moment vanish. However, the field quantity (26) has the following properties:

$$\begin{aligned} r \to \infty &\quad : \quad d_g \to 0\,, \\ r \to 0 &\quad : \quad d_g \to \infty\,. \end{aligned} \tag{28}$$

Thus, d_g is more intense in the vicinity of the particle, and vanishes at spatial infinity. In view of Equations (27) and (28), d_g may be interpreted as the gravitational COM distribution function. It is clearly related to the intensity of the gravitational field. The analyses of the space-time configurations below support this interpretation, as we will see.

3.2. *The Schwarzschild-de Sitter Space-Time*

The line element of the Schwarzschild-de Sitter space-time is given by

$$ds^2 = -\alpha^2\,dt^2 + \frac{1}{\alpha^2}dr^2 + r^2 d\theta^2 + r^2\sin^2\theta d\phi^2\,, \tag{29}$$

where

$$\alpha^2 = 1 - \frac{2m}{r} - \frac{r^2}{R^2}\,, \tag{30}$$

$R = \sqrt{3/\Lambda}$ and Λ is the cosmological constant. Here we are considering the speed of light $c = 1$. The Schwarzschild-de Sitter space-time has been considered in the TEGR in Reference [41]. The set of tetrad fields adapted to stationary observers in space-time is given by

$$e_{a\mu} = \begin{pmatrix} -\alpha & 0 & 0 & 0 \\ 0 & \alpha^{-1}\sin\theta\,\cos\phi & r\cos\theta\,\cos\phi & -r\sin\theta\,\sin\phi \\ 0 & \alpha^{-1}\sin\theta\,\sin\phi & r\cos\theta\,\sin\phi & r\sin\theta\,\cos\phi \\ 0 & \alpha^{-1}\cos\theta & -r\sin\theta & 0 \end{pmatrix} . \tag{31}$$

After long but simple calculations we find that the components of Equation (18) read

$$\begin{aligned} -M^{(0)(1)} &= 4k\sin\theta\left(\frac{1}{\alpha}-1\right) r\sin\theta\cos\phi\,, \\ -M^{(0)(2)} &= 4k\sin\theta\left(\frac{1}{\alpha}-1\right) r\sin\theta\sin\phi\,, \\ -M^{(0)(3)} &= 4k\sin\theta\left(\frac{1}{\alpha}-1\right) r\cos\theta\,. \end{aligned} \tag{32}$$

We identify $x = r\sin\theta\cos\phi$, $y = r\sin\theta\sin\phi$, $z = r\cos\theta$ as usual, and write Equation (17) as

$$\begin{aligned} L^{(0)(1)} &= \int d^3x\, 4k\sin\theta\left(\frac{1}{\alpha}-1\right) x\,, \\ L^{(0)(2)} &= \int d^3x\, 4k\sin\theta\left(\frac{1}{\alpha}-1\right) y\,, \\ L^{(0)(3)} &= \int d^3x\, 4k\sin\theta\left(\frac{1}{\alpha}-1\right) z\,, \end{aligned} \tag{33}$$

where $d^3x = dr\,d\theta\,d\phi$. Integration in the angular variables implies the vanishing of the three integrals $L^{(0)(i)}$, i.e., the total centre of mass vanishes, as expected. The equations above may be written exactly as Equation (27) provided we identify

$$d_g = 4k\,\sin\theta\left(\frac{1}{\alpha}-1\right) \equiv 4k\,\sin\theta\, f(r)\,. \tag{34}$$

The analysis of the expression above leads to interesting results. Let r_1 and r_2 denote the two horizons of the Schwarzschild-de Sitter space-time, $\alpha(r_1) = 0$ and $\alpha(r_2) = 0$, so that $r_1 < r_2$. The radius r_1 is close to the Schwarzschild radius, $r_1 \approx \frac{2m}{r}$, and $r_2 \approx R$. We have

$$\begin{aligned} r \to r_1 &\quad : \quad f(r) \to \infty\,, \\ r \to r_2 &\quad : \quad f(r) \to \infty\,. \end{aligned} \tag{35}$$

The function $f(r)$ is defined by Equation (34). The minimum of $f(r)$ is given by

$$\frac{df}{dr} = -\frac{1}{\alpha^2}\frac{d\alpha}{dr} = 0\,,$$

and takes place at $r_{min} = (mR^2)^{1/3}$. Thus, d_g is intense close to both r_1 and r_2, i.e., close to the Schwarzschild and cosmological horizons.

The radial position r_{min} is related to the inertial accelerations of an observer. In order to understand this feature, we evaluate the translational (non-gravitational) accelerations of a frame given by Equation (19). We find

$$\begin{aligned} \phi_{(0)(1)} &= \frac{d\alpha}{dr}\sin\theta\cos\phi\,, \\ \phi_{(0)(2)} &= \frac{d\alpha}{dr}\sin\theta\sin\phi\,, \\ \phi_{(0)(3)} &= \frac{d\alpha}{dr}\cos\theta\,. \end{aligned} \tag{36}$$

We define the inertial acceleration vector Φ as

$$\Phi(r) = (\phi_{(0)(1)}, \phi_{(0)(2)}, \phi_{(0)(3)}) \equiv \phi(r)\hat{r} = \frac{d\alpha}{dr}\hat{r}\,, \tag{37}$$

where $\hat{r} = (\sin\theta\cos\phi, \sin\theta\sin\phi, \cos\theta)$. Since

$$\frac{d\alpha}{dr} = \frac{1}{\alpha}\left(\frac{m}{r^2} - \frac{r}{R^2}\right),$$

we see that

$$\begin{aligned} r_1 < r < r_{min}: \quad & \frac{d\alpha}{dr} > 0 \quad \rightarrow \quad \phi(r) > 0\,, \\ r_{min} < r < r_2: \quad & \frac{d\alpha}{dr} < 0 \quad \rightarrow \quad \phi(r) < 0\,. \end{aligned} \tag{38}$$

Thus, given that the inertial acceleration $\phi(r) > 0$ is repulsive in the region $r_1 < r < r_{min}$, the gravitational acceleration is attractive in this interval, as expected. By means of a similar argument, we see that the gravitational acceleration is repulsive in the region $r_{min} < r < r_2$, as expected.

In view of the analysis above, we may interpret d_g given by Equation (34) as the gravitational COM distribution function, in similarity to Equation (26), and therefore one may understand the gravitational repulsion as attraction to a region of intense gravitational COM distribution function, which, in the present case, is the region in the vicinity of the cosmological horizon. If dark energy is indeed related to the existence of a cosmological constant, then it is natural that it is concentrated close to the radius $r_2 \approx R = \sqrt{3/\Lambda}$ in the context of a simple Schwarzschild-de Sitter model.

The function d_g plays the role of a gravitational COM density. However, mathematically it is not a density. The integrands in Equations (27) and (33) are in fact densities, but not d_g alone. In Newtonian mechanics, d_g in Equations (27) and (33) plays the role of mass density.

3.3. *Dark Matter Simulated by Non-Local Gravity*

A non-local formulation of general relativity, based on a geometrical framework similar to the one established by Equations (2)–(4) has been developed by Hehl, Mashhoon and collaborators [42–44]. One interesting consequence of this development is an extension of Newtonian gravity that may play a relevant role in the dynamics of galaxies, and might provide an explanation that is expected to come from dark matter models of gravity. We restrict the considerations to a simplified space-time with spherical symmetry, so that Equations (29), (31), (33) and (34) remain valid.

The Newtonian approximation is established by

$$\alpha^2 = -g_{00} \simeq 1 + \frac{2\Phi_g}{c^2}\,, \tag{39}$$

where Φ_g is the Newtonian potential, and $2\Phi_g/c^2 << 1$.

It follows that

$$f(r) = \frac{1}{\alpha} - 1 \simeq -\frac{\Phi_g}{c^2}\,. \tag{40}$$

The Newtonian potential that arises in the non-local formulation of gravity is given by [43,44]

$$\Phi_g \simeq -\frac{GM}{r} + \frac{GM}{\lambda}\ln\left(\frac{r}{\lambda}\right), \tag{41}$$

where λ is a constant length, and is taken to be $\lambda \approx 1kpc = 3260$ light-years. Consequently, the influence of the second term on the right hand side of Equation (41) in the solar system is negligible. Therefore, we find

$$f(r) = \frac{1}{\alpha} - 1 \simeq \frac{m}{r} - \frac{m}{\lambda}\ln\left(\frac{r}{\lambda}\right). \tag{42}$$

In the expression above, $m = GM/c^2$. For values of r within a galaxy, $r < \lambda$ and thus $-(m/\lambda)\ln(r/\lambda)$ is positive, and decreases as $1/r$ with increasing values of r, a result that shows that the gravitational field is sufficiently intense at the borders of a galaxy to explain the rotation curves of spiral galaxies. The function $d_g = 4k \sin\theta f(r)$ may again be understood as the gravitational COM distribution function of the spherically symmetric space-time.

3.4. Arbitrary Space-Time with Axial Symmetry

The analysis of a space-time that is not spherically symmetric allows to obtain the generalization of Equations (27) and (33). Let us consider an arbitrary space-time with axial symmetry. It is described by following line element,

$$ds^2 = g_{00}dt^2 + g_{11}dr^2 + g_{22}d\theta^2 + g_{33}d\phi^2 + 2g_{03}d\phi\, dt\,, \tag{43}$$

where all metric components depend on r and θ, but not on ϕ : $g_{\mu\nu} = g_{\mu\nu}(r,\theta)$. The determinant $e = \sqrt{-g}$ is $e = [g_{11}g_{22}\,\delta]^{1/2}$, where

$$\delta = g_{03}g_{03} - g_{00}g_{33}\,.$$

The inverse metric components are $g^{00} = -g_{33}/\delta$, $g^{03} = g_{03}/\delta$ and $g^{33} = -g_{00}/\delta$.

The set of tetrad fields in spherical coordinates that is adapted to static observers in space-time is given by

$$e_{a\mu} = \begin{pmatrix} -A & 0 & 0 & -C \\ 0 & \sqrt{g_{11}}\,\sin\theta\cos\phi & \sqrt{g_{22}}\cos\theta\cos\phi & -D\,r\sin\theta\sin\phi \\ 0 & \sqrt{g_{11}}\,\sin\theta\sin\phi & \sqrt{g_{22}}\cos\theta\sin\phi & D\,r\sin\theta\cos\phi \\ 0 & \sqrt{g_{11}}\,\cos\theta & -\sqrt{g_{22}}\sin\theta & 0 \end{pmatrix}. \tag{44}$$

The functions A, C and D are defined such that Equation (44) yields (43). They read

$$\begin{aligned} A(r,\theta) &= (-g_{00})^{1/2}\,, \\ C(r,\theta) &= -\frac{g_{03}}{(-g_{00})^{1/2}}\,, \\ D(r,\theta) &= \frac{1}{(r\sin\theta)}\left[\frac{\delta}{(-g_{00})}\right]^{1/2}. \end{aligned} \tag{45}$$

After simple calculations, we find that Equations (17) and (18) yield

$$\begin{aligned} L^{(0)(1)} &= \int d^3x\, d_{g1}\,(r\sin\theta\cos\phi)\,, \\ L^{(0)(2)} &= \int d^3x\, d_{g2}\,(r\sin\theta\cos\phi)\,, \\ L^{(0)(3)} &= \int d^3x\, d_{g3}\,(r\cos\theta)\,, \end{aligned} \tag{46}$$

where now we have

$$\begin{aligned} d_{g1} = d_{g2} &= 2k\Big\{-\frac{1}{r}\partial_1\left[\frac{g_{22}\delta}{(-g_{00})}\right]^{1/2} - \frac{1}{r\sin\theta}\partial_2\left[\left(\frac{g_{11}\delta}{(-g_{00})}\right)^{1/2}\cos\theta\right] \\ &\quad + \frac{1}{r\sin\theta}(g_{11}g_{22})^{1/2}\Big\}\,, \\ d_{g3} &= 2k\Big\{-\frac{1}{r}\partial_1\left[\frac{g_{22}\delta}{(-g_{00})}\right]^{1/2} + \frac{1}{r\cos\theta}\partial_2\left[\left(\frac{g_{11}\delta}{(-g_{00})}\right)^{1/2}\sin\theta\right]\Big\}. \end{aligned} \tag{47}$$

In the flat space-time, the quantities above vanish. It is not difficult to see that if the metric tensor components above represent the exterior gravitational field of a typical rotating source, the expressions above are not divergent. Note that $L^{(0)(1)}$ and $L^{(0)(2)}$ vanish due to integration in ϕ, as a consequence of the axial symmetry, but $L^{(0)(3)}$ is non-vanishing in general.

In the equations above we obtain $d_{g1} = d_{g2}$ because of the axial symmetry of the space-time. We see that, in general, we may have three different COM distribution functions, one for each direction in the three-dimensional space, in contrast to the situation in classical mechanics, where there is a single mass density in the definition of centre of mass.

4. Conclusions

In this article we have investigated the definition of centre of mass of the gravitational field, in the realm of the teleparallel equivalent of general relativity. The analysis of the gravitational centre of mass density leads to the concept of COM distribution function. We may understand the latter as a quantity that provides a description of the intensity of the gravitational field in space-time. The emergence of this quantity justifies the analysis of the centre of mass density of arbitrary configurations of the gravitational field, including gravitational wave configurations. We have applied this definition to the space-time endowed with a positive cosmological constant. At the speculative level, dark energy might be a consequence of the existence of a positive cosmological constant that induces a strong gravitational acceleration very far from our present location in the universe. In the simple model established by the Schwarzschild-de Sitter space-time, dark energy is roughly located in the region beyond $r = r_{min} = (mR^2)^{1/3}$, according to Equation (38).

The centre of mass moment naturally arises in the Hamiltonian formulation of the teleparallel equivalent of general relativity, and its definition is obtained from the primary constraints of the theory—Equation (13). It is given by Equations (17) and (18). The analysis led us to interpret the quantity d_g in the integrand of Equations (27), (33) and (46) as the gravitational COM distribution function. Although d_g plays the role of a density, mathematically it is not a density. It vanishes when the gravitational field is turned off. The expressions of $L^{(0)(i)}$ given by Equations (27) and (33) do remind the standard expression of centre of mass in classical mechanics. The distribution function d_g in the three-dimensional space is related to the intensity of the gravitational field. In the space-time of a point massive particle, d_g is intense (and in fact diverges) in the vicinity of the particle, and in the Schwarzschild-de Sitter space-time d_g is positive definite and diverges at both the Schwarzschild and cosmological horizons, which are precisely the regions where the gravitational field is more intense.

In relativistic field theory or in the Newtonian approximation of general relativity, energy, momentum and angular momentum are frame dependent field quantities, and so they are, in general, in the present context. In particular, the gravitational COM moment is evaluated in the frame adapted to an arbitrary observer in space-time. The gravitational centre of mass given by Equations (17) and (18) is invariant under coordinate transformations of the three-dimensional space, and under time reparametrizations. It transforms covariantly under global SO(3,1) transformations, provided the tetrad fields transform as $\tilde{e}^a{}_\mu = \Lambda^a{}_b e^b{}_\mu$, where $\Lambda^a{}_b$ are matrices of the SO(3,1) group. However, definition (17) is not covariant under local SO(3,1) transformations. In relativistic field theory, the COM definition is also not covariant under local SO(3,1) transformations.

We conclude that repulsion, in the Schwarzschild-de Sitter space-time, is in fact attraction to a region of intense gravitational COM distribution function. We have seen that in the region $r < r_{min} = (mR^2)^{1/3}$ the gravitational acceleration is attractive, and is repulsive in the dark energy region $r > r_{min}$. We expect the present analysis to be useful in the investigation of realistic cosmological models endowed with a positive cosmological constant.

Acknowledgments: I am grateful to B. Mashhoon for enlightening comments and for pointing out relevant references.

References

1. Arnowitt, R.; Deser, S.; Misner, C.W. *Gravitation: An Introduction to Current Research*; Witten, L., Ed.; Wiley: New York, NY, USA, 1962.
2. Weinberg, S. *Gravitation and Cosmology*; Wiley: New York, NY, USA, 1972.

3. Dixon, W.G. Dynamics of Extended Bodies in General Relativity. I. Momentum and Angular Momentum. *Proc. R. Soc. Lond. A* **1970**, *314*, 499–527.
4. Dixon, W.G. Dynamics of Extended Bodies in General Relativity. II. Moments of the Charge-Current Vector. *Proc. R. Soc. Lond. A* **1970**, *319*, 509–547.
5. Dixon, W.G. The definition of multipole moments for extended bodies. *Gen. Relativ. Gravit.* **1973**, *4*, 199–209.
6. Ehlers, J.; Rudolph, E. Dynamics of extended bodies in general relativity center-of-mass description and quasirigidity. *Gen. Relativ. Gravit.* **1977**, *8*, 197–217.
7. Regge, T.; Teitelboim, C. Role of surface integrals in the Hamiltonian formulation of general relativity. *Ann. Phys.* **1974**, *88*, 286–318.
8. Hanson, A.; Regge, T.; Teitelboim, C. *Constrained hamiltonian systems*; Accademia Nazionale dei Lincei: Roma, Italy, 1976.
9. Beign, R.; ÓMurchadha, N. The Poincaré group as the symmetry group of canonical general relativity. *Ann. Phys.* **1987**, *174*, 463–498.
10. Baskaran, D.; Lau, S.R.; Petrov, A.N. Center of mass integral in canonical general relativity. *Ann. Phys.* **2003**, *307*, 90–131.
11. Nester, J.M.; Ho, F.H.; Chen, C.M. Quasilocal Center-of-Mass for Teleparallel Gravity. 2004, arXiv:gr-qc/0403101.
12. Nester, J.M.; Meng, F.F.; Chen, C.M. Quasilocal Center-of-Mass. *J. Korean Phys. Soc.* **2004**, *45*, S22–S25.
13. Maluf, J.W. The teleparallel equivalent of general relativity. *Ann. Phys.* **2013**, *525*, 339–357.
14. Aldrovandi, R.; Pereira, J.G. *Teleparallel Gravity: An Introduction*; Springer: Heidelberg, Germany, 2013.
15. Blagojevic, M.; Hehl, F.W. *Gauge Theories of Gravitation*; Imperial College: London, UK, 2013.
16. Einstein, A. Riemannsche Geometrie unter Aufrechterhaltung des Begriffes des Fernparallelismus (Riemannian Geometry with Maintaining the Notion of Distant Parallelism). In *Sitzungsberichte der Preussischen Akademie der Wissenshcaften*; Verlag der Akademie der Wissenschaften: Berlin, Germany, 1928; pp. 217–221.
17. Einstein, A. Unified Field Theory based on Riemannian Metrics and Distant Parallelism. *Math. Ann.* **1930**, *102*, 685–697.
18. Cho, Y.M. Einstein Lagrangian as the translational Yang-Mills Lagrangian. *Phys. Rev. D* **1976**, *14*, 2521–2525;
19. Cho, Y.M. Gauge theory of Poincaré symmetry. *Phys. Revs. D* **1976**, *14*, 3335–3341.
20. Hayashi, K.; Shirafuji, T. New general relativity. *Phys. Rev. D* **1979**, *19*, 3524–3554.
21. Hayashi, K.; Shirafuji, T. Addendum to "New general relativity". *Phys. Rev. D* **1981**, *24*, 3312–3315.
22. Hehl, F.W. Four lectures on poincaré Gauge field theory. In *Cosmology and Gravitation: Spin, Torsion, Rotation and Supergravity*; Bergmann, P.G., de Sabbata, V., Eds.; Plenum Press: New York, NY, USA, 1980.
23. Nitsch, J. The macroscopic limit of the poincaré gauge field theory on gravitation. In *Cosmology and Gravitation: Spin, Torsion, Rotation and Supergravity*; Bergmann, P.G., de Sabbata, V., Eds.; Plenum Press: New York, NY, USA, 1980.
24. Schweizer, M.; Straumann, N.; Wipf, A. Post-Newtonian Generation of Gravitational Waves in a Theory with Torsion. *Gen. Relativ. Gravit.* **1980**, *12*, 951–961.
25. Nester, J.M. Positive energy via the teleparallel Hamiltonian. *Int. J. Mod. Phys. A* **1989**, *4*, 1755–1772.
26. Wiesendanger, C. Translational gauge invariance and classical gravitodynamics. *Class. Quantum Gravity* **1995**, *12*, 585–603.
27. Cai, Y.-F.; Capozziello, S.; de Laurentis, M.; Saridakis, E.N. $f(T)$ Teleparallel Gravity and Cosmology. 2015, arXiv:1511.07586.
28. Maluf, J.W.; da Rocha-Neto, J.F. Hamiltonian formulation of general relativity in the teleparallel geometry. *Phys. Rev D* **2001**, *64*, 084014.
29. Da Rocha-Neto, J.F.; Maluf, J.W.; Ulhoa, S.C. Hamiltonian formulation of unimodular gravity in the teleparallel geometry. *Phys. Rev. D* **2010**, *82*, 124035.
30. Maluf, J.W.; Ulhoa, S.C.; Faria, F.F.; da Rocha-Neto, J.F. The angular momentum of the gravitational field and the Poincaré group. *Class. Quantum Gravity* **2006**, *23*, 6245–6256.
31. Maluf, J.W. Hamiltonian formulation of the teleparallel description of general relativity. *J. Math. Phys.* **1994**, *35*, 335–343.
32. Maluf, J.W. The gravitational energy-momentum tensor and the gravitational pressure. *Ann. Phys. (Berlin)* **2005**, *14*, 723–732.

33. Da Rocha-Neto, J.F.; Maluf, J.W. The angular momentum of plane-fronted gravitational waves in the teleparallel equivalent of general relativity. *Gen. Relativ. Gravit.* **2014**, *46*, 1667.
34. Mashhoon, B.; Muench, U. Length measurement in accelerated systems. *Ann. Phys. (Berlin)* **2002**, *11*, 532–547.
35. Mashhoon, B. Vacuum electrodynamics of accelerated systems: Nonlocal Maxwell's equations. *Ann. Phys. (Berlin)* **2003**, *12*, 586–598.
36. Maluf, J.W.; Faria, F.F.; Ulhoa, S.C. On reference frames in spacetime and gravitational energy in freely falling frames. *Class. Quantum Gravity* **2007**, *24*, 2743–2754.
37. Maluf, J.W.; Faria, F.F. On the construction of Fermi-Walker transported frames. *Ann. Phys. (Berlin)* **2008**, *17*, 326–335.
38. Maluf, J.W. Repulsive gravity near naked singularities and point massive particles. *Gen. Relativ. Gravit.* **2014**, *46*, 1734.
39. Parker, E.P. Distributional geometry. *J. Math. Phys.* **1979**, *20*, 1423–1426.
40. Katanaev, M.O. Point massive particle in General Relativity. *Gen. Relativ. Gravit.* **2013**, *45*, 1861–1875.
41. Ulhoa, S.C.; da Rocha-Neto, J.F.; Maluf, J.W. The gravitational energy problem for cosmological models in teleparallel gravity. *Int. J. Mod. Phys. D* **2010**, *19*, 1925–1935.
42. Hehl, F.W.; Mashhoon, B. Formal framework for a nonlocal generalization of Einstein's theory of gravitation. *Phys. Rev. D* **2009**, *79*, 064028.
43. Blome, H.J.; Chicone, C.; Hehl, F.W.; Mashhoon, B. Nonlocal modification of Newtonian gravity. *Phys. Rev. D* **2010**, *81*, 065020 .
44. Hehl, F.W.; Mashhoon, B. Nonlocal gravity simulates dark matter. *Phys. Lett. B* **2009**, *673*, 279–282.

7

General Relativity and Cosmology: Unsolved Questions and Future Directions

Ivan Debono [1,*,†] **and George F. Smoot** [1,2,3,†]

[1] Paris Centre for Cosmological Physics, APC, AstroParticule et Cosmologie, Université Paris Diderot, CNRS/IN2P3, CEA/lrfu, Observatoire de Paris, Sorbonne Paris Cité, 10, rue Alice Domon et Léonie Duquet, 75205 Paris CEDEX 13, France

[2] Physics Department and Lawrence Berkeley National Laboratory, University of California, Berkeley, 94720 CA, USA; gfsmoot@lbl.gov

[3] Helmut and Anna Pao Sohmen Professor-at-Large, Hong Kong University of Science and Technology, Clear Water Bay, Kowloon, 999077 Hong Kong, China

* Correspondence: ivan.debono@apc.univ-paris7.fr.

† These authors contributed equally to this work.

Academic Editors: Lorenzo Iorio and Elias C. Vagenas

Abstract: For the last 100 years, General Relativity (GR) has taken over the gravitational theory mantle held by Newtonian Gravity for the previous 200 years. This article reviews the status of GR in terms of its self-consistency, completeness, and the evidence provided by observations, which have allowed GR to remain the champion of gravitational theories against several other classes of competing theories. We pay particular attention to the role of GR and gravity in cosmology, one of the areas in which one gravity dominates and new phenomena and effects challenge the orthodoxy. We also review other areas where there are likely conflicts pointing to the need to replace or revise GR to represent correctly observations and consistent theoretical framework. Observations have long been key both to the theoretical liveliness and viability of GR. We conclude with a discussion of the likely developments over the next 100 years.

Keywords: General Relativity; gravitation; cosmology; Concordance Model; dark energy; dark matter; inflation; large-scale structure

---

1. Perspective

Scientists have been fascinated by General Relativity ever since it was developed. It has been described as poetic, beautiful, elegant, and, at times, as impossible to understand.

General Relativity is often described as a simple theory. It is hard to define simplicity in science. One can always construct an entire theory encapsulated in one equation. Richard Feynman famously demonstrated this in a thought experiment where he rewrote all the laws of physics as $\vec{U} = 0$, where each element of $\vec{U}$ contained the hidden structure [1]. His point was that simplicity does not automatically bring truth.

An examination of the mathematical structure of General Relativity gives us a more sober definition of "simplicity". Under certain assumptions about the structure of physical theories, and of the properties of the gravitational field, General Relativity is the only theory that describes gravity. Alternative theories introduce additional interactions and fields.

General Relativity is also unique among theories of fundamental interactions in the Standard Model. Like electromagnetism, but unlike the strong and weak interactions, its domain of validity covers the entire range of length scales from zero to infinity. However, unlike the other forces, gravity as described by General Relativity acts on all particles. This implies that the theory does not fail below the Planck scale. All gravitational phenomena, from infinitesimal scales to distances beyond the observable universe, may be modelled by General Relativity. We may therefore formulate a mathematically rigorous description of General Relativity: it is the most complete theory of gravity ever developed.

All gravitational phenomena that have ever been observed can be modelled by General Relativity. It describes everything from falling apples, to the orbit of planets, the bending of light, the dynamics of galaxy clusters, and even black holes and gravitational waves. The domain of validity of the theory covers a wide range of energy levels and scales. That is why it has survived so long, and that is why it survives today, one hundred years after it was formulated, in an age in which the amount of data and knowledge increases by orders of magnitude every few years.

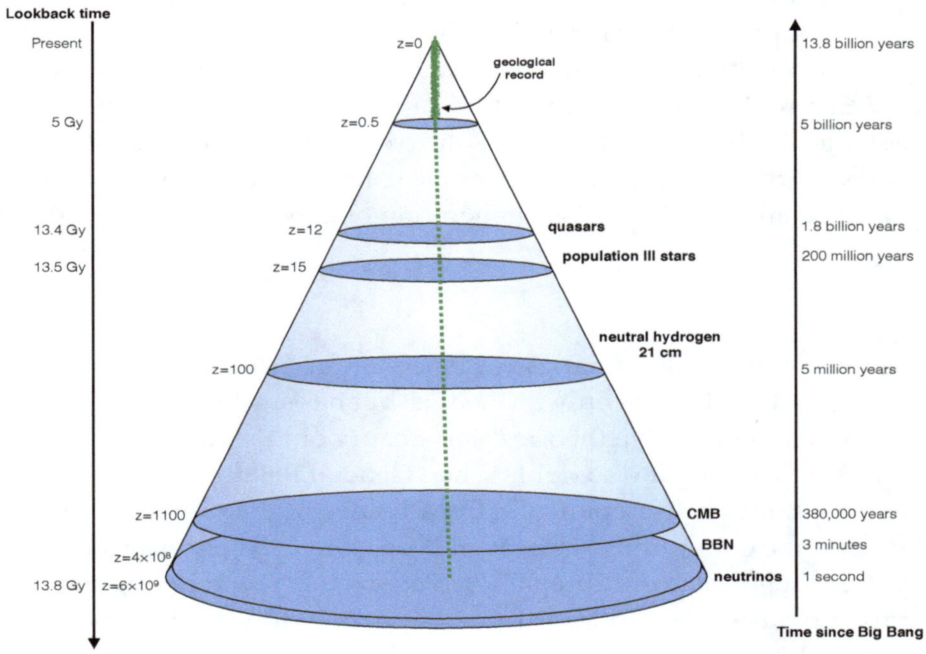

Figure 1. How we observe the universe. The lookback time is the difference between the age of the universe now, and the age of the universe when photons from an object were emitted. The more distant an object, the farther in its past we are observing its light. This distance in both space and time is expressed by the cosmological redshift z. We obtain most of our astrophysical information from the surface of our past light cone, because it is carried by photons. The only information from within the cone come from local experiments and observations, such as geological records. The green dotted line is the world-line of the atoms and nuclei providing the material for our geological data. Local experiments are carried out along this bundle of world-lines. They provide a useful test of physical constants. One example is the observation of the Oklo phenomenon [2]. The earliest information we have collected so far comes from the cosmic microwave background (CMB). Earlier than the CMB time-like slice is the cosmic neutrino background. We observe Big Bang nucleosynthesis (BBN) indirectly, through observations of the abundances of chemical elements.

Why, then, are we still testing General Relativity? Why do we still develop, discuss, test and fine-tune alternative theories? Because there are some very fundamental open questions in physics, particularly in cosmology. Moreover, the big questions in cosmology happen to be the ones that are not answered by General Relativity: the accelerated expansion of the universe, the presence of a mysterious form of matter which cannot be observed directly, and the initial conditions in the early universe.

The theoretical completeness described above is both a necessary and aesthetic feature of a fundamental theory. However, it creates experimental difficulties, for it compels us to test the theory at extreme scales, where experimental errors may be large enough to allow several alternative theories.

At extremely small scales below the Planck length, classical mechanics should break down. This compels us to question whether General Relativity is still accurate at these scales, whether it needs to be modified, and whether a quantum description of gravity can be formulated. At the other end of the scale, at cosmological distances, we may question whether General Relativity is valid, given that the universe cannot be modelled sufficiently accurately by General Relativity without invoking either a cosmological constant, or some additional, unknown component of the universe. Finally, we may question the accuracy of our solutions to the equations of General Relativity, which depend on some approximation scheme. These approximations provided analytical solutions which enabled most of the early progress in General Relativistic cosmology and astrophysics. However, one century after the formulation of the theory, we now have a flood of data from increasingly accurate observations (as shown in Figure 1), coupled with computing power which was hitherto unheard of. Tests of the higher-order effects predicted by General Relativity and some of its competitors are now within reach.

The purpose of this review is to examine the motivation for the development of alternative theories throughout the history of GR, to give an overview of the state of the art in General Relativistic cosmology, and to look ahead. In the next few decades, some of the open questions in cosmology may well be answered by a new generation of experiments, and GR may be challenged by alternative theories.

2. A Brief History

Let us start this review by breaking our own rule about unscientific adjectives. General Relativity is a beautiful theory of gravity. It has not only thrilled us, but has survived 100 years of challenges, both by experimental tests and by alternative theories. The beauty of the theory was clear at the beginning, but the initial focus was on whether it was right. When General Relativity provided an explanation for the 43 seconds of arc per century discrepancy in the advance of the perihelion of Mercury [3], it got the attention of the scientific community. However, it was the prediction and the observation of the bending of light by the Sun [4] that confirmed GR's place as the new reigning theory of gravity [5].

The setting at the Royal Society under the portrait of Newton for the report of the eclipse light bending observations led by Arthur Eddington, and reported by the great writer Aldous Huxley, was perfect to describe to the world the ascendancy of a new theory replacing Newton's gravity (see, e.g., [6]). From this point onwards, the scientific community started to take General Relativity seriously, and theorists worked hard to understand this new theory, beguilingly simple but hard to apply, and to advance its predictions.

Shortly after its publication, GR quickly became the framework for astrophysics, and for the Standard or Concordance Model of cosmology. However, it was still challenged by alternative theories. Initially, the alternatives were motivated by theoretical considerations. This early period led to a fuller understanding of GR and its predictions. Some of the predictions, such as black holes and gravitational waves, divided the scientific community. Did they exist as physical objects, or just as mathematical artifacts of the theory?

By the time GR turned 50, the model of cosmology had been established, GR had been tested, and things had started to stagnate. However, advances in observations led to new discoveries, which in turn led to renewed challenges.

First came the missing mass in the universe. Could GR be modified to account for it? Then came the theories about the very early universe, and the behaviour of the quantum-scale, tiny initial universe. Finally, twenty years ago, came the confirmation of cosmic acceleration. This had a twofold effect. On one hand, it spurred the development of a whole range of alternative theories of gravity. On the other hand, it confirmed GR like never before, for General Relativity, with a cosmological constant, can account for the observations perfectly.

In 2015, on the 100th birthday of General Relativity, gravitational waves were observed for the first time. This had been the last major untested prediction of General Relativity. It was a remarkable achievement, and in many ways it heralds a new age of astrophysical observations. The experimental capabilities and the computing power have finally caught up with the theory. Cosmology and astrophysics have now entered the era of Big Data, and much of the theoretical effort is now driven by data. However, the foundation for almost the entire scientific endeavour is still this theory of chronogeometrodynamics, developed 100 years ago when today's instruments and computers were still a distant dream.

From Aristotle to Einstein

General Relativity is the basis for the Standard Model of physical cosmology, and here we shall discuss the development of General Relativity (GR). The history of cosmology and GR are intertwined. We shall discuss why the theory has been so successful, and the criteria that must be satisfied by any alternative theory of physics, and by cosmological models.

Cosmology, in its broadest definition, is the study of the cosmos. It aims to provide an accurate description of the universe. Throughout much of the history of science, the development of cosmology was hampered by the lack of a universal physical theory. Observational tools were extremely limited, and there was no mathematical formulation for physical laws. The cosmos was described in metaphysical, rather than physical terms.

Discussions on the history of physics often refer to Karl Popper's concept of 'Falsifizierbarkeit' (falsifiability) [7]. In this formalism, scientific discovery proceeds by successive falsifications of theories. A falsifiable theory that covers observations, and that has not yet been proven false can be regarded as provisionally acceptable. Yet we know that in reality it is not quite as straightforward. A theory that is considered to be correct acquires this status by accumulation of evidence rather than by a single falsification of a previous theory [8]. This is especially true in cosmology, where the selection of theoretical models often depends on the outcome of statistical calculations.

The scientific revolution which brought about the development of a precise mathematical language for physical theories heralded the scientific age of cosmology. Physical laws, tested here on Earth and later in the Solar System, could be applied to the 'entire universe', and could thus provide a precise physical description of the cosmos. Modern cosmology is based upon this epistemological framework. Cosmology depends upon a fundamental premise. As a science, it must deal strictly with what can be observed, but the observable universe forms only a fraction of the whole cosmos. One is forced to make the fundamental but unverifiable assumption that the portion of the universe which can be observed is representative of the whole, and that the laws of physics are the same throughout the whole universe [9]. Once we make this assumption, we can construct a model of the universe based on a description of its observable part.

Any cosmological model which assumes the universality of physical laws must be based upon some physical theory. Since cosmology aims to describe the universe on the largest possible scales, it must be based upon an application long-range physical interactions. Since the theory of gravitation is the physical theory at the basis of standard cosmology, and is also at the centre of the big questions facing modern cosmology, we shall give an overview of the development of theories of gravitation.

The development of physical theories of gravity was far from smooth, nor did it always conform to Popper's scheme. Before the logical tools (mathematics) for the phenomenological description (physics) were invented, progress was rather haphazard.

According to Popper's scheme, this development should be driven by the search for ever more general principles. Yet Aristotelian theory, to take one example, considered itself to be general enough—its claimed region of validity was the entire universe, except that rising smoke, floating feathers, falling apples and orbiting celestial spheres each had their own rules.

The real revolution came when it was realised that the behaviour of all bodies could be described by a single rule—a universal theory of gravitation. This theory is a description of the long range forces that electrically neutral bodies exert on one another because of their matter content.

Whether they choose to or not, scientists will always stand on the shoulders of giants. No theory is invented in a scientific vacuum. This goes all the way back to the cosmology of the Euro-Mediterranean Ancient World, codified in the Aristotelian teachings of the 4th century B.C. This Hellenic "natural philosophy" provided qualitative rather than quantitative descriptions for what we would call today the free parameters of the theory [10]. It stands to reason—the instruments had not yet been invented that could test the theory of gravity to within numerical accuracy. Without accurate timekeeping instruments, processes could at best be described as "slower than" or "faster than". However, instruments to measure the movement of the celestial bodies, such as sundials, quadrants and astrolabes, were invented and improved upon, and measurements were carried out [11]. Astronomy flourished.

There is a certain logic to the development of physical theories from the Ancient World, to the Middle Ages, and right up to the Renaissance [12–14]. The basic tenet of the physics of Aristotle is that actions follow logically from causes. He distinguished between natural and violent motion. Natural motion implies falling at a speed proportional to the weight of the object and inversely proportional to the density of the medium. Violent motion happens whenever there is a force acting on an object, and the speed of the object is proportional to this force. Strato of Lampsacus replaced Aristotle's explanation of 'unnatural' motion with one that is very close to the modern notion of inertia. He identified natural motion as a form of acceleration, and demonstrated experimentally that falling bodies accelerate. In the 14th century, Jean Buridan came up with the notion of impetus, where the initial force imparts motion to the object, which gradually diminishes as gravity and air resistance act against this initial force. Concurrently, Nicole Oresme was using a crude early form of graph to describe motion, and unwittingly showing the complicated notions of differentiation and integration in pictorial form[15,16].

The cosmological observations, limited to the innermost five planets of the Solar System (Mercury, Venus, Mars, Jupiter, and Saturn) and the sphere of stars, seemed to confirm the Aristotelian-Ptolemaic theory. Celestial bodies moved in regular patterns made up of repeating circles. Small discrepancies were explained by circles within circles.

The fact that the theories were based on these regular patterns is no accident. Patterns are the keyword in all of physics. Human beings are wired to recognise patterns. We can only build theories because we recognise patterns in the universe. This characteristic of valid theories has been called sloppiness. The patterns fall within some hyper-ribbon of stability in the theory [17].

The revolution in physics came with the development of mathematical, quantitative, models to describe physical reality. Starting in the 1580 Galileo carried out a series of observations in which he subjected kinematics to rigorous experiment, and showed that naturally-falling objects really do accelerate. Crucially, he showed that the composition of the body has no effect whatsoever on this acceleration. He also realised that for violent motion, the speed is constant in the absence of friction. Galileo also took rigorous observations of astronomical objects. In 1610 he made the first observation of Jupiter's satellites, and the first observation of the phases of Venus, which is impossible according to the Ptolemaic geocentric model. His observations were important in putting to rest the Aristotelian theory of perfect and unchanging heavens.

By the time Newton came along, telescopes had been invented. Galileo had observed moons orbiting the Solar System planets, and hundreds of stars invisible to the naked eye. His 1610 treatise, aptly called *Siderus Nuncius* ("Starry Message", or "Astronomical Report" in modern language) [18], was the first scientific work based on observations through a telescope. Mechanical clocks had been invented. The sphere of observed data had expanded [19]. Calculus provided the tool to make sense of this new flood of data. Thus, physicists of Newton's generation found a very different scientific environment than the one in which Galileo had started off.

In 1687, Isaac Newton published in his "Mathematical Principles of Natural Philosophy", known by its abbreviated Latin title as *Principia* [20]. This was a significant milestone in physics. Newton's model of gravitation was, in his own words, a "universal" law. It applied to all bodies in the universe, whether it was cannonballs on Earth, or planets orbiting the Sun. For more than two centuries, Newton's theory, was the standard physical description of gravity. There was no other attempt to find a different theory for the gravitational force, although the intervening years between Newtonian gravity and Relativity produced some important physical concepts such as de Maupertuis's "Principle of Least Action" [21], further developed by Euler [22], Lagrange [23] and Hamilton [24,25]. The path of each particle is assigned a number called an action, which is the integral of the Lagrangian. In classical mechanics, the action principle is equivalent to Newton's Laws. Lagrangian field theory is an important cornerstone of modern physics. The Lagrangian of any physical interaction, when subjected to an action principle, give us field equations and conservation laws for the theory. It is an expression of the symmetries in physical laws.

Newtonian gravity was the great success story of nineteenth century physics, the golden age of mathematical astronomy. It allowed astronomers to calculate the position of planets and asteroids with ever greater precision, and to confirm their calculations by observation. Thus the size of the known universe grew. Evidence started to accumulate suggesting that there might be other galaxies in the universe besides our own. In 1845, the planet Neptune was discovered, after Urbain le Verrier suggested pointing telescopes in a region of the Solar System which he predicted by Newtonian calculations [26,27]. The search was motivated in the first place by an anomaly in the orbit of Uranus which could not be otherwise explained using Newtonian theory [28]. The discovery of Neptune showed that Newtonian theory was valid even in the very farthest limits of the Solar System.

There was another anomaly which could not be explained—the excessive perihelion precession of Mercury by 43 arcseconds per century, confirmed by le Verrier himself. Urbain le Verrier thus holds the distinction of being one of the few experimentalists to have confirmed Newton's theory and then disproved it. Astronomers attempted to explain this perihelion anomaly using Newtonian mechanics, which led them to speculate on the existence of Vulcan, a hypothetical planet whose orbit was even closer to the Sun [29].

The first doubts on Newtonian theory began to take shape just at the time when theorists were examining the full implications of the theory for complex, multi-body dynamical systems such as the Solar System. In 1890, Henri Poincaré published his magnum opus on the three-body problem [30], a masterpiece of celestial mechanics. At the time, Poincaré was working on another open question in physics: the aether. This led him to formulate a theory which was very close to Special Relativity [31], but which did not quite fit with Maxwell's electromagnetism [32], and was ultimately flawed.

By the end of the 19th century, the necessary mathematical tools were in place which would enable the development of Special and then General Relativity. There is an intimate connection between physics and the language of mathematics which is often overlooked. The former, especially in modern times, depends on the latter. Could Aristotle have developed General Relativity? No. Because he had not the mathematical language. Equations and mathematical formulations are relatively recent in the history of physics. Even Newton, for all his fame as a mathematical genius, never wrote the equation $F = -GMm/r^2$. He wrote a series of statements implying this law in (Latin) words: "*Gravitatem, quæ Planetam unumquemque respicit, ese reciprocæ ut quadratum distantiæ locorum ab ipsius centro*", and so on. It is hard to imagine how human beings could manipulate tensors and solve the field equations of Relativity in anything but numbers and symbols. Theories and physics do not happen in a cultural and scientific vacuum. They are human creations, and they depend intimately on tools for the transmission and communication of human knowledge.

The physical theory of gravity—the laws that govern gravitational interactions—remained unchanged until Einstein's time. In 1905, Einstein published his Theory of Special Relativity (SR) [33]. Soon after, he turned to the problem of including gravitation within four-dimensional spacetime [34–37].

Newton's formulation of the gravitational laws is expressed by the equations:

$$\frac{d^2x^i}{dt^2} = -\frac{\partial \Phi}{\partial x^i}, \tag{1}$$

$$\triangle \Phi = 4\pi G\rho\,, \tag{2}$$

where Φ is the gravitational potential, G is the universal gravitational constant, ρ is the mass density, and $\triangle = \nabla^2$ is the Laplace operator. These equations cannot be incorporated into Special Relativity as they stand. The equation of motion (1) for a particle is in three-dimensional form, so it must be modified into a four-dimensional vector equation for $d^2x^\mu / d\tau^2$. Similarly, the field Equation (2) is not Lorentz-invariant, since the three-dimensional Laplacian operator instead of the four-dimensional d'Alembertian $\Box = \partial^\mu \partial_\mu$ means that the gravitational potential Φ responds instantaneously to changes in the density ρ at arbitrarily large distances. The conclusion is that Newtonian gravitational fields propagate with infinite velocity. In other words, instantaneous action in Newtonian theory implies action at a distance when reconsidered in the light of Special Relativity. This violates one of the postulates of SR. How do we reconcile gravity and Special Relativity?

3. The Development of General Relativity

3.1. From Special to General Relativity

The simplest relativistic generalisation of Newtonian gravity is obtained by representing the gravitational field by a scalar Φ. Since matter is described in Relativity by the stress-energy tensor $T_{\mu\nu}$, the only scalar with dimensions of mass density (which corresponds to ρ) is T^μ_μ. A consistent scalar relativistic theory of gravity would thus have the field equation

$$\Box \Phi = 4\pi G T^\mu_\mu\,. \tag{3}$$

However, when the equation of motion from this theory are applied to a static, spherically symmetric field Φ, such as that of the sun, acting on an orbiting planet, they would result in a negative precession, or retardation of the perihelion. Experimental evidence since the time of Le Verrier and his observation of the orbit of Mercury [38] clearly shows that planets experience a prograde precession of the perihelion. Moreover, in the limit of a zero rest-mass particle, such as a photon, the equations of motion show that the particle experiences no geodesic deviation. The existence of an energy-momentum tensor due to an electromagnetic field would also be impossible, since $(T_{\text{electromagnetic}})^\mu_\mu = 0$. The theory therefore allows neither gravitational redshift, nor deviation of light by matter, both of which are clearly observable phenomena [39]. Another route to generalisation could be to represent the gravitational field by a vector field Φ_μ, analogous to electromagnetism. Following through with this strategy, the "Coulomb" law in this theory gives a repulsion between two massive particles, which clearly contradicts observations. The theory also predicts that gravitational waves should carry negative energy, and, like the scalar theory, predicts no deviation of light. Like the scalar theory, then, the vector theory must be discarded.

What about a flat-space tensor theory? The gravitational field in this theory is described by a symmetric tensor $h_{\mu\nu} = h_{\nu\mu}$. The choice of the Lagrangian in this theory is dictated by the requirement that $h_{\mu\nu}$ be a Lorentz-covariant, massless, spin-two field.

In the 1930s, Wolfgang Pauli and Markus Fierz [40] were the first to write down this Lagrangian and investigate the resulting theory. The predictions of the theory for deviation of light agree with those of General Relativity, and are consistent with observations. Since the field equations and gauge properties are identical to those of Einsein's linearised theory, the predictions for the properties of gravitational waves, including positive energy, agree with those obtained using the linearised theory in General Relativity. However, the theory differs from General Relativity in its predicted value for the perihelion precession, which is $\frac{4}{3}$ of that given by GR. This disagrees with the value obtained from observations of Mercury's orbit.

The theory has an even worse deficiency. If two gravitating bodies (that is, not test particles) are considered, and the field equations are applied to them, then the theory predicts that gravitating bodies cannot be affected by gravity, since they all move along straight lines in a global Lorentz reference frame. This holds for bodies made of arbitrary stress-energy, and since all bodies gravitate, then one must conclude that no body can be accelerated by gravity, which is a obvious self-inconsistency in the theory.

The only way in which a consistent theory of gravity can be constructed within Special Relativity is to consider the geometry of spacetime as the gravitational field itself. In other words, *all matter* moves in an effective Riemann space of metric $g^{\mu\nu} \equiv \eta^{\mu\nu} + h^{\mu\nu}$, where $\eta^{\mu\nu}$ is the Minkowski metric. The requirement of consistency leads us to universal coupling, which implies the Equivalence Principle.

The existence of curved spacetime can be deduced from purely physical arguments. In 1911, before he had fully developed General Relativity, Einstein [34] showed that a photon must be affected by a gravitational field, using conservation of energy applied to Newtonian gravitation theory. Schild [41–43] showed by a simple thought experiment, formulated within Special Relativity, that a consistent theory of gravity cannot be constructed within this framework. His argument is based upon a gravitational redshift experiment carried out in the field of the Earth, using a global Lorentz frame tied to the Earth's centre. Successive pulses of light rising to the same height should experience a redshift, and therefore the pulse rate at the top should be slower than that at the bottom. But light rays are drawn at 45 degrees in Minkowski spacetime diagrams, so that top and bottom time intervals are equal, which is impossible if redshift occurs. Hence the spacetime must be curved. One therefore concludes that in the presence of gravity, Special Relativity cannot be valid over any sufficiently extended region.

General Relativity may be understood as a generalisation of Special Relativity over extended regions. Since Special Relativity can comfortably be described using tensor calculus, it was only natural to extend the flat Minkowski spacetime of Special Relativity to the curved spacetime of General Relativity. This was a physical application of Riemannian geometry [44,45], which had been developed in the second half of the 19th century. The idea of tensor calculus on curved manifolds was already mathematically well-established. Einstein's innovation lay in identifying the Einstein tensor, itself related to the Riemann curvature tensor, as the "gravitational field" in the theory.

Einstein had been working on the problem for some years, starting in 1907. He arrived at the final, correct form in 1915 [46,47]. He was well-aware of the significance of his publication, and he gave it the succinct title of "The Field Equations of Gravitation" (*Feldgleichungen der Gravitation*). The correct field equations for the theory contained in this publication served as the starting point or subsequent derivations.

3.2. The Formalism of General Relativity

General Relativity is based on two independent but mutually supporting postulates.

The first postulate is sometimes referred to collectively as the *Einstein Equivalence Principle*:

- *The Strong Equivalence Principle:* The laws of physics take the same form in a freely-falling reference frame as in Special Relativity
- *The Weak Equivalence Principle:* An observer in freefall should experience no gravitational field. That is to say, an observer cannot determine from a local experiment whether the his laboratory is being accelerated by a rocket of static at the surface of a gravitating body. Gravity is erased up to tidal forces, which are determined by the size of the laboratory and its distance to the centre of the gravitational attraction.

The Equivalence Principle allows us to construct the metric and the equation of motion by transforming from a freely-falling to an accelerating frame. It can be mathematically expressed by the assuming that all matter fields are minimally coupled to a single metric tensor $g_{\mu\nu}$. The distance between two points in 4-dimensional spacetime, called events, is:

$$ds^2 = c^2\,d\tau^2 = g_{\mu\nu}\,dx^\mu\,dx^\nu\,. \tag{4}$$

Throughout the text, we follow the Einstein summation convention for repeated indices, so that $c_i x^i = \sum_{i=1}^{n} c_i x^i$ for $i = 1, \ldots, n$. Greek indices are used for space and time components, while Latin indices are spatial ones only. We use the following metric signature: $(-+++)$.

The metric defines lengths and times measured by laboratory rods and clocks. This metric implies that the action for any matter field ψ is of the form

$$S_{\text{matter}}[\psi, g_{\mu\nu}]\,, \tag{5}$$

which gives us three important results. First, it implies the universality of freefall. Second, it implies that all non-gravitational constants are spacetime independent. Third, it implies that the laws of physics are isotropic. This equation defines how matter behaves in a given curved geometry, how light rays propagate, how stars, planets and galaxies move, and gives us verifiable observational consequences.

The second postulate is related to the dynamics of the gravitational interaction. This is assumed to be governed by the Einstein-Hilbert action:

$$S_{\text{gravity}} = \frac{c^3}{16\pi G}\int d^4x\sqrt{-g_*}R_* \tag{6}$$

where $g^*_{\mu\nu}$ is a massless spin-2 field called the Einstein metric. General Relativity identifies the Einstein metric with the physical metric, that is: $g_{\mu\nu} = g^*_{\mu\nu}$. This implements the Strong Equivalence Principle.

The Einstein-Hilbert action defines the dynamics of gravity itself. Relativity is thus a geometrical approach to fundamental interactions. These are realised though continuous classical fields which are inseparably connected to the geometrical structures of spacetime, such as the metric, affine connection, and curvature.

The General Relativistic equation of motion is simply parallel transport on curved spacetime. It is given by

$$\frac{d^2x^\mu}{d\tau^2} + \Gamma^\mu_{\alpha\beta}\frac{dx^\alpha}{d\tau}\frac{dx^\beta}{d\tau} = 0\,, \tag{7}$$

where x^μ is some set of coordinates for a point in spacetime. $\Gamma^\mu_{\alpha\beta}$ are the components of the affine connection (or metric connection). The fundamental theorem of Riemannian geometry states that the affine connection can be expressed entirely in terms of the metric:

$$\Gamma^\alpha_{\lambda\nu} = \frac{1}{2}g^{\alpha\nu}(g_{\mu\nu,\lambda} + g_{\lambda\nu,\mu} - g_{\mu\lambda,\nu})\,, \tag{8}$$

where the comma denotes a derivative, i.e., $g_{\mu\nu,\lambda} = \frac{\partial g_{\mu\nu}}{\partial x^\lambda}$.

We need to construct invariant quantities in GR (quantities that are the same for all observers). To achieve this, we need to contract the covariant A_μ and contravariant A^μ components of a vector or tensor A by using the metric to raise or lower indices: $A_\mu = g_{\mu\nu}A^\mu$. Thus the equation of motion (7) can be made covariant by recasting it as the covariant derivative of the 4-velocity $U^\mu = \gamma(c, \mathbf{v})$:

$$\frac{D_\mu U^\mu}{d\tau} = 0\,, \tag{9}$$

where the covariant derivative is defined as

$$D_\mu A^\mu = dA^\mu + \Gamma^\mu_{\alpha\beta}A^\alpha\,dx^\beta\,. \tag{10}$$

The quantity γ is the Lorentz factor:

$$\gamma = \frac{1}{\sqrt{1 - v^2/c^2}} \,. \tag{11}$$

The transformation from SR to GR is then carried out by mapping the Minkowski metric to a general metric: $\eta \to g$ and by mapping $\partial \to D$.

In GR, freely-falling bodies travel along a geodesic. Geometrically, this is the shortest distance between two points in spacetime. The path length along a geodesic is given by

$$S = \int (g_{\mu\nu}\, \mathrm{d}x^{\mu}\, \mathrm{d}x^{\nu})^{1/2} \,. \tag{12}$$

In cosmology, it is essential for us to be able to describe spacetime which is not "empty". In the presence of a perfect fluid (an inviscid fluid with density ρ and isotropic pressure p), the energy and momentum of spacetime is described by the energy-momentum tensor (or stress-energy tensor)

$$T^{\mu\nu} = \left(\rho + \frac{p}{c^2}\right) U^{\mu} U^{\nu} - p g^{\mu\nu} \,. \tag{13}$$

Classical energy and momentum conservation are generalized in GR as the four conservation laws

$$D_{\mu} T^{\mu\nu} = 0 \,. \tag{14}$$

In other words, the stress-energy tensor has a vanishing covariant divergence. In the absence of a component possessing pressure or density, or both, the energy-momentum tensor is zero.

The central notion in General Relativity is that gravitation can be described by a metric. The Einstein equations give us the relation between the metric and the matter and energy in the universe:

$$G^{\mu\nu} = -\frac{8\pi G}{c^4} T^{\mu\nu}. \tag{15}$$

The left-hand side of this equation is a function of the metric: $G^{\mu\nu}$ is the Einstein tensor, defined as:

$$G^{\mu\nu} = R^{\mu\nu} - \frac{1}{2} g^{\mu\nu} R \,, \tag{16}$$

where $R^{\mu\nu}$ is the Ricci tensor, which depends on the metric and its derivatives, and the Ricci scalar R is the contraction of the Ricci tensor ($R = g_{\mu\nu} R^{\mu\nu}$). The right-hand side of Equation (15) is a function of the energy: G is Newton's constant, and $T^{\mu\nu}$ is the energy-momentum tensor.

Einstein's Relativity has three main distinguishing characteristics:

- it agrees with experiment
- it describes gravity entirely in terms of geometry
- it is free of any "prior geometry"

These characteristics are lacking in most of the other theories [48,49]. Apart from the issue of agreement with experiment, Einstein's theory is unique in its physical simplicity.

Every other theory introduces auxiliary gravitational fields, or involves prior geometry. Prior geometry is any aspect of the geometry of spacetime which is fixed immutably, that is, it cannot be changed by changing the distribution of gravitating sources.

A rigorous mathematical definition of the unique simplicity of General Relativity is given by Lovelock's theorem [50–52]. This is a generalisation of an earlier theorem by Élie Cartan [53], and may be formulated as follows:

> *In 4 spacetime dimensions, the only divergence-free symmetric rank-2 tensor constructed solely from the metric g and its derivatives up to second differential order, and preserving diffeomorphism invariance, is the Einstein tensor plus a cosmological term.*

In simple terms, the theorem states that GR emerges as the unique theory of gravity if the conditions of the theorem are followed. In fact, Lovelock's theorem provides a useful scheme for classifying alternatives to General Relativity.

Einstein described both the demand for "no prior geometry" and for a "geometric, coordinate-independent formulation of physics" by the single phrase "general covariance", but the two concepts are not quite the same.

While many physical theories can be formulated in a generally covariant way, General Relativity is actually based on the "no prior geometry" demand. This distinction was not always made, especially in the first decades after Einstein's publications [54,55]. Erich Kretschmann's famous objection in 1917 [56] concerned this point, since he regarded general covariance merely as formal feature that any theory could have, not as a special feature belonging to GR.

3.3. Newtonian Nostalgia: The First Wave of Alternative Theories

Newton's theories had predicted observations of Solar System objects, comets and asteroids, with astounding precision. Why should they be tampered with? The first wave of alternative theories were driven more by theoretical considerations than by observations. Equations (1) and (2) can be generalized so that they are consistent with the postulates of Special and General Relativity. Several generalisations of this kind were attempted in the first few decades following the development of GR, motivated by lingering resistance to any deviation from Newtonian gravity.

One early theory, involving prior geometry, was formulated by Nordstrøm in 1913 [57]. In this theory, the physical metric of spacetime g is generated by a background flat spacetime metric η, and by a scalar gravitational field ϕ. Stress-energy generates ϕ:

$$\eta^{\alpha\beta}\phi_{,\alpha\beta} = -4\pi\phi\eta^{\alpha\beta}T_{\alpha\beta} \tag{17}$$

and g is constructed from ϕ and η:

$$g_{\alpha\beta} = \phi^2\eta_{\alpha\beta}. \tag{18}$$

Prior geometry cannot be removed by rewriting Nordstrøm's equations in a form devoid of η and ϕ [58]. Mass only influences one degree of freedom in the spacetime geometry, while the other degrees of freedom are fixed *a priori*. This prior geometry, if it existed, could be detected by physical experiments.

In the 1920s, Alfred North Whitehead [59] formulated a two-tensor theory of gravity in which the prior geometry is quite different from later theories such as Ni's [48]. Whitehead's theory is remarkable in that it agrees with Einstein's in its predictions for the four standard tests (bending of light, gravitational redshift, perihelion shift, and time delay). It was accepted as a viable alternative for Einstein's theory until Clifford Martin Will [60] showed that it predicts velocity-independent anisotropies in the Cavendish constant (the gravitational constant G in Newtonian theory). This would produce time-dependent Earth tides which are clearly contradicted by everyday observations. Any valid theory of gravity must not only agree with relativistic experiments, but also with past experiments in the Newtonian regime.

One theory which disagrees violently with non-relativistic experiments is due to George David Birkhoff [61]. It was developed in the 1940s, and it predicts the same redshift, perihelion shift, deflection and time-delay as General Relativity, but it requires that the pressure inside gravitating bodies should be equal to the total density of mass-energy ($p = \rho$). This means that sound waves travel with the speed of light. This clearly contradicts everyday experiments.

Most of the early alternative theories were abandoned either because they were contradicted by observations, or because of internal inconsistencies in the theories themselves. One notable exception is Dicke-Brans-Jordan theory, sometimes called Brans-Dicke, or Jordan-Fierz-Brans-Dicke theory [62,63], developed in the 1960s by Robert H. Dicke and Carl H. Brans following earlier work by Pascual Jordan and Markus Fierz. The different names arise from the fact that the theory is a special case of Jordan's, with $\eta = -1$. An alternative mathematical representation of the theory is given by [64].

This theory introduced auxiliary gravitational fields. Brans and Dicke took the equivalence principle as the starting point of their theory, and thus they describe gravity in terms of spacetime curvature, but their gravitational field, unlike Einstein's, is a scalar-tensor combination. In this way it overcomes the difficulties associated with tensor or scalar-only theories mentioned earlier. The trace of the energy-momentum tensor $(T_{\rm M})_{\mu\nu}$ (representing matter) and a coupling constant λ generate the long-range scalar field ϕ via the equation

$$\Box^2\phi = 4\pi\lambda(T_{\rm M})^{\mu}_{\mu}. \tag{19}$$

The scalar field ϕ fixes the value of G, which is therefore not a constant, but simply the coupling strength of matter to gravity. The gravitational field equations relate the curvature to the energy-momentum tensors of the scalar field and matter:

$$R_{\mu\nu} - \tfrac{1}{2}g_{\mu\nu}R = -\frac{8\pi}{c^4\phi}\left[(T_{\rm M})_{\mu\nu} + (T_{\Phi})_{\mu\nu}\right], \tag{20}$$

where $(T_{\rm M})_{\mu\nu}$ is the energy-momentum tensor of matter and $(T_{\Phi})_{\mu\nu}$ is the energy-momentum tensor of the scalar field ϕ. For historical reasons, it is usual to write the coupling constant as

$$\lambda = \frac{2}{3+2\omega}, \tag{21}$$

where ω is the dimensionless 'Dicke coupling constant'. In the limit $\omega \to \infty$, we have $\lambda \to 0$, so ϕ is not affected by the matter distribution, and can be set to a constant $\phi = 1/G$. Hence Dicke-Brans-Jordan theory reduces to Einstein's theory in the limit $\omega \to \infty$.

The equivalence principle is satisfied in this theory since the special-relativistic laws are valid in the local Lorentz frames of the metric g of spacetime. The scalar field does not exert any direct influence on matter. It only enters the field equations that determine the geometry of spacetime. On a conceptual level, Brans-Dicke theory can be seen as more fully Machian than Einstein's theory. Einstein himself attempted to incorporate Mach's Principle into his theory, but in Einstein's General Relativity, the inertial mass of an object will always be independent of the mass distribution in the universe. In Brans-Dicke theory, the long-range scalar field is an indirectly coupling field, so it does not directly influence matter, but the Einstein tensor is determined partly by the energy-momentum tensor, and partly by the long-range scalar field.

Dicke-Brans-Jordan theory is self-consistent and complete, but experimental evidence based on Solar System tests, shows that $\omega \geq 600$ [65], as a conservative estimate. Some calculations raise this limit even higher, with $\omega \gtrsim 10^4$ [66]. The Cassini mission set a comparable limit of $\omega > 40{,}000$ [67]. Recent cosmological data from the *Planck* probe show that $\omega \geq 890$ [68,69]. This is consistent with the Solar System bounds. Future cosmological experiments and data from large-scale structure could provide even better constraints [70].

Brans-Dicke theory is a special case of general scalar-tensor theories with $\omega(\phi) =$ constant, where ϕ is a value depending on the cosmological epoch. In these theories, the function $\omega(\phi)$ could be such that the theory is very different from GR in the early universe or in future epochs, but very close to GR in the present. In fact, it has been shown that GR is a natural attractor for such scalar-tensor theories, since cosmological evolution naturally drives the fields towards large values of ω [71,72].

3.4. Self-Consistency, Completeness, and Agreement with Experiment

Any viable theory must satisfy three fundamental criteria: self-consistency, completeness, and agreement with past experiment.

To be self-consistent, a theory must not contain any internal contradictions. The spin-two field theory of gravity [40] is equivalent to linearised General Relativity but it is internally inconsistent since it predicts that gravitating bodies should have their motion unaffected by gravity. When one tries to remedy this inconsistency, the resulting theory is nothing but General Relativity. Another self-inconsistent theory is due to Paul Kustaainheimo [73,74]. It predicts zero gravitational redshift when the wave version of light (Maxwell theory) is used, and nonzero redshift when the particle version (photon) is used.

To be complete, a theory must be able to analyse the outcome of any experiment. This means that it must be compatible with other physical theories which describe any other forces that are present in experiments. This can only be achieved if the theory is derived from first principles, since the theoretical postulates must be as general as possible if the theory is to cover the widest range of phenomena.

A viable theory must agree with past experiment, which includes experiments in the Newtonian regime, and the standard tests of General Relativity. Its results must agree with those obtained from Newtonian theory in the weak field limit, and with GR in relativistic situations. It also means that the theory must agree with cosmological observations.

The experimental criterion also works the other way. Any alternative to General Relativity that claims to have a smaller set of limiting cases must be experimentally distinguishable, perhaps by future experiment. At some point, the divergence between GR and other theories must manifest itself physically, in the form of predictions which can be verified by experiment. This is perhaps the greatest challenge of current alternatives to GR.

3.5. Metric Theories and Quantum Gravity

Most theories of gravity incorporate two principles: *spacetime possesses a metric; and that metric satisfies the equivalence principle*. Such theories are called metric theories. There are some exceptions.

Soon after the publication of the theory of General Relativity, it became apparent that its formulation is incompatible with a Quantum Mechanical description of the gravitational field. It was Einstein himself who pointed out that quantum effects must lead to a modification of General Relativity [75]. Back then, the first successful applications of Quantum Mechanics to electromagnetism were starting to give useful results. These developments led to the question of whether General Relativity can be quantized.

This is a difficult question. First, Einstein's field equations are much more complicated than Maxwell's equations, and in fact are nonlinear. The physical reason for this is that the gravitational field is coupled to itself—the stress-energy tensor acts as the source for spacetime curvature, which in general contributes to the stress-energy tensor. This means that the equations seem to violate the superposition principle, which requires the existence of a linear vector space (see, e.g., [76,77]). This is the mathematical expression of wave-particle duality—a central tenet of Quantum Theory.

Second, to quantize the gravitational field we would have to quantize spacetime itself. The physical meaning of this is not completely clear.

Finally, there are experimental problems. Maxwell's equations predict electromagnetic radiation, which was first observed by Hertz [78]. Quantization of the field results in being able to observe individual photons, and these were first seen in the photoelectric effect predicted by Einstein [79]. Similarly, Einstein's equations for the gravitational field predict gravitational radiation [75], so there should be, in principle, the possibility of observing individual gravitons, which are the quanta of the field. The direct observation of gravitational waves was finally achieved in September 2015 by the LIGO instrument [80]. The detection of individual gravitons is more difficult and is beyond the capability of current experiments.

To develop a quantum theory of General Relativity, the fundamental interactions in GR would have to follow quantum rules. In Quantum Theory, particle interactions are described by gauge theories, so GR would have to follow the gauge principle. Although the gauge principle was first recognized in electromagnetism, modern gauge theory, formulated initially by Chen Ning Yang and Robert Mills [81,82], emerged entirely within the framework of the quantum field programme. As more particles were discovered after the 1940s, various possible couplings between those elementary particles were being proposed. It was therefore necessary to have some principle to choose a unique form out of the many possibilities suggested. The principle suggested by Yang and Mills in 1954 is based on the concept of gauge invariance, and is hence called the gauge principle.

3.6. The Gauge Approach and Non-Metric Theories

The idea of gauge invariance, and the term itself, originated earlier, from the following consideration due to Hermann Weyl in 1918 [83,84]. In addition to the requirement of General Relativity that coordinate systems have to be defined only locally, so likewise the standard of length, or scale, should only be defined locally. It is therefore necessary to set up a separate unit of length at every spacetime point. Weyl called such a system of unit-standards a gauge system (analogous to the standard width, or "gauge", of a railway track).

The gauge principle therefore may be formulated as follows: If a physical system is invariant with respect to some global (spacetime independent) group of continuous transformations G, then it remains invariant when that group is considered locally (spacetime dependent), that is $G \mapsto G(x)$. Partial derivatives are replaced by covariant ones, which depend on some new vector field.

In Weyl's view, a gauge system is as necessary for describing physical events as a coordinate system. Since physical events are independent of our choice of descriptive framework, Weyl maintained that gauge invariance, just like general covariance, must be satisfied by any physical theory.

In Euclidean geometry, we know that translation of a vector preserves its length and direction. In Riemannian geometry, the Christoffel connection [85] (or affine connection) guarantees length preservation, but a vector's orientation is path dependent. However, the angle between two vectors, following the same path, is preserved under translation. Weyl wondered why the remnant of planar geometry, length preservation, persisted in Riemannian geometry. After all, our measuring standards (rigid rods and clocks), are known only at one point in spacetime. To measure lengths at another point, we must carry our measuring tools along with us. Weyl maintained that only the *relative* lengths of any two vectors at the same point, and the angle between them, are preserved under parallel transport. The length of any single vector is arbitrary. To encode this mathematically, Weyl made the following substitution:

$$g_{\mu\nu}(x) \mapsto \lambda(x) g_{\mu\nu}(x), \tag{22}$$

where the conformal factor $\lambda(x)$ is an arbitrary, positive, smooth function of position. Weyl required that in addition to GR's coordinate invariance, formulae must remain invariant under the substitution (Equation (22)). He called this a *gauge transformation*. The scale therefore becomes a local property of the metric.

Weyl's theory enabled him to unify gravity and electromagnetism, the only two forces known at the time. However, Weyl's original scale invariance was abandoned soon after it was proposed, since its physical implications seemed to contradict experiments. In particular, if two identical clocks C_1 and C_2 are transported on two different paths, which both end at the same point Q, the time-like vectors l_1 and l_2 given by C_1 and C_2 at Q would be different in the presence of an electromagnetic field. Therefore the two clock rates would differ. As Einstein (probably the only expert who could keep an eye on Weyl's theory at the time) pointed out, this concept meant that spectral lines with definite frequencies could not exist, since the frequency of the spectral lines of atomic clocks would depend on the atom's location, both past and present. However, we know the atomic spectral lines to be definite, and independent of spacetime position [86–89].

Despite its initial failures, Weyl's idea of a local gauge symmetry survived, and acquired new meaning with the development of Quantum Mechanics. According to Quantum Mechanics, interactions are realized through quantum (that is, non-continuous) fields which underlie the local coupling and propagation of field quanta, but which have nothing to do with the geometry of spacetime.

The question is whether General Relativity can be formulated as a gauge theory. This question has been discussed by ever since it was first posed [90–96].

If features of General Relativity could be recovered from a gauging argument, then that would show that the two formulations are not inconsistent. The first to succeed in this was Kibble [91], who elaborated on an earlier, unsuccessful attempt by Utiyama [90]. Kibble arrived at a set of gravitational field equations, although not the Einstein equations, constructing a slightly more general theory, known as "spin-torsion" theory. The inclusion of torsion in Einstein's General Relativity had long been theorized. In fact the necessary modifications to General Relativity were first suggested by Élie Cartan in the 1920s [97–100], who identified torsion as a possible physical field.

The connection between torsion and quantum spin was only made later [91,101,102], once it became clear that the stress-energy tensor for a massive fermion field must be asymmetric [103,104]. The Einstein-Cartan (1920s) and the Kibble-Sciama (late 1950s) developments occurred independently. For historical reasons, spin-torsion theories are sometimes referred to as Einstein-Cartan-Kibble-Sciama (ECKS) theories, but Einstein-Cartan Theory (ECT) is the term more commonly employed.

The Einstein-Cartan Theory of gravity is a modification of GR allowing spacetime to have torsion in addition to curvature, and, more importantly, relating torsion to the density of intrinsic angular momentum. This modification was put forward by Cartan before the discovery of quantum spin, so the physical motivation was anything but quantum theoretic. Cartan was influenced by the works of the Cosserat brothers [105] who considered a rotation stress tensor in a generalized continuous medium besides a force tensor.

Cartan assumed the linear connection to be metric and derived, from a variational principle, a set of gravitational field equations. However, Cartan required, without justification, that the covariant divergence of the energy-momentum tensor be zero, which led to algebraic constraint equations, thus severely restricting the geometry. This probably discouraged Cartan from pursuing his theory. It is now known that the conservation laws in relativistic theories of gravitation follow from the Bianchi identities and in the presence of torsion, the divergence of the energy-momentum tensor need not vanish.

In simple mathematical terms, a non-zero torsion tensor means that

$$T^{\mu}_{\nu\sigma} = \Gamma^{\mu}_{\nu\sigma} - \Gamma^{\mu}_{\sigma\nu} \neq 0 \,. \tag{23}$$

Geometrically, it means that an infinitesimal geodesic parallelogram forms a non-closed loop. Torsion is therefore a local property of the metric. The Lagrangian action of Einstein-Cartan theory takes the usual Einstein-Hilbert form:

$$S = \int \mathrm{d}^4x\sqrt{-g}\left(-\frac{g^{\mu\nu}R_{\mu\nu}(\Gamma)}{16\pi G} + \mathcal{L}_m\right), \tag{24}$$

where Γ is a general affine connection and $\mathcal{L}_m$ is the matter Lagrangian. The theory differs from GR in the structure of Γ, leading to a field theory with additional interactions.

Torsion vanishes in the absence of spin and the Einstein-Cartan field equation is then the classical Einstein field equation. In particular, there is no difference between the Einstein and Einstein-Cartan theories in empty space. Since practically all tests of relativistic theory are based on free space experiments, the two theories are, to all effects, indistinguishable via the standard tests of GR. The inclusion of torsion only results in a slight change in the energy-momentum tensor. Cartan's theory holds the distinction of being complete, self-consistent and in agreement with experiment, but of being a non-metric theory of gravitation. The link between torsion and quantum spin means that it could be possible to study the divergence between the GR and ECKS theories at the quantum level. Such experiments have recently been proposed [106].

Kibble's theory contains some features which were criticized [107]. It is now accepted that torsion is an inevitable feature of a gauge theory based upon the Poincaré group. Classical GR must be modified by the introduction of a spin-torsion interaction if it is to be viewed as a gauge theory. The gauge principle alone fails to provide a conceptual framework for GR as a theory of gravity.

In the 1990s, Anthony Lasenby, Chris Doran and Stephen Gull proposed an alternative formulation of General Relativity which is derived from gauge principles alone [108–113]. Their treatment is very different from earlier ones where only infinitesimal translations are considered [91,107]. There are a few other theories similar in their approach to that of Lasenby, Doran and Gull (e.g., [114,115]).

4. Why Consider Alternative Theories?

The motivation for considering alternatives to GR comes mainly from theoretical arguments, like scale invariance of the gravitational theory, additional scalar fields that emerge from string theories, Dark Matter, dark energy or inflation, or additional degrees of freedom that arise in the framework of brane-world theories.

In Table 1, we draw up a list of some of the more well-known alternatives to General Relativity. This list is far from exhaustive, but it serves to highlight the major elements which differentiate these theories. There are several works containing a more detailed listing and discussion of the various alternative theories (e.g., [39,116,117]).

Table 1. A "comparative morphology" of some of the major alternatives to General Relativity, in approximate chronological order. We have only listed the theories of particular historical significance. The current landscape, in which cosmologists seek to explain Dark Matter, dark energy, and inflation, offers far more theories. It is generally easier to incorporate the non-gravitational laws of physics within metric theories, since other theories would result in greater complexity, rendering calculations difficult. The only way in which metric theories significantly differ from each other is in their laws for the generation of the metric. Abbreviations: Tensor (T), V (Vector), S (Scalar), P (Potential), Dy (Dynamic), Einstein Equivalence Principle (EEP), i.e., uniqueness of freefall, Local Lorentz Invariance (LLI), Local Position Invariance (LPI), param (Parameter), ftn (Function).

Theory	Metric	Other Fields	Free Elements	Status
Newton 1687 [20]	Nonmetric	P	None	Nonrelativistic, implicit action at a distance
Poincaré 1890s–1900s [31,118]				Fails; does not mesh with electromagnetism
Nordstrøm 1913 [57]	Minkowski	S	None	Fails to predict observed light detection
General Relativity 1915 [46]	Dy	None	None	Viable
Whitehead 1922 [59]				Violates LLI; contradiction by everyday observation of tides
Cartan 1922–1925 [98]	ST			Still viable; introduces matter spin
Kaluza-Klein 1920s [119,120]	T	S	Extra dimensions	Violates Equivalence Principle
Birkhoff 1943 [61]	T			Fails Newtonian test; demands speed of sound equal to speed of light
Milne 1948 [121]	Machian background			Incomplete; no gravitational redshift prediction; contradicts cosmological observations.
Thiry 1948 [122]	ST			Unlikely; extremely constrained by results on γ^{PPN}
Belifante-Swihart 1957 [123]	Nonmetric	T	K param	Violates EEP; contradicted by Dicke–Braginsky experiments
Brans–Dicke 1961 [63]	Generic S	Dy	S	Viable for $\omega > 500$
Ni 1972 [48]	Minkowski	T, V, S	1 param, 3 ftns	Violates LPI; predicts preferred-frame effects
Will-Nordtvedt 1972 [124]	Dy T	V		Viable but can only be significant at high energy regimes
Barker 1978 [125]	ST			Unlikely; severely constrained.
Rosen 1973 [126,127]	Fixed	T	None	Contradicted by binary pulsar data
Rastall 1976 [128]	Minkowski	S, V	None	Contradicted by gravitational wave data
$f(R)$ models 1970s [129,130]	$n+1$ST	S	Free ftn	Consistent with Solar System tests; viable but severely constrained
MOND 1983 [131–133]	Nonmetric	P	Free ftn	Nonrelativistic theory
DGP 2000 [134]	ST/Quantum			Appears to be contradicted by BAOs, CMB and Supernovae Ia unless DE added
TeVeS 2004 [135]	T,V,S	Dy S	Free ftn	Highly unstable [136]; ruled out by SDSS data [137]

5. From General Relativity to Standard Cosmology

When Einstein published his seminal GR papers it became almost immediately apparent that the theory could be applied to the whole universe, under certain assumptions, to obtain a relativistic cosmological description. If the content of the universe is known, then the energy-momentum tensor can be constructed, and the metric derived using Einstein's equations. Einstein himself was the first to apply GR to cosmology in 1917 [138]. The first expanding-universe solutions to the relativistic field equations, describing a universe with positive, zero and negative curvature, were discovered by Alexander Friedmann [139,140]. This occurred before Edwin Hubble's observations and the empirical confirmation, in 1929, that the redshift of a galaxy is proportional to its distance. Hubble formulated the law which bears his name: $v = H_0 r$, where H_0 is the constant of proportionality [141]. The problem of an expanding universe was independently followed up during the 1930s by Georges Lemaître [142], and by Howard P. Robertson [143–145] and Arthur Geoffrey Walker [146].

These exact solutions define what came to be known as the Friedmann-Lemaître-Robertson-Walker (FLRW) metric, also referred to as the FRW, RW, or FL metric. This metric starts with the assumption of spatial homogeneity and isotropy, allowing for time-dependence of the spatial component of the metric. Indeed, it is the only metric which can exist on homogeneous and isotropic spacetime. The assumption of homogeneity and isotropy, known as the Cosmological Principle, follows from the Copernican Principle, which states that we are not privileged observers in the universe. This is no longer true below a certain observational scale of around 100 Mpc (sometimes called the "End of Greatness"), but it does simplify the description of the distribution of mass in the universe.

The FLRW metric describes a homogeneous, isotropic universe, with matter and energy uniformly distributed as a perfect fluid. Using the definition of the metric in Equation (4), it is written as:

$$-\,\mathrm{d}s^2 = c\,\mathrm{d}\tau^2 - R^2(t)[\,\mathrm{d}r^2 + S_k^0(r)(\,\mathrm{d}\theta^2 + \sin^2\theta\,\mathrm{d}\phi^2)]\,, \tag{25}$$

where r is a time independent comoving distance, θ and ϕ are the transverse polar coordinates, and t is the cosmic or physical time. $R(t)$ is the scale factor of the universe. The function $S_k^0(r)$ is defined as:

$$S_k^0(r) = \begin{cases} \sin(r) & (k = +1) \\ r & (k = 0) \\ \sinh(r) & (k = -1) \end{cases} \tag{26}$$

where k is the geometric curvature of spacetime, the values 0, +1, and −1 indicating flat, positively curved, and negatively curved spacetime, respectively.

Another common form of the metric defines the comoving distance as $S_k^0(r) \to r$, so that

$$-\,\mathrm{d}s^2 = c\,\mathrm{d}t^2 - R^2(t)\left[\frac{\mathrm{d}r^2}{1-kr^2} + r^2(\,\mathrm{d}\theta^2 + \sin^2\theta\,\mathrm{d}\phi^2)\right], \tag{27}$$

where t is again the physical time, and r, θ and ϕ are the spatial comoving coordinates, which label the points of the 3-dimensional constant-time hypersurface.

The dimensionless scale factor $a(t)$ is defined as

$$a(t) \equiv \frac{R(t)}{R_0}\,, \tag{28}$$

where R_0 is the present scale factor (i.e., $a = 1$ at present). The scale factor is therefore a function of time, so it can be abbreviated to a. The metric can then be written in a dimensionless form:

$$-\,\mathrm{d}s^2 = c^2\,\mathrm{d}\tau^2 = c^2\,\mathrm{d}t^2 - a^2\left[\mathrm{d}r^2 + S_k^2(r)(\,\mathrm{d}\theta^2 + \sin^2\theta\,\mathrm{d}\phi^2)\right], \tag{29}$$

where $S_k(r)$ can be redefined as

$$S_k(r) = \begin{cases} R_0 \sin(r/R_0) & (k=+1) \\ r & (k=0) \\ R_0 \sinh(r/R_0) & (k=-1)\,. \end{cases} \tag{30}$$

Equivalently, using the definition in Equation (27),

$$-\,\mathrm{d}s^2 = c^2\,\mathrm{d}t^2 - a^2\left[\frac{\mathrm{d}r^2}{1-k(r/R_0)^2} + r^2(\,\mathrm{d}\theta^2 + \sin^2\theta\,\mathrm{d}\phi^2)\right]. \tag{31}$$

The comoving distance is distance between two points measured along a path defined at the present cosmological time. It means that for objects moving with the Hubble flow, the comoving distance remains constant in time. The proper distance, on the other hand, is dynamic and changes in time. At the current age of the universe, therefore, the proper and comoving distances are numerically equal, but they differ in the past and in the future. The comoving distance from an observer to a distant object such as a galaxy can be computed by the following formula:

$$\chi = \int_{t_e}^{t} c\,\frac{\mathrm{d}t'}{a(t')} \tag{32}$$

where $a(t')$ is the scale factor, t_e is the time of emission of photons from the distant object, and t is the present time.

The comoving distance defines the comoving horizon, or particle horizon. This is the maximum distance from which particles could have travelled to the observer since the beginning of the universe. It represents the boundary between the observable and the unobservable regions of the universe.

If we take the time at the Big Bang as $t = 0$, we can define a quantity called the conformal time η at a time t as:

$$\eta = \int_0^t \frac{\mathrm{d}t'}{a(t')}\,. \tag{33}$$

This is useful, because the particle horizon for photons is then simply the conformal time multiplied by the speed of light c. The conformal time is not the same as the age of the universe. In fact it is much larger. It is rather the amount of time it would take a photon to travel from the furthest observable regions of the universe to us. Because the universe is expanding, the conformal time is continuously increasing.

The concept of particle horizons is important. It defines causal contact. The only objects not in causal contact are those for which there is no event in the history of the universe that could have sent a beam of light to both. This is at the origin of some of the big questions about the universe associated with the Big Bang model, which gave rise to the Inflationary paradigm (see [147]). We shall discuss this later.

5.1. *Cosmological Expansion and Evolution Histories*

The FLRW metric relates the spacetime interval ds to the cosmic time t and the comoving coordinates through the scale factor $R(t)$. The scale factor is the key quantity of any cosmological model, since it describes the evolution of the universe. The notion of distance is fairly straightforward in Euclidean geometry. In General Relativity, however, where we work with generally curved spacetime, the meaning of distance is no longer unique. The separation between events in spacetime depends on the definition of the distance being used.

By combining the GR field equation (Equation (15)) and the definition of the metric (Equation (31)), we obtain two independent Einstein equations, known as the Friedmann equations:

$$\left(\frac{\dot{a}}{a}\right)^2 + \frac{kc^2}{a^2} = \frac{8\pi G}{3}\rho \tag{34}$$

and

$$2\left(\frac{\ddot{a}}{a}\right) + \left(\frac{\dot{a}}{a}\right) + \frac{kc^2}{a^2} = -\frac{8\pi G}{c^2}p\,. \tag{35}$$

The Friedmann equations relate the total density ρ of the universe, including all contributions, to its global geometry. There exists a critical density ρ_c for which $k = 0$. By rearranging the Friedmann equation and using the definition of the Hubble parameter we then obtain

$$\rho_c(t) = \frac{3H^2(t)}{8\pi G}\,. \tag{36}$$

A universe whose density is above this value will have a positive curvature, that is, it will be spatially closed (k = +1); one whose density is less than or equal to this value will be spatially open ($k = 0$ or $k = -1$).

A dimensionless density parameter for any fluid component of the universe (i.e., a component for whose gravitational field is produced entirely by the mass, momentum, and stress density) can be defined by

$$\Omega(t) = \frac{\rho(t)}{\rho_c(t)} = \frac{8\pi G\rho(t)}{3H^2(t)}\,. \tag{37}$$

The current value of the density parameter is denoted Ω_0.

Subtracting Equation (34) from Equation (35) yields the acceleration equation:

$$\frac{\ddot{a}}{a} = -4\pi G\left(\frac{\rho}{3} + \frac{p}{c^2}\right)\,. \tag{38}$$

The geodesic Equation (12) allows us to compute the evolution in time of the energy and momentum of the various components particles which make up the universe. From this evolution, we can construct the fluid equation, or continuity equation, which describes the relation between the density and pressure:

$$\dot{\rho} + 3\frac{\dot{a}}{a}\left(\rho + \frac{p}{c^2}\right) = 0\,. \tag{39}$$

This is valid for any fluid component of the universe, such as baryonic and nonbaryonic matter, or radiation.

The foundations of the Concordance Model of cosmology depend on General Relativity. Any modification to the theory that changes the Einstein equations will have solutions that differ from the Friedmann equations.

The FLRW universe contains different mass-energy components which are assumed to evolve independently. This is physically valid at late cosmological times, when the components are decoupled, so the density evolutions are distinct. In Table 2, we give the equation of state and the evolution of the density and scale factor for different components of the universe. The quantities in this table are explained in detail in the following sections.

Table 2. The evolution of the various cosmological components. The quantities are the equation of state $w \equiv p/\rho c^2$, the density ρ, the pressure p, and the scale factor $a(t)$.

Component	w	$\rho = a^{3(1+w)}$	$a(t) = t^{2/3(1+w)}$
Radiation (photons and relativistic neutrinos)	$1/3$	$\sim a^{-4}$	$\sim t^{1/2}$
Dust (includes CDM, baryons and non-relativistic neutrinos)	0	$\sim a^{-3}$	$\sim t^{2/3}$
Curvature	$-1/3$	$\sim a^{-2} \to a^{-4}$	t
Cosmic strings	$-1/3$	$\sim a^{-2} \to a^{-4}$	t
Domain walls	$-2/3$	a^{-1}	$\sim t^2$
Inflation	$\to -1$	$\frac{1}{2}\dot{\phi}^2 + V(\phi)$	$\sim e^{Ht}$
Vacuum energy	-1	constant	$\sim e^{Ht}$

5.2. Matter (Dust)

Matter which is pressureless is referred to as "dust". This is a useful approximation for cosmological structures which do not interact, such as individual galaxies. Substituting $p_m = 0$ in the equation of state for dust shows that the density of this component scales as:

$$\rho_m(a) = \frac{\rho_{m,0}}{a^3}, \tag{40}$$

where $\rho_{m,0}$ is the current density. Assuming spatial flatness, the time evolution of the scale factor is then

$$a(t) = \left(\frac{t}{t_0}\right)^{2/3}, \tag{41}$$

which gives us

$$H(t) = \frac{2}{3t}. \tag{42}$$

This is known as the Einstein-de Sitter (EdS) solution, and it describes the evolution of H in a constant-curvature homogeneous universe with a pressureless fluid as the only component. It was first described by Einstein and Willem de Sitter in 1932 [148].

5.3. Radiation

In the early universe, the energy content was dominated by photons and relativistic particles (especially neutrinos). The expansion of the universe dilutes the radiation fluid, and the wavelength is increased by the expansion so that the energy decreases. From thermodynamics,

$$E_{\rm rad} = \rho_{\rm rad} c^2 = \alpha T^4, \tag{43}$$

where T is the radiation temperature and α is the Stefan-Boltzmann constant. The equation of state for radiation can then be derived from the fluid Equation (39):

$$\rho_{\rm rad}(a) = \frac{\rho_{\rm rad,0}}{a^4} \quad ; \quad p_{\rm rad} = \frac{\rho_{\rm rad} c^2}{3}. \tag{44}$$

Combining this with the Friedmann equations, and assuming flatness ($k = 0$), we obtain the time dependence of the scale factor and the Hubble parameter:

$$a(t) = \left(\frac{t}{t_0}\right)^{1/2} \quad ; \quad H(t) = \frac{1}{2t}. \tag{45}$$

6. The Components and Geometry of the Universe and Cosmic Expansion

How do we relate the expansion of the universe to its contents? The total density of the universe in terms of its constituent components can be written as the sum of the densities of these components at any given time or scale factor:

$$\rho = \rho_m + \rho_{\text{rad}} + \rho_{\text{DE}} , \tag{46}$$

where the subscript "DE" denotes another component of the universe, called Dark Energy.

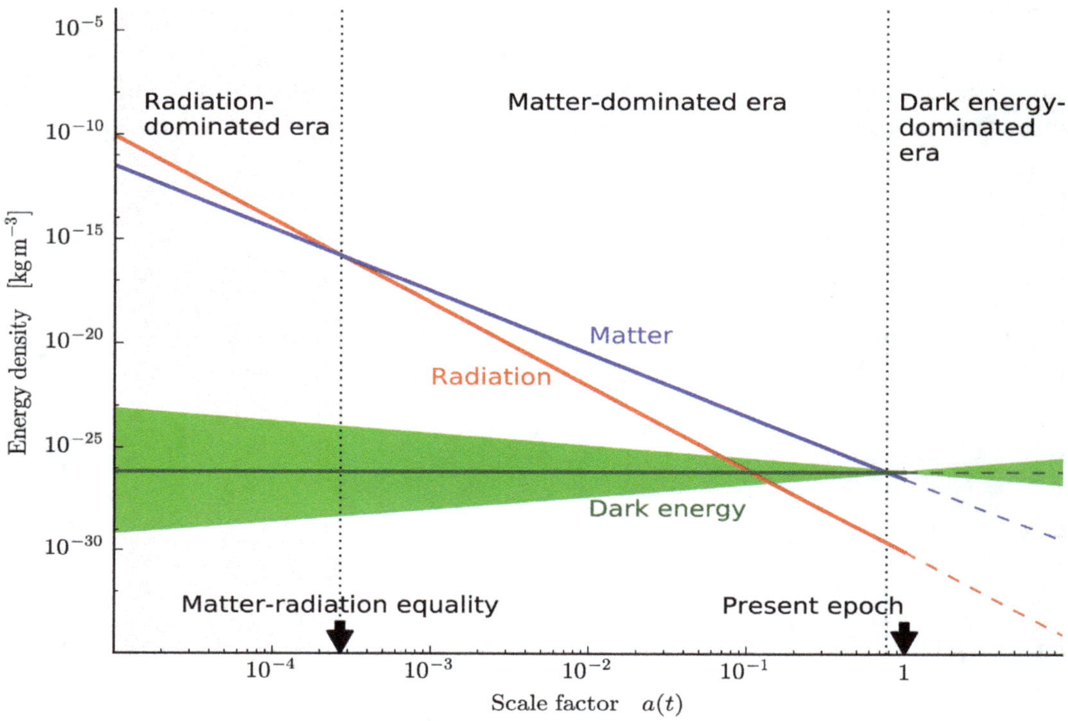

Figure 2. The density evolution of the main components of the universe. The early universe was radiation-dominated, until the temperature dropped enough for matter density to being to dominate. The energy density of dark energy is constant if its equation of state parameter $w = 1$. Because the matter energy density drops as the scale factor increased, dark energy began to dominate in the recent past. At the present time ($a(t) = 1$), we live in a universe dominated by dark energy. For dark energy, the green band represents an equation of state parameter $w = -1 \pm 0.2$, showing how a small change in the value of this parameter can give very different evolution histories for dark energy. If the Concordance Model is correct, the universe will be completely dominated by dark energy in future epochs (shown by the dashed lines). The matter density will keep decreasing as the universe expands. Our Milky Way will merge with the Andromeda Galaxy, and eventually, the entire Local Group will coalesce into one galaxy. The luminosities of galaxies will begin to decrease as the stars run out of fuel and the supply of gas for star formation is exhausted. In the very far future, this galaxy will be in the only one in our Hubble patch, as all the other galaxies will pass behind the cosmological horizon. The night sky, save for the stars in the Local Group, will be very dark indeed. Stellar remnants will either escape galaxies or fall into the central supermassive black hole. Eventually, baryonic matter may disappear altogether as all nucleons including protons decay, or all matter may decay into iron. In either scenario, the universe will end up being dominated by black holes, which will evaporate by Hawking radiation. The end result is a Dark Era with an almost empty universe, and the entire universe in an extremely low energy state, with a possible heat death as entropy production ceases (see, e.g., [149,150]) What happens after that is speculative.

The total dimensionless density can then be written:

$$\Omega = \Omega_{\rm m} + \Omega_{\rm rad} + \Omega_{\rm DE}\,, \tag{47}$$

where we have dropped the subscript for clarity, i.e., $\Omega_{m,0} = \Omega_{\rm m}$, etc. The Friedmann Equation (34) can now be rewritten using the equations of state for the different components:

$$H^2(a) = \frac{8\pi G}{3}\left(\rho_m a^{-3} + \rho_{\rm rad} a^{-4} + \rho_{\rm DE} e^{-3\int_a^1 [1+w(a')]\,{\rm d}\ln a'}\right) - \frac{kc^2}{a^2}. \tag{48}$$

This can be rearranged to give:

$$H^2(a) = H_0^2\left[\Omega_{\rm m} a^{-3} + \Omega_{\rm rad} a^{-4} + \Omega_{\rm DE} e^{-3\int_a^1 [1+w(a')]\,{\rm d}\ln a'} + (1-\Omega)a^{-2}\right], \tag{49}$$

or, in terms of redshift:

$$H^2(z) = H_0^2\left[\Omega_{\rm m}(1+z)^3 + \Omega_{\rm rad}(1+z)a^4 + \Omega_{\rm DE} e^{-3\int_0^z [1+w(z')]/(1+z')\,{\rm d}\ln z'} + (1-\Omega)(1+z)^2\right]. \tag{50}$$

The term $1-\Omega$ is sometimes replaced by Ω_k, the density due to the intrinsic geometry of spacetime. Equation (50) is of central importance since it relates the redshift of an object to the global density components and geometry of the universe.

The density evolution of the various components of the universe is shown in Figure 2.

7. The Hot Big Bang

In the Standard Model, it is generally accepted that the universe arose from an initial singularity, often termed the "Big Bang", which occurred some 13.8 billion years ago (as measured by *Planck* [151]). This is not discussed here, but it should be noted that there are several proposals for the mechanism of this singularity. During this epoch, we are dealing with Planck scale physics, so most of these mechanisms involve quantum gravity. Other proposals (such as some superstring and braneworld theories) do away with the need for an initial singularity altogether.

7.1. The Cosmic Microwave Background

The radiation density $\rho_{\rm rad} \propto a^{-4}$, so the temperature evolution of the universe from an initial T_0 is:

$$T = \frac{T_0}{a}\,. \tag{51}$$

In other words, the universe cools down as it expands. Conversely, this means at early times, when the scale factor was close to zero, the temperature was very high (hence the term "Hot Big Bang"). The radiation left from the early hot universe, cooled by expansion, is known as the Cosmic Microwave Background, or CMB.

The properties of atomic and nuclear processes in an expanding universe provided the first clue for the existence of a hot Big Bang. This was a remarkable achievement of the Big Bang model, because it provided an explanation for the observed abundances of chemical elements in terms of nucleosynthesis. The processes that created nuclei and atoms could only have been possible in an early universe in thermal equilibrium, with black-body spectrum which cooled down as the universe expanded. This allowed Ralph Alpher, Robert Herman, Hans Bethe and George Gamow to predict the existence and temperature of the CMB in 1948 [152–155]. The universe therefore has a thermal as well as an expansion history. Hence the 'Hot Big Bang'.

The first direct evidence for the Hot Big Bang came two decades later, with the observation of the CMB by Arno Penzias and Robert Woodrow Wilson in 1964 [156].

The confirmation of the thermal history of the universe, together with the discovery of charge parity violation in 1964 [157], provided clues about baryogenesis and the observed matter-antimatter imbalance in the universe. This inspired the first proposals for a mechanism for baryogenesis by Andrei Sakharov in1967 [158], followed by electroweak symmetry breaking by Vadim Kuzmin in 1970 [159].

This is a remarkable demonstration of the success of the Concordance Model. The cosmological model fits very well with the predictions of particle physics, which in turn can be tested by cosmological observations. The Concordance Model of the structure and evolution of the universe requires a mechanism for baryogenesis as well as an explanation for Dark Matter and dark energy. The challenge for physical theories beyond (or within) the Standard Model is to explain the preference of matter over antimatter, and to explain the magnitude of this asymmetry. Cosmological observations can be used to address these challenges [160].

The CMB is an extremely isotropic source of microwave radiation, with a spectrum corresponding to a perfect blackbody at a temperature $T_0 = 2.7260 \pm 0.0013\,\mathrm{K}$ [161]. Using the current temperature and $E_{\mathrm{rad}} = \rho_{\mathrm{rad}} c^2 = \alpha T^4$, the radiation density today is given by:

$$\Omega_{\mathrm{rad}} = 2.47 \times 10^{-5} h^{-2}\,. \tag{52}$$

At some time in the early universe, the ambient radiation temperature corresponded to the ionisation potential of hydrogen, which is 13.6 eV. During this epoch, the universe was filled with a sea of highly energetic particles and photons—a hot ionised plasma. The particles were mainly electrons and protons. Other fundamental particles (quarks) existed earlier when the ambient energy corresponded to their rest mass. At some point, as the universe expanded and cooled, the energy of the photons was no longer sufficient to ionise the hydrogen, and within a relatively short time, all of the electrons and protons combined to form neutral hydrogen. The photons were then free to move through the universe. This process is known as decoupling and it occurred at a temperature of $\sim$2500 K, when the universe was approximately 380, 000 years old [162]. It is these decoupled photons which make up the CMB. The surface on the sky from which these photons originate is known as the surface of last scattering.

7.2. Matter-Radiation Equality

At the present epoch, neglecting dark energy, the universe is dominated by matter. This component is characterised by the fact that the matter particles can be treated in a non-relativistic regime, whereas photons and relativistic neutrinos both behave like radiation. The total contribution to the energy density from non-relativistic components (matter) and relativistic components (radiation and relativistic neutrinos) can be written as Ω_{NR} and $\Omega_{\mathrm{R}} = \Omega_{\mathrm{rad}} + \Omega_{\nu}$, respectively. Using the fact that $\Omega_{\mathrm{m}} = \Omega_{\mathrm{m},0} a^{-3}$, the ratio of the contributions of the components is a function of the scale factor a:

$$\frac{\Omega_{\mathrm{R}}}{\Omega_{\mathrm{NR}}} = \frac{\Omega_{\mathrm{rad}} + \Omega_{\nu}}{\Omega_{\mathrm{m}}} = \frac{4.15 \times 10^{-5}}{\Omega_{\mathrm{m},0} h^2 a^{-3}}\,, \tag{53}$$

where we explicitly use the subscript 0 for the present-day values.

Then there must exist a scale factor for which the ratio is unity. This is given by:

$$a_{\mathrm{eq}} = \frac{4.15 \times 10^{-5}}{\Omega_{\mathrm{m},0} h^2}\,, \tag{54}$$

or, in terms of redshift,

$$1 + z_{\mathrm{eq}} = 2.4 \times 10^4 \Omega_{\mathrm{m},0} h^2\,. \tag{55}$$

The epoch at which the matter energy density equals the radiation energy density is called matter-radiation equality, and it has a special role in large-scale structure formation.

7.3. Neutrinos

Neutrinos have particular properties which give rise to a distinct evolution history. They are known to exist from the Standard Model of particle physics, and the Hot Big Bang model predicts the amount of neutrinos in the universe. Neutrinos can be thought of as "dark" matter because of their very small reaction cross-section, which implies negligible self-interaction. However, they are not *cold* Dark Matter. They are simply extremely light particle that can stream out of high-density regions. They therefore cause the suppression of perturbations on scales smaller than the free-streaming scale. Unlike photons and baryons, cosmic neutrinos have not been observed. However, particle physics allows us to chart the history of this particle during nucleosynthesis, and to relate the neutrino temperature to the photon temperature today [163–165].

The scale on which perturbations are damped by neutrinos is determined by the comoving distance that a neutrino can travel in one Hubble time at equality. For a neutrino mass $\sim 1\,\text{eV}$, the average velocity, T_ν / m_ν is of order unity at equality. This leads to a suppression of power on all scales smaller than k_{eq}. Note that this phenomenon depends on the individual neutrino mass, rather than the total neutrino mass. A lighter neutrino can free-stream out of larger scales, so the suppression begins at lower k for the lighter neutrino species. Heavier neutrinos constitute more of the total neutrino density, and so suppress small-scale power more than lighter neutrino species, which means that we need at least two parameters to model massive neutrino phenomenology to sufficient accuracy: the neutrino mass fraction Ω_ν, or some expression of this quantity in terms of the total neutrino mass $\sum m_\nu$, and the number of massive neutrino species N_ν.

Neutrinos introduce a redshift and scale dependence in the transfer function. We know that the perturbation modes of a certain wavelength λ can grow if they are greater than the Jeans wavelength. Above the Jeans scale, perturbations grow at the same rate independently of the scale. For the baryonic and cold Dark Matter components, the time and scale dependence of the power spectrum can therefore be separated at low redshifts. This is not the case with massive neutrinos, which introduce a new length scale given by the size of the comoving Jeans length when the neutrinos become non-relativistic. In terms of the comoving wavenumber k_{nr}, this scale is given by:

$$k_{\text{nr}} = 0.026 \left(\frac{m_\nu}{1\,\text{eV}}\right)^{1/2} \Omega_{\text{m}}^{1/2}\, h\,\text{Mpc}^{-1} \tag{56}$$

for three neutrinos of equal mass, each with mass m_ν. The growth of Fourier modes with $k > k_{\text{nr}}$ is suppressed because of neutrino free-streaming. From the equation above, it is evident that the free-streaming scale varies with the cosmological epoch (since there is a dependence on Ω_{m}), and therefore the scale and time dependence of the power spectrum cannot be separated.

Neutrinos are fermions, with a Fermi-Dirac distribution with assumed zero chemical potential. When they decoupled from the plasma, their distribution remained Fermi-Dirac, with their temperature falling as a^{-1}. This decoupling occurred slightly before the annihilation of electrons and positrons, which occurred when the cosmic temperature was of the order of the electron mass ($T \approx m_e$). Neutrinos decoupled when the cosmic plasma had a temperature of around $1\,\text{MeV}$. The energy associated with this annihilation was therefore not inherited by the neutrinos, and the entropy was completely transferred to the entropy of the photon background. Thus:

$$(S_{\text{e}} + S_\gamma)_{\text{before}} = (S_\gamma)_{\text{after}}\,, \tag{57}$$

where S_{e} and S_γ are respectively the entropy of the electron-positron pairs and the photon background, and 'before' and 'after' refer to the annihilation time.

The entropy per particle species, ignoring constant factors, is $S \propto gT^3$, where g is the statistical weight of the species. For bosons, $g = 1$ and for fermions, $g = 7/8$ per spin state. According to the Standard Model, the neutrino has one spin degree of freedom, each neutrino has an antiparticle, and there are three generations of neutrinos, also called "families" or "species" (μ, τ and electron neutrinos). This means that the degeneracy factor of neutrinos is equal to 6. Before annihilation, the fermions are electrons (2 spin states), positrons (2), neutrinos and antineutrinos (6 spin states). The bosons are photons (2 spin states). We therefore have $g_{\text{before}} = 4(7/8) + 2 = 11/2$, while after annihilation $g = 2$ because only photons remain. Applying entropy conservation and counting relativistic degrees of freedom, the ratio of neutrino and photon temperatures below m_e is therefore:

$$\frac{T_\nu}{T_\gamma} = \left(\frac{4}{11}\right)^{1/3}, \tag{58}$$

so that the present neutrino temperature is

$$T_{\nu,0} = \left(\frac{4}{11}\right)^{1/3} T_{\text{CMB}} = 1.945\,\text{K}. \tag{59}$$

The number density of neutrinos is then

$$n_\nu = \frac{6\zeta(3)}{11\pi^2} T^3_{\text{CMB}}, \tag{60}$$

where $\zeta(3) \approx 1.202$, which gives $n_\nu \approx 112\,\text{cm}^{-3}$ at the present epoch [166]. In the early universe, neutrinos are relativistic and behave like radiation. So they contribute to the total radiation energy density ρ_{rad}, which includes the photon energy density ρ_γ:

$$\rho_{\text{rad}} = \left[1 + \left(\frac{7}{8}\right)\left(\frac{4}{11}\right)^{4/3} N_{\text{eff}}\right] \rho_\gamma, \tag{61}$$

where N_{eff} is the effective number of neutrino species. At late times, when massive neutrinos become non-relativistic, their contribution to the mass density is $m_\nu n_\nu$, giving

$$\Omega_\nu = \frac{\rho_\nu}{\rho_c} \approx \frac{\sum_i^{N_\nu} m_{\nu,i}}{93.14\,\text{eV}\,h^2}, \tag{62}$$

where $m_{\nu,i}$ is the mass of individual neutrino species and N_ν is the number of massive neutrino species. This expression relates the total neutrino mass $\sum m_\nu$ to the neutrino fraction Ω_ν.

It can be seen from the above that this equation can be modified through a change in the effective number of neutrino species by many factors: a non-zero initial chemical potential, or a sizeable neutrino-antineutrino asymmetry, or even a fourth, 'sterile' neutrino [166–168]. The Standard Model predicts a value of $N_{\text{eff}} = 3.046$ for the effective number of neutrino species. This accounts for the three neutrino families together with relativistic degrees of freedom, since neutrinos are not completely decoupled at electron-positron annihilation and are subsequently slightly heated [169]. Any significant deviation from this value could be a signature of hidden physical effects, possibly requiring a modification of General Relativity [170].

Neutrino oscillation experiments do not, at present, determine absolute neutrino mass scales, since they only measure the difference in the squares of the masses between neutrino mass eigenstates. Cosmological observations, on the other hand, can constrain the neutrino mass fraction, and can distinguish between different mass hierarchies [166].

Observations of neutrino flavour oscillations in atmospheric and solar neutrinos, provide evidence of a difference between the masses of the different species or flavours, as well as for a non-zero mass. For three neutrino mass eigenstates m_1, m_2 and m_3, the squared mass differences are [171]:

$$\begin{aligned} |\Delta m^2_{21}| &= m^2_2 - m^2_1| \cong 7.5 \times 10^{-5}\,\mathrm{eV}^2 \\ |\Delta m^2_{31}| &= |m^2_3 - m^2_1| \cong 2.5 \times 10^{-3}\,\mathrm{eV}^2 \\ \frac{|\Delta m^2_{21}|}{|\Delta m^2_{31}|} &\cong 0.03\,. \end{aligned} \tag{63}$$

The ambiguity in the sign of the mass differences Δm allows for two possible mass hierarchies: the normal hierarchy given by the scheme $m_3 \gg m_2 > m_1$, or the inverted hierarchy $m_2 > m_1 \gg m_3$. Given Equation (63), constraining the total neutrino mass to a small enough maximum value could exclude an inverted hierarchy. Conversely, a total neutrino mass $m_\nu \sim 2$ eV is only possible with a degenerate neutrino mass scheme. Hence the interest in finding cosmological neutrino mass bounds.

The fact that cosmological constraints could be stronger than constraints from particle accelerators was noticed quite early (see [172]). The 'closure limit' gives us $m_\nu < 90$ eV. This was first derived in the late 1960s and 1970s [173–176]. Since then, cosmological neutrino bounds have improved significantly, with different methods being used e.g., luminous red galaxies [177], CMB anisotropies [151,178], or weak lensing [179–181].

Joint *Planck* CMB and BAO observations give us $m_\nu < 0.23$ eV, but various data combinations can change this figure, and strong priors on the value of the Hubble constant can provide tighter constraints [151].

8. Inflation: The Second Wave of Alternative Theories

In the late 1970s, General Relativity had been largely accepted by the scientific community. But a series of cosmological considerations led to renewed interest in alternative theories. These were not so much attempt to solve problems in the theory itself, but to find explanations for observations that were not explained by the theory.

General Relativity applied to the universe gave us the Hot Big Bang model: a universe expanding out of an initial highly energetic, dense state. The Hot Big Bang model was successful in explaining many interlinked phenomena which were subsequently confirmed by observation: the Hubble Law and the expansion of the universe, the thermal history of the universe, primordial nucleosynthesis, the existence of the cosmic microwave background, the relation between the temperature and scale factor, and finally the blackbody nature of the CMB. The remarkable fact is that these phenomena occur on extremely different scales, and are observed via different physical processes, and yet they all fit neatly within one model.

However, there are some observations which the Hot Big Bang model fails to explain. These cosmological problems are linked to the primordial universe, the most obvious being the following (for details see [182–186], and references therein):

- The Horizon Problem
- The Flatness Problem
- The Monopole Problem

The horizon problem arises from the structure of spacetime. In the standard cosmological model described by the FLRW equations, different regions of the universe observed today could have not been in causal contact with because of the great distances between them which are greater than the distance that could have been traversed by light since the Big Bang. The transfer of information (i.e., any physical interaction) or energy can occur, at most, at the speed of light, but these regions have the same temperature and other physical characteristics. In particular, we observe causally-disconnected regions of the CMB to be in thermal equilibrium. How could this have happened?

The horizon problem was first identified in the late 1960s. This led to to early attempts to model chaotic solutions to Einstein's field equations near the initial singularity [187,188]. In the late 1970s, Alexei Starobinsky noted that quantum corrections to General Relativity should be important in the very early universe. These corrections would lead to a modification of gravity, which induces an inflationary phase [189]. Starobinsky's was the first model of inflation.

The flatness problem is one the so-called "coincidence problems" of modern cosmology. By the 1960s, observations had determined that the density of matter in the universe is comparable to the critical density necessary for a flat universe. So the contribution of curvature had to be of the same order of magnitude as the contribution of matter throughout the history of the universe. This represents a fine-tuning problem. Observations of the CMB have confirmed that the universe is spatially flat to within a few percent. Why is the global geometry of the universe so flat?

The magnetic monopole problem arises from the Hot Big Bang model. Grand Unified Theories predict the production of a large number of magnetic monopoles [190,191] in the early, extremely hot universe. Why have none ever been observed? If they exist at all, they are much more rare than the Big Bang theory predicts. This was noted by Zel'dovich and others in the late 1970s [192,193].

Hot Big Bang Plus Inflation

This gave rise to the idea of a model in which the early universe undergoes a period of exponential accelerated expansion. This theory, called "inflation", was first formulated by Alan Guth in the 1980s [194] while he was trying to investigate why no magnetic monopoles are observed. It was realised that inflation solves the horizon and flatness problems, as well as explaining the absence of relic monopoles. Better still, it explains the origin of structure in the universe.

In the Standard ΛCDM Model, the initial perturbations from which structure evolved are assumed to have been seeded by the inflationary potential. Reconstructing the primordial power spectrum is no easy task, and poses two main problems. Observationally, we want to extract the amplitude and scale variation from the data. Theoretically, we seek to explain the origin of the perturbations. At present, the leading theoretical paradigm for the primordial fluctuations is inflation, which provides initial conditions for both large-scale structure and the cosmic microwave background radiation. The theory of inflation offers a plethora of models, each of which predicts a certain power spectrum of primordial fluctuations $\mathcal{P}(k)$. Since the inflationary paradigm is linked to the theoretical description of the primordial power spectrum, it is necessary to briefly explain some of the main concepts here (for the full details, see [183,185]).

The precise definition of inflation is any period during which the scale factor of the universe is accelerating, that is, $\ddot{a} > 0$. This expression is equivalent to other definitions of inflation:

$$\frac{\mathrm{d}}{\mathrm{d}t}\frac{H^{-1}}{a} < 0 \implies \epsilon \equiv -\frac{\dot{H}}{H^2} < 1 \iff \frac{\mathrm{d}^2 a}{\mathrm{d}t^2} > 0 \iff \rho + 3p < 0 \,. \tag{64}$$

The first expression above has a remarkable physical interpretation. It means that the observable universe becomes smaller during inflation.

The basic theory of inflation states that from the initial Big Bang singularity to approximately 10^{-37} s, there existed a set of highly energetic scalar fields. By definition, Ω is driven towards 1 during inflation. Inflationary theories assume that gravity is described by GR, which means that the component driving inflation must satisfy $\rho + 3P < 0$. If for example, the universe was dominated during the inflationary phase by a scalar field (or set of fields) ϕ with a self-interaction potential $V(\phi)$. It is the form of this potential which differentiates the various inflationary theories. Most theories assume a "Mexican hat" potential, with a single field, while chaotic inflation assumes a simple power law potential with a slowly varying field [195]. The action for this potential is then [196]

$$S = -\int \mathrm{d}^4x\sqrt{-g}\left[\frac{m_{\mathrm{Pl}}^2 R}{16\pi} - \frac{1}{2}(\nabla\phi)^2 + V(\phi)\right] , \tag{65}$$

where $m_{\rm Pl}$ is the Planck mass. As the universe cooled, the scalar field became trapped in a false vacuum, so its energy density became constant. The potential energy, however, is nonzero, so the pressure is negative. The scale factor during inflation has the de Sitter form:

$$a(t) = e^{(\Lambda_{\rm I}/3)^{1/2} t}, \tag{66}$$

where $\Lambda_{\rm I}$ represents the energy density of the inflationary field (sometimes called the inflaton).

Since the energy density of the inflaton field was very high, the associated magnitude of the negative pressure would have been very large as well. The scale factor is thought to have increased during inflation by $\sim e^{65}$, and any point in the universe which found itself in a false vacuum state would have undergone inflation. The accelerated expansion lasted until the field rolled down to a minimum, when it decays into the familiar particles of the Standard Model, and the universe can then be described by an FLRW model.

The inflationary paradigm provides an explanation for the origin of structure and for the observed geometry of the universe, in addition to solving the aforementioned cosmological problems.

First, inflation solves the flatness problem. Using Equation (66), the evolution of Ω during inflation can be written as:

$$|\Omega(t) - 1| \propto e^{-(4\Lambda_{\rm I}/3)^{1/2} t}, \tag{67}$$

so that $|\Omega - 1|$ is driven very close to 0 as t increases. This explains why the universe is flat. It also means that this value has not deviated significantly from its initial value right after expansion. We can therefore safely assume spatial flatness throughout the history of the universe. Given the observational difficulties, this provides a theoretical motivation for taking the idea of a large $\Omega_{\rm DE}$ seriously.

Second, inflation solves the horizon problem (Figure 3). Regions of the universe which are causally disconnected today evolved out of the same causally-connected region in the early universe. The observed uniformity of the CMB is no longer a problem.

Third, inflation explains why we have never observed magnetic monopoles. Due to rapid expansion of the universe during inflation, they become so rare in any given volume of space that we would be very unlikely to ever encounter one. Nor would they have sufficient density to alter the gravity and thereby the normal expansion of the universe following inflation. The problem of magnetic monopoles motivated the Guth's development of his theory in 1981 [194]. The solution of the monopole problem, and problems related to other relics, was an early success of the inflationary paradigm, and inspired similar theories [197,198].

Fourth, the inflationary scenario provides a natural explanation for the origin of structure, providing a link between quantum mechanics and relativistic cosmological paradigm. This was realised soon after the development of the theory of inflation, and the details were worked out in the early 1980s [199–204].

An initially smooth background needs seed fluctuations around which gravitational collapse can occur. The inflationary scenario attributes their origin to quantum fluctuations in the inflaton field potential, so that the universe is not perfectly symmetric. Different points in the universe inflate from slightly different points on the potential, separated by $\delta\phi$. Inflation for these two points ends at different times, separated by $\delta t = \delta\phi/\dot{\phi}$. This induces a density fluctuation $\delta = H\delta t$ (see [184]). Since all the points undergoing inflation are part of the same potential field, the initial fluctuations are nearly scale invariant. This means that the density amplitude on the horizon scale will also be constant:

$$\delta_{\rm H} = H\delta t = \frac{H^2}{2\pi\dot{\phi}} = \text{constant}. \tag{68}$$

In summary, inflation solves the three cosmological problems listed above:

- The Horizon Problem. Solution: the entire universe evolved out of the same causally-connected region.

- The Flatness Problem. Solution: any initial curvature is diluted by the inflationary epoch and driven to zero.
- The Monopole Problem. Solution: the rapid expansion of the universe drastically reduces the predicted density of magnetic monopoles, if they exist.

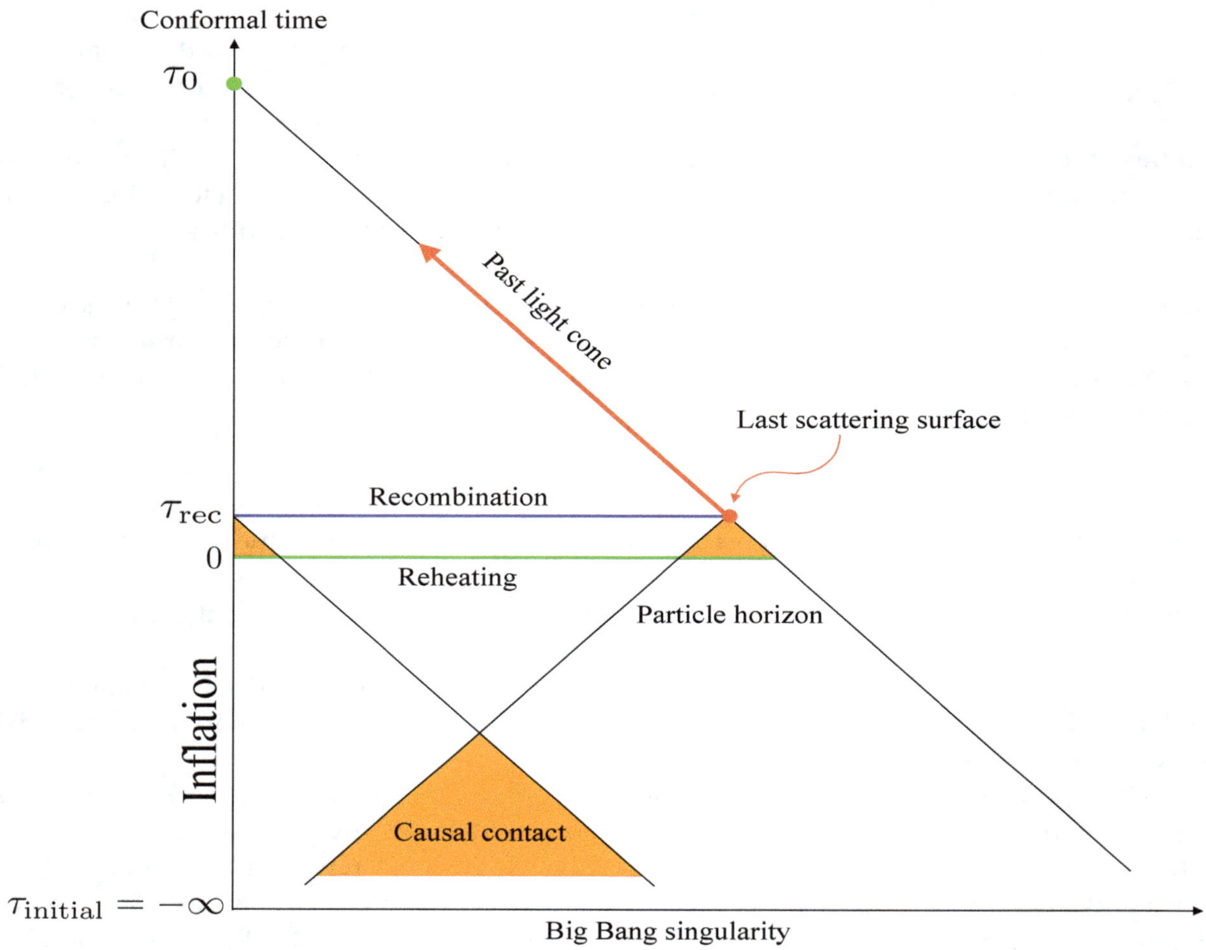

Figure 3. How inflation solves the horizon problem. The light cones on the causal diagram of an inflationary FLRW model are at $\pm 45°$. The worldlines of comoving matter are vertical on this kind of diagram. The particle horizons are horizontal lines. Here we have shown the particle horizon for the CMB. Without inflation, conformal time would only go back to τ_0, and different regions of the CMB which we observe today along our past light cone would never have been in causal contact. Because of inflation, conformal time is extended to the Big Bang singularity, so these regions would have been in causal contact at some point in our past light cone.

How long did inflation last? The answer is given by looking closely at Equation (67) above. A convenient measure of expansion is the so-called e-fold number, defined as:

$$N \equiv \ln\left(\frac{a_f}{a_i}\right) = \int_{t_i}^{t_f} H \, dt \,. \tag{69}$$

Here, a_i and a_f are the values of the scale factor at the beginning and end of inflation, while t_i and t_f are the corresponding proper times. The scale factor a is only physically meaningful up to a normalisation constant, so the e-fold number is defined with respect to some chosen origin. The reason is that in cosmology, what is fixed is not the initial condition, but the current expansion—we cannot measure any H, but we measure H_0 then extrapolate backwards.

We can search for the minimum duration of inflation required to solve the horizon problem. At the very least, we require that the observable universe today fits in the comoving Hubble radius at the beginning of inflation:

$$(a_0 H_0)^{-1} < (a_i H_i)^{-1}. \tag{70}$$

The condition is the same for the horizon and flatness problems.

If we assume that the universe was radiation-dominated since the end of inflation (giving us $H \propto a^{-2}$), and ignore the relatively recent matter- and dark energy-dominated epochs, we obtain

$$\frac{a_\mathrm{f}}{a_\mathrm{i}} = \frac{a_0}{a_\mathrm{f}}. \tag{71}$$

In the general case, the condition becomes:

$$\frac{a_\mathrm{f}}{a_\mathrm{i}} \geq \frac{a_0}{a_\mathrm{f}}, \tag{72}$$

or in terms of the number of *e*-folds,

$$N_\mathrm{f} - N_\mathrm{i} \geq N_0 - N_\mathrm{f}. \tag{73}$$

In other words, there should be as much expansion during inflation as after inflation.

The solution to the horizon problem is the same as the solution to the flatness problem. Taking into account the present energy density of the universe, we need a minimum of about 50 to 60 *e*-folds. This already gives us a useful criterion for realistic inflation models. The most recent *Planck* results show a preference for a higher number of *e*-folds: $78 < N < 157$ [205].

Most models of inflation are slow-roll models, in which the Hubble rate varies slowly [185,206,207]. This model was first developed by Andrei Linde in 1982 [208]. It solved a major problem in Guth's early theory. Instead of tunnelling out of a false vacuum state, inflation occurrs by a scalar field rolling down a potential energy gradient. When the field rolls very slowly compared to the expansion of the universe, inflation occurs. Hence the name "slow-roll inflation". However, when the gradient becomes steeper, inflation ends and reheating can occur. It is beyond the scope of this review to go into the detail of the theory, but it is necessary for us to briefly refer to the link between this theory and the spectral index of primordial fluctuations, which is an important observational parameter in the Concordance Model of cosmology.

To quantify slow roll, cosmologists typically use two parameters ϵ and η which vanish in the limit that ϕ becomes constant. The first parameter is defined as:

$$\epsilon \equiv \frac{\mathrm{d}}{\mathrm{d}t}\left(\frac{1}{H}\right) = \frac{-\dot{H}}{aH^2}, \tag{74}$$

which is always positive, since H is always decreasing. The second complementary variable which defines how slowly the field is rolling is:

$$\eta \equiv \frac{1}{aH\dot{\phi}^{(0)}}\left[3aH\dot{\phi}^{(0)} + a^2 V'\right], \tag{75}$$

where $\phi^{(0)}$ is the zero-order field, and V is the potential.

The scalar spectral index can be defined in terms of some function, usually a polynomial, involving the two slow-roll parameters ϵ and η. As an example we shall give two such parameterisations: $n = 1 - 4\epsilon - 2\eta$ [209] and $n = 1 - 6\epsilon + 2\eta$ [206]. The rate of change of n can also be expressed in terms of inflationary parameters: $\mathrm{d}n/\,\mathrm{d}\ln k = 16\epsilon\eta + 24\epsilon^2 + 2\xi^2$ [210], where

$$\xi^2 \equiv m_{\mathrm{Pl}}^4 \frac{V'(\mathrm{d}^3/\,\mathrm{d}\phi^3)}{V^2}, \tag{76}$$

m_{Pl} being the reduced Planck mass (4.342×10^{-6} g).

Therefore, by extracting the values of ϵ and η from the data, using methods such as weak lensing, we can directly probe the potential of of the inflaton field. Likewise for the tilt or spectral index of the primordial power spectrum. Slow-roll inflation predicts that the spectral index of primordial fluctuations should be slightly less than 1. The reason for this is simple. For inflation to end, the Hubble parameter H has to change in time. This time-dependence changes the conditions at the time when each fluctuation mode exits the Hubble horizon and therefore gets translated into a scale-dependence.

Inflation accounts for the observed spatial flatness of the universe, and the absence of magnetic monopoles. These predictions have been confirmed by various probes, most notably by precision measurements of CMB anisotropies, starting with the Cosmic Background Explorer (COBE) in 1992 [211–213], then with the Wilkinson Microwave Anisotropy Probe (WMAP) which ran for nine years from 2001 to 2010. WMAP data placed tight constraints on the predicted burst of growth in the very early universe, providing compelling evidence that the large-scale fluctuations are slightly more intense than the small-scale ones, which is a subtle prediction of many inflation models [162,214–219]. Significantly, WMAP found evidence that the scalar spectral index is less than 1 (a 2σ deviation), implying a deviation from scale invariance for the primordial power spectrum. As explained above, this is a major prediction of inflation, and this observation reinforced the evidence in favour of the theory. Conclusive proof of a scale-dependent primordial power spectrum (a 5σ deviation from $n_s = 1$) was provided by the *Planck* CMB anisotropy probe in 2013 [220,221] and confirmed in 2015 [151,205].

One current experimental challenge is to observe the B-modes of polarisation of the CMB caused by primordial gravitational waves produced by inflation. Their detection by the BICEP2 experiment was announced in early 2014. However, more accurate modelling of the signal over the next few months, which allowed the observation to be explained by polarised dust emission in our Galaxy, decreased the statistical confidence of the initial result [222]. This was confirmed by *Planck* data in 2016 [223]. Upcoming large-scale structure surveys, such as the *Euclid* satellite mission, or 21-cm radiation surveys such as the Square Kilometre Array, may measure the power spectrum with greater precision than current CMB probes, and could provide further evidence in favour of the inflationary paradigm [224–226].

9. The First Unknown Component: Dark Matter

The first evidence for Dark Matter came from astronomy rather than cosmology. Newtonian physics and General Relativity both provide very precise rules for the dynamics of galaxies: the mass determines the rotation velocity. Starting in the 1920s, stronomers noticed that amount of visible matter in galaxies did not match the observed rotation curves. These curves relate the tangential velocity of the constituent stars (or gas) about the centre of the galaxy to their radial distance. Observations of the velocities of globular clusters about galaxies showed that at large radii the velocities are approximately constant, implying that the amount of mass in the galaxies is much higher than the visible mass.

The first suggestion of the existence of hidden matter, motivated by stellar velocities, was made by Jacobus Kapteyn in 1922 [227]. Radio astronomy pioneer Jan Oort also hypothesized the existence of Dark Matter in 1932 [228]. Oort was studying stellar motions in the local galactic neighbourhood and found that the mass in the galactic plane must be greater than what was observed. This measurement was later determined to be erroneous.

In 1933, Fritz Zwicky, who studied galactic clusters while working at the California Institute of Technology, made a similar inference [229]. Zwicky applied the virial theorem to the Coma galaxy cluster and obtained evidence of unseen mass that he called *dunkle Materie* in German, or "Dark Matter". Zwicky estimated its mass based on the motions of galaxies near its edge and compared that to an estimate based on its brightness and number of galaxies. He estimated that the cluster had about 400 times more mass than was visually observable. The gravity effect of the visible galaxies was far too small for such fast orbits, thus mass must be hidden from view. Based on these conclusions, Zwicky inferred that some unseen matter provided the mass and associated gravitation attraction to hold the cluster together.

In 1937, Zwicky made the bold assertion that galaxies would be unbound without some form of invisible matter [230]. Zwicky's estimates were off by more than an order of magnitude, mainly due to an obsolete value of the Hubble constant. The same calculation today shows a smaller fraction, using greater values for luminous mass. However, Zwicky did correctly infer that the bulk of the matter was dark.

More evidence started to accumulate for the existence of some non-emitting component which was now being called Dark Matter [231]. In 1959, Kahn and Woltjer [232] pointed out that the motion of Andromeda towards us implied that there must be Dark Matter in our Local Group of galaxies. Dynamical evidence for massive Dark Matter halos around individual galaxies came later, starting in the 1970s, when rotation curve data from multiple galaxies confirmed the Dark Matter halo hypothesis [233–236]. Like baryonic matter, Dark Matter is a fluid with vanishingly small pressure. Unlike baryonic matter, it has no interaction with photons, making it both dark and transparent. It also has a vanishingly small self-interaction beside gravity. One result of this is that the Dark Matter halos surrounding galaxies are rounder than the galaxies themselves [237] .

In the last few decades, cosmology has contributed one important piece of information: the amount of Dark Matter. The observed value of the matter density in the universe is $\Omega_{\mathrm{m}} = 0.3089 \pm 0.0062$. However, the density of baryonic matter is $\Omega_b = 0.0486 \pm 0.0010$ [151]. The missing mass is made up of Dark Matter.

The name "Dark Matter" is an indication of its nonbaryonic nature: it cannot be observed by emission of photons, so observers need to find a way around this problem. Current evidence for the existence of Dark Matter comes from a variety of sources besides galactic dynamics [231]. The two most important ones are CMB anisotropies and gravitational lensing. In addition, Big Bang nucleosynthesis provides evidence that some of the Dark Matter may be baryonic. The inventory of observed baryons in the local universe falls short of the total anticipated abundance from Big Bang nucleosynthesis, implying that most of the baryons in the universe are unseen [238].

Anisotropies in the CMB are related to anisotropies in the baryonic density field by the Sachs-Wolfe effect [239]. This means that the baryon density field variation at the time of decoupling can be linked to CMB anisotropies. If all matter were made of baryons, the amplitude of the density fluctuations should have reached $\delta \sim 10^{-2}$ at the present epoch. However, we observe structures with $\delta \gg 1$ at the present epoch (e.g., galaxies and galaxy clusters). The discrepancy can only be explained by the presence of additional matter, which created potential wells for the baryons to fall into after decoupling. These potential wells would have had to be formed by a weakly interacting fluid that decoupled well before baryons and began to cluster much earlier. Such a fluid would only interact via the gravitational and possibly the weak nuclear force. As the baryons accumulated in the potential wells, their pressure would have built up, leading to oscillations in the baryon fluid, termed "baryon acoustic oscillations" (BAO) [240,241]. These oscillations leave an imprint on the CMB power spectrum, which has been confirmed observationally, and which constrains the mass density, leading to a further confirmation of the existence of this missing mass.

The phenomenon of gravitational lensing includes cosmic shear, weak lensing, cosmic magnification. Although the theory of cosmic shear had been worked out from the 1960s to the early 1990s [242], the first detection had to await the development of instruments sensitive enough to make the required observations, and image analysis software to accurately correct for unwanted effects when measuring the shapes of galaxies. In March 2000, four groups independently announced the first discovery of cosmic shear [243–246]. Since then, cosmic shear has established itself as an important technique in observational cosmology.

Gravitational lensing shows that the amount of lensing of galaxies around galaxy clusters is too high to be caused by the visible matter. Apart from the stars themselves, a galaxy cluster also has a gas component, but X-ray observations show that this is still not enough to account for the extra mass. The cluster must therefore have a non-emitting halo of Dark Matter around it.

Various Dark Matter candidates have been proposed [247–251]. However, all these candidates have one common characteristic: a very small reaction cross-section, making them extremely difficult to detect directly [252–255]. Experiments have, however, placed limits on the mass of Weakly Interacting Massive Particles (WIMPs), which are the current best candidate for Dark Matter (together with axions). WIMPs are an entire new class of fundamental particle outside of the Standard Model that result from supersymmetry [256,257]. These results show that even the lightest Dark Matter particle should have a mass which is not below $\sim$10 MeV. We also know that $\Omega_{\rm CDM} = 0.2589 \pm 0.0057$ [151]. The conclusion is that $\Omega_\nu \ll \Omega_{\rm m}$, implying that hot Dark Matter (i.e., neutrinos) cannot account for the Dark Matter density $\Omega_{\rm CDM}$.

An alternative to Dark Matter is to explain the missing mass by means of a modification of gravity at large distances or more specifically at small accelerations. In 1983, Morderhai Milgrom proposed a phenomenological modification of Newton's law which fits galaxy rotation curves [131–133]. The theory, known as Modified Newtonian Dynamics (MOND) automatically recovers the Tully-Fischer law. The theory modifies the acceleration of a particle below a small acceleration $a_0 \sim 10^{-10}\ {\rm ms}^{-2}$, which therefore enters the theory as a universal constant. The gravitational acceleration at large distances then reads $a = \sqrt{GMa_0}/r$ at large distances, instead of the Newtonian law $a = \sqrt{GM/r^2}$.

There are two main difficulties with MOND. First, it does not explain how galaxy clusters can be bound without the presence of some hidden mass [258,259]. Second, attempts to derive MOND from a consistent relativistic field theory have failed. One such attempt is the Tensor-Vector-Scalar Theory (TeVeS) [135] is more successful and actually relativistic but not apparently necessary since it still requires dark energy and Dark Matter. Many models are unstable [260], or require actions which depend on the mass M of the galaxy, thereby giving a different theory for each galaxy. Moreover, modified gravity theories have serious difficulties reproducing the CMB power spectrum and the evolution of large-scale structure [261–263].

The greatest challenge to modified gravity theories, and also the clearest direct evidence of Dark Matter, comes from observations of a pair of colliding galaxy clusters known as the Bullet Cluster [264] in which the stars and Dark Matter separate from the substantial mass of ionised gas. The Dark Matter follows the less substantial stars and not the more massive gas.

The modifications of gravity proposed as alternative to the Dark Matter paradigm illustrate the need for tests of GR at large distances and low accelerations. They also illustrate the problems faced by models which favour goodness of fit over parsimony. Modified gravity theories can give an excellent phenomenological fit through an adjustment of the values of the extra parameters, but there is no universal principle to determine these values. This requirement for simplicity and predictivity is met by General Relativity.

10. The Second Unknown: Dark Energy and the New Wave Alternative Theories

The current motivation for alternative theories seems to be the search for an explanation of the observed accelerating expansion of the universe. Let us consider the justification for the dark energy paradigm within the inflationary ΛCDM model, and the process which led to its acceptance by the scientific community (see [265]).

Round about the time that GR was developed, the universe was thought to be static. There was no compelling reason to think otherwise. Einstein realised that his equations implied a non-static universe, so in 1917 he revised his field equations of GR to read [138]:

$$G_{\mu\nu} - \Lambda g_{\mu\nu} = G_{\mu\nu} - 8\pi G\rho_\Lambda g_{\mu\nu} = 8\pi G T_{\mu\nu} \tag{77}$$

where $\rho_\Lambda = \Lambda/8\pi G$ is proportional to the cosmological constant Λ.

It can be seen from this equation that Einstein did not consider the cosmological constant to be part of the stress-energy term. One could, of course, put $\rho_\Lambda g_{\mu\nu}$ on the right-hand side of the equation and count it as part of the source term of the stress-energy tensor and simply consider ρ_Λ to be the vacuum energy. This is not just a semantic distinction. When ρ_Λ takes part in the dynamics of the universe, then the field equation is properly written with ρ_Λ, or its generalisation, as part of the stress-energy tensor:

$$G_{\mu\nu} = 8\pi G(T_{\mu\nu} + \rho_\Lambda g_{\mu\nu})\,. \tag{78}$$

The equation describing gravity is then unchanged from its original form—there is no new physical theory. Instead, there is a new component in the content of the universe.

This component must satisfy Special Relativity (that is, an observer can choose coordinates so that the metric tensor has Minkowskian form). An observer moving in spacetime in such a way that the universe is observed to be homogeneous and isotropic would measure the stress-energy tensor to be

$$T_{\mu\nu} = \begin{pmatrix} \rho & 0 & 0 & 0 \\ 0 & p & 0 & 0 \\ 0 & 0 & p & 0 \\ 0 & 0 & 0 & p \end{pmatrix}. \tag{79}$$

This means that the new component in the stress-energy tensor looks like an ideal fluid with negative pressure:

$$p_\Lambda = -\rho_\Lambda. \tag{80}$$

In modern concordance cosmology, this component is usually termed "dark energy". If the equation of state parameter of dark energy is constant, i.e., $w(z) = -1$, then its energy density will be constant regardless of the expansion of the universe.

Einstein inserted the cosmological constant because he felt that the non-static universe predicted by the formalism of GR was incorrect, given the data available in 1917 [138]. At the time, observations of the universe were limited primarily to stars in our own galaxy, with observed low velocities, so there was solid observational evidence justifying the assumption that the universe was static. Einstein's goal was to obtain a universe that satisfied Mach's principle of the relativity of inertia. However, observational evidence started to accumulate for another paradigm. In 1917, Vesto Slipher [266] published his measurements of the spectra of spiral nebulae, which showed that most were shifted towards the red. The breakthrough came when the linear redshift-distance relation was formulated by Hubble [141], who showed that the universe was expanding. Einstein then dropped his support for the cosmological constant.

In the FLRW cosmological model, the expansion history of the universe is determined by the mass density of the different components, whose sum is normalised to unity:

$$\Omega_{m,0} + \Omega_{\text{rad},0} + \Omega_{X,0} + \Omega_{k,0} = 1\,, \tag{81}$$

where the 0 subscript indicates the present epoch. We use the term Ω_X to show that this equation does not assume anything about the nature of the additional energy component (dark energy). In fact we could have used Ω_Λ or Ω_{DE} in the current Concordance Model.

Big bang nucleosynthesis and observations of large scale structure provide a good determination of the mass content of the universe, allowing Ω_{m} and Ω_{rad} to be fixed. However, observations in the 1980s and 1990s started to show inconsistencies with the cosmological model at the time, which was a matter-dominated, expanding universe with a present-epoch Hubble constant of $H_0 \simeq 0.7\,\text{kms}^{-1}$ Mpc and $\Omega_\Lambda = 0$ [267,268]. This was the so-called "age problem", where the predicted age of the universe seemed to be younger than the age of the oldest stars. Angular-diameter distances to the last scattering surface at $z = 1100$ measured from the CMB are in fact 1.7 times smaller than those predicted by an isotropic and homogeneous universe containing only pressureless matter (see [269]). Since the inflationary scenario, which by then was well established, predicts a flat $\Omega_{\text{total}} = 1$ universe, there was a problem with the cosmological model.

It was realised that one of the three assumptions of the cosmological model had to be wrong. Either the universe contains exotic matter with a negative pressure, or standard General Relativity is wrong, or the universe is not homogeneous and isotropic. The solution could also lie in some combination of the three. Most of the research since the late 1990s has followed the first approach, and the term "Concordance Model" refers to an FLRW universe, following General Relativistic cosmology, containing dark energy.

Within the FLRW framework, two main proposals were put forward: one was ΛCDM, in which there is a contribution to the energy density from a term similar to the cosmological constant (or the cosmological constant itself), and the other was $\nu +$ CDM, where the missing mass came from massive neutrinos ($m_\nu \simeq 7$eV) (e.g., [270,271]).

The first strong evidence of dark energy came in 1998 and 1999, when observations of the luminosities of type Ia supernovae indicated that the expansion of the universe is accelerating [272,273]. Concurrently, other observations constrained the neutrino mass to $m_\nu \ll 7$eV, thus discounting the $\nu +$ CDM model and confirming ΛCDM as the Concordance Model (e.g., [265,274]). It is not clear when the term "dark energy" was first used, but it seems to have been around 1998. The term is analogous to "Dark Matter", which had been in use for some time [275].

Since then, numerous observations have confirmed cosmic acceleration, including supernovae, the cosmic microwave background, large-scale structure and baryon acoustic oscillations (see, e.g., [276–278]). The values of the present epoch matter and radiation components are well established:

$$\Omega_{m,0} \equiv \frac{8\pi G \rho_{m,0}}{3H_0^2} \sim 0.3, \quad \Omega_{\mathrm{rad},0} \equiv \frac{8\pi G \rho_{\mathrm{rad},0}}{3H_0^2} \sim 1 \times 10^{-4}, \tag{82}$$

where H_0 is the present value of the Hubble parameter $H(a = 1)$.

The data also indicate that the universe is currently nearly spatially flat:

$$|\Omega_K| \ll 1. \tag{83}$$

This is normally taken to imply that the spatial curvature $K = 0$, since

$$\Omega_{k,0} = 0 \equiv \frac{-K}{a_0^2 H_0^2} \sim 0. \tag{84}$$

Thus it also justifies the inflationary paradigm. However, inflation only tells us that $\Omega_K \to 0$, so that the curvature may have had a nonzero value in the past. In the present universe, however, the distinction is negligible. In any case, Equation (81) implies that there has to be a nonzero Λ (a constant term added to the Einstein equation) such that

$$\Omega_{\Lambda,0} \equiv \frac{\Lambda}{3H_0^2} \sim 0.7. \tag{85}$$

Inserting these values into the Friedmann equation leads to the dramatic conclusion that the expansion of the universe is accelerating:

$$\ddot{a}_0 = H_0^2 \left(\Omega_\Lambda - \frac{1}{2}\Omega_\mathrm{m} - \Omega_\mathrm{rad} \right) > 0, \tag{86}$$

where a_0 is the present value of the scale factor $a(t)$.

Note that this conclusion only holds if the universe is homogeneous and isotropic (i.e., a Friedmann-Lemaître model). In such a universe, the distance to a given redshift z and the time elapsed since that redshift are tightly related via the only free function, $a(t)$. If the universe is isotropic around us, but not homogeneous, that is, a non-Copernican Tolman-Bondi-Lemaître model [279], then this relation would be lost and present data might not imply acceleration. A Copernican model where this relation again breaks down is the inhomogeneous universe, where the acceleration can be produced via nonlinear averaging—the backreaction of inhomogeneities.

Dark energy is a fluid component whose equation of state is:

$$p_{\mathrm{DE}} = wc^2 \rho_{\mathrm{DE}} \,. \tag{87}$$

This is the equation of state in its most general form, since w can be any function of redshift, scale factor or cosmic time, with the constraint that $w \leq 0$ (i.e., the fluid has a negative pressure). Assuming that $w = w(a)$, we have the following density-scale relation:

$$\rho_{\mathrm{DE}}(a) = \rho_{\mathrm{DE},0} \mathrm{e}^{-3\int_a^1 [1+w(a')]\,\mathrm{d}(\ln a')} \,. \tag{88}$$

It can be seen that in the special case of a constant $w = -1$, the fluid equation implies that the density is constant.

If dark energy is a cosmological constant, it still leaves the question of its physical nature. Its observed value of $\Lambda \approx 3 \times 10^{-122} c^3/\hbar G$ is so small that it is hard to interpret as the vacuum energy.

One possibility is that the final value of the cosmological constant is zero, and that cosmic acceleration is due to the potential energy of a scalar field, with some sort of mechanism to dynamically relax it to a small value. This notion leads to models of dark energy which invoke a slowly-rolling cosmological scalar field to source accelerated expansion, smilar to cosmological inflation.

How can a scalar field drive cosmic acceleration? The action of a scalar field minimally coupled to Einstein gravity is

$$S = \int \mathrm{d}^4x \sqrt{-g} \left(\frac{m_{\mathrm{Pl}}^2 R}{2} - \frac{1}{2}(\partial\phi)^2 - V(\phi) \right) \,. \tag{89}$$

The stress-energy tensor for the scalar field is given by

$$T^{\phi}_{\mu\nu} = \partial_\mu \phi \partial_\nu \phi - g_{\mu\nu} \left(\frac{1}{2}(\partial\phi)^2 + V(\phi) \right) \,. \tag{90}$$

For a homogeneous field such that $\phi = \phi(t)$, a cosmological scalar acts like a perfect fluid with equation of state $w = P/\rho$ given by

$$w_\phi = \frac{\frac{1}{2}\dot{\phi}^2 - V(\phi)}{\frac{1}{2}\dot{\phi}^2 + V(\phi)} \,. \tag{91}$$

The observed expansion leads us to a value of $w_\phi \simeq -1$, which requires a very slowly-rolling field: $\dot{\phi}^2 \ll V(\phi)$. There has also been a lot of interest in constructing quintessence models which can produce an equation of state of the "phantom" type ($w_\phi < -1$).

The equation of state of dark energy that has the potential to distinguish between dark energy candidates. The most important distinction that can be made between different dark energy models is whether the energy density of this component is constant, filling space homogeneously, or whether it is some form of quintessence field whose energy density can vary in time and space. There is a multitude of alternative models, such as $f(R)$ [280–282], or Chameleon Models [283,284]. It is therefore useful to consider the redshift evolution of w, so that w is an arbitrary function of redshift z. There are a number of different parameterisations of $w(z)$ ([285], and references therein).

The most common parameterisation is sometimes termed the Chevallier-Linder-Polarski (or CPL) parameterisation [285,286]:

$$w(z) = w_0 - \frac{\mathrm{d}w}{\mathrm{d}a}(1-a) \,, \tag{92}$$

where the scale factor $a = 1/(1+z)$. If we define

$$w_a = -\frac{\mathrm{d}w}{a\,\mathrm{d}\ln a} \,, \tag{93}$$

then the equation becomes

$$w(a) = w_0 + w_a(1 - a)\,, \tag{94}$$

which is the most commonly used form.

It has been shown that this parameterisation is stable and robust over large redshift ranges [287]. A wide range of functional forms of $w(a)$ can be parameterised by the $w_0 - w_a$ combination. However, there are some dark energy models which it cannot reproduce (see [288–291]).

The problem with the dark energy paradigm, stated simply, is that the parameters are not constrained well enough to rule out certain models. We have fairly good bounds on the dark energy density, but the dark energy equation of state is still poorly constrained. Even for a constant w model, corresponding to ΛCDM, the bounds are such that a time-varying $w(a)$ could mimic a constant w, thereby disguising underlying physics (see [292–294]).

11. The Evolution of Large-Scale Structure

After the epoch of matter-radiation equality, and before the onset of dark energy domination, the mass-energy content of the universe became dominated by matter. From an initially smooth background (as evidenced by CMB observations), structures have evolved to a scale of more than 100 Mpc, with the term "large scale structure" being used to refer to objects modelled on this scale. At this scale, the mass-energy inhomogeneities can be modelled as perturbations on a homogeneous and isotropic unperturbed background spacetime. Below this scale we observe galaxy clusters, individual galaxies, and stars. The model of structure formation must be accurate enough to provide an good description of the universe on a wide range of scales.

The standard model for the formation of structure assumes that at some early time there existed small fluctuations, which grew by gravitational instability. The origins of these fluctuations are unclear, but they are thought to arise from quantum fluctuations of the primordial universe, uncorrelated and with Gaussian amplitudes, which were then amplified during a later inflationary phase [185]. The assumption that the amplitudes of the relative density contrasts is much smaller than unity means that we can think of the primordial fluctuations as small perturbations on a homogeneous and isotropic background density. This ensures that we can describe them using linear theory.

Heuristically, the mechanism of structure formation can be understood in terms of gravitational self-collapse. Matter collapses gravitationally around initial mass overdensities. This increases the relative density of that region, causing further collapse of more matter, and amplifying the effect. The linear theory of structure formation needs to be relativistic, because the perturbations on any length scale are comparable or larger than the horizon size at sufficiently early times. The horizon size is defined as the distance ct which light can travel in time t since the Big Bang. Dissipative effects and pressure also affect structure formation, as explained below (for details of the theory, see [295,296]).

The relative density is the density ρ at a particular point in space $\mathbf{x}$ relative to the mean $\overline{\rho}$ at some time parameterised by the scale factor a, and can be expressed as a dimensionless density contrast:

$$\delta(\mathrm{x}, a) = \frac{\rho(\mathrm{x}, a) - \overline{\rho}(a)}{\overline{\rho}(a)}\,. \tag{95}$$

This quantity can be understood as the dimensionless density perturbation of some background matter distribution.

There are two types of density perturbations that can occur within a matter-radiation fluid. If the fluid could be compressed adiabatically in space, the perturbations have a constant matter-to-radiation ratio everywhere. Since the energy density of radiation is proportional to T^4, and the number density is proportional to T^3, the energy densities of radiation and matter are related by:

$$\delta_{\mathrm{rad}} = \frac{4}{3}\delta_{\mathrm{m}}\,. \tag{96}$$

Isocurvature perturbations occur when the entropy density is perturbed, but not the energy density. Since the total energy density remains constant, there is no change in the spatial curvature and

$$\rho_{\rm rad}\delta_{\rm rad} = \rho_m \delta_{\rm m} \,. \tag{97}$$

Perturbations can occur at different scales, or 'modes'. The latter term is used when the amount of perturbation on a particular scale is expressed using Fourier analysis. The Fourier transform pair of $\delta(\mathbf{x})$ is:

$$\begin{aligned} \hat{\delta}(\mathbf{k}) &= \int \mathrm{d}^3 x \delta(\mathbf{x}) \mathrm{e}^{i\mathbf{k}.\mathbf{x}} \quad ; \\ \delta(\mathbf{x}) &= \int \frac{\mathrm{d}^3 k}{(2\pi)^3} \hat{\delta}(\mathbf{k}) \mathrm{e}^{-i\mathbf{k}.\mathbf{x}}, \end{aligned} \tag{98}$$

with each mode assumed to evolve independently. In the Einstein-de Sitter regime, linear adiabatic perturbations scale with time as follows:

$$\delta \propto \begin{cases} a(t)^2 & \text{(radiation domination)} \\ a(t) & \text{(matter domination)} \end{cases} \tag{99}$$

while isocurvature perturbations are initially constant and then decline:

$$\delta \propto \begin{cases} \text{constant} & \text{(radiation domination)} \\ a(t)^{-1} & \text{(matter domination)} \,. \end{cases} \tag{100}$$

In both cases, the overall shape of the spectrum of the perturbations over all modes is preserved, while the amplitude changes with time. The evolution described above is affected on small scales by a number of processes.

11.1. Evolution on Small Scales

During the radiation-dominated epoch the growth of certain modes is suppressed. This behaviour can be modelled in terms of the horizon scale $\lambda_H(a)$, which is the distance ct that light could have travelled since the initial singularity (a comoving horizon size). A mode k is said to enter the horizon when $\lambda = \lambda_H(a)$, where $\lambda = (2\pi)/k$. If $\lambda < \lambda_H(a_{\rm eq})$ then a mode enters the horizon during the radiation-dominated epoch. The time scale for collapse of matter during this epoch is larger than the typical expansion time scale ($t \sim 1/H(a)$) due to the relatively rapid expansion $\rho_{\rm rad} \propto a^{-4}$. The growth of these modes is therefore suppressed. After the epoch of matter-radiation equality ($a = a_{\rm eq}$), these perturbations can then start to collapse gravitationally. We can define the suppression factor for a particular mode as the factor by which the amplitude is reduced had it not entered the horizon:

$$f_{\rm sup} = \left(\frac{a_{\rm enter}}{a_{\rm eq}}\right)^2 = \left(\frac{k_0}{k}\right)^2 \tag{101}$$

where the mode evolves as $\propto a^2$ until it enters the horizon at $a_{\rm enter}$ and is suppressed until $a_{\rm eq}$, when its evolution resumes as $\propto a$. The second equality in the above equation comes from applying an Einstein-de Sitter approximation where $k_0 = 1/\lambda_H(a_{\rm eq})$ (see [297]).

Pressure opposes gravitational collapse for modes with a wavelength less than the Jeans length [298], sometimes called the free-streaming scale, defined as

$$\lambda_J = c_s \sqrt{\frac{\pi}{G\rho}} \,. \tag{102}$$

During the radiation-dominated epoch, the sound speed $c_s = c/\sqrt{3}$ and the Jeans length is always close to the horizon size. The Jeans length then reaches a maximum at $a = a_{\rm eq}$ and then begins to decrease as the sound speed declines. This means that on scales larger than the comoving horizon size, perturbations are only affected by gravity, and the spectrum starts to turn over at this point (where the effects of pressure begin to dominate). The comoving horizon size at $z_{\rm eq}$ is given by:

$$R_0 r_H(z_{\rm eq}) \approx \frac{16.0}{\Omega_{\rm m} h^2}{\rm Mpc}\,. \tag{103}$$

Another important scale occurs where photon diffusion erases perturbations in the matter-radiation fluid. This process is termed Silk damping [299]. The scale at which it occurs is characterised by the distance travelled by the photon in a random walk by the time of last scattering:

$$\lambda_S \approx 16.3(1+z)^{-5/4}(\Omega_b^2 \Omega_{\rm m} h^6)^{-1/4}{\rm Gpc}\,. \tag{104}$$

All of the effects mentioned above are particularly important where the behaviour of massive neutrinos is concerned. Heuristically we can understand the complexity of their behaviour by considering them as a component whose equation of state changes as the universe evolves. From a component which behaves like photons (since the particles have a very small mass and relativistic speeds), massive neutrinos lose energy and start behaving like baryonic matter [300,301].

11.2. Growth oF Perturbations in the Presence of Dark Energy

All of the above effects were described in an Einstein-de Sitter universe. In a universe with a smooth non-clustering dark energy component below the horizon scale, the matter perturbation fields evolves according to:

$$\begin{aligned} \ddot{\delta} + 2H\dot{\delta} - (3/2)H^2\Omega_{\rm m}\delta &= 0 \\ \delta'' + (2-q)a^{-1}\delta' - (3/2)\Omega_{\rm m}a^{-2}\delta &= 0, \end{aligned} \tag{105}$$

where a dot denotes a time derivative and a dash denotes a derivative with respect to a. The term q is the deceleration parameter. This can be interpreted in the following way: the perturbations grow according to a source term which involves the amount of matter ($\Omega_{\rm m}$) but the growth is suppressed by the friction term due to the expansion of the universe. The latter is also known as the Hubble drag.

If we define the growth as the ratio of the amplitude of a perturbation at a time a to some initial amplitude, i.e.,

$$D(a) = \frac{\delta(a)}{\delta(a_{\rm initial})}, \tag{106}$$

the equation becomes, for a general dark energy scenario where $w = w(a)$ (see [302])

$$D'' + \frac{3}{2}\left(1 - \frac{w(a)}{1+X(a)}\right)\frac{D'(a)}{a} - \frac{3}{2}\left(\frac{X(a)}{1+X(a)}\right)\frac{D}{a^2} = 0\,, \tag{107}$$

where

$$X(a) = \frac{\Omega_{\rm m}}{\Omega_{\rm DE}} {\rm e}^{-3\int_a^1 {\rm d}\ln a' w(a')} \tag{108}$$

is the ratio of the matter density to the dark energy density. For large X (i.e., $\Omega_{\rm m} \sim 1$ where $\Omega_{\rm DE} \sim 1 - \Omega_{\rm m}$) we recover the matter-dominated behaviour ($D \sim a$). To parameterise deviations from this behaviour we define the "normalised growth" as $G = D/a$. The evolution equation is then:

$$G'' + \left[\frac{7}{2} - \frac{3}{2}\left(\frac{w(a)}{1+X(a)}\right)\right]\frac{G''}{a} + \frac{3}{2}\left(\frac{1-w(a)}{1+X(a)}\right)\frac{G}{a^2} = 0\,. \tag{109}$$

This equation allows us to physically interpret the effects of dark energy. In the presence of dark energy, the Hubble drag term is increased, so that growth is suppressed in a universe with an accelerating expansion. This is similar to the suppression due to radiation dominance.

11.3. The Power Spectrum of Matter

In an FLRW universe, the homogeneity and isotropy assumption means that any statistical properties must also be homogeneous and isotropic. The implication for the matter perturbation field is that its Fourier modes must be uncorrelated (due to homogeneity). Usually, we assume that the mode amplitudes are Gaussian. This assumption is well motivated since the theory for the seed fluctuations assumes that they have a quantum origin. Due to the central limit theorem, the sum of a sufficiently large number of mode amplitudes will tend towards a Gaussian distribution (see, e.g., [303,304]).

Such a field, with uncorrelated modes, and a Gaussian distribution of mode amplitudes is called a Gaussian random field, and can be entirely described by its two-point correlation function:

$$\langle \delta(\mathbf{x})\delta^*(\mathbf{y})\rangle = C_{\delta\delta}(|\mathbf{x}-\mathbf{y}|)\,. \tag{110}$$

The angled brackets denote an ensemble average (an average over a multitude of realisations). The value of δ at a given point in the universe will have a different value in each realisation, with a variance $\langle\delta^2\rangle$. Since we can only observe one realisation of our universe (in other words, at most only a finite region in this one universe), we apply the ergodic principle: The average over a sufficiently large volume is equal to the ensemble average.

In Fourier space, the correlation function can be written as:

$$\langle \hat{\delta}(\mathbf{k})\hat{\delta}^*(\mathbf{k}')\rangle = \int \mathrm{d}^3x e^{i\mathbf{k}.\mathbf{x}} \int \mathrm{d}^3x' e^{-i\mathbf{k}'.\mathbf{x}'}\langle \delta(\mathbf{x})\delta^*(\mathbf{x}')\rangle. \tag{111}$$

Replacing $\mathbf{x}' = \mathbf{x}+\mathbf{y}$, and substituting Equation (110), this can be written as:

$$\begin{aligned}
\langle \hat{\delta}(\mathbf{k})\hat{\delta}^*(\mathbf{k}')\rangle &= \int \mathrm{d}^3x e^{i\mathbf{k}.\mathbf{x}} \int \mathrm{d}^3y e^{-i\mathbf{k}'.(\mathbf{x}+\mathbf{y})} C_{\delta\delta}(|\mathbf{y}|) &(112)\\
&= (2\pi)^3\delta_D(\mathbf{k}-\mathbf{k}') \int \mathrm{d}^3y e^{-i\mathbf{k}.(\mathbf{y}) C_{\delta\delta}(|\mathbf{y}|)} &(113)\\
&= (2\pi)^3\delta_D(\mathbf{k}-\mathbf{k}') P_\delta(|\mathbf{k}|)\,. &(114)
\end{aligned}$$

The power spectrum has been defined as the Fourier transform of the correlation function:

$$P_\delta(|\mathbf{k}|) = \int \mathrm{d}^3y e^{i\mathbf{k}.(\mathbf{y})} C_{\delta\delta}(|\mathbf{y}|)\,. \tag{115}$$

The standard convention in cosmology is to abbreviate $P_\delta(|\mathbf{k}|)$ to $P(k)$, where $k = |\mathbf{k}|$. The power spectrum can be expressed in dimensionless form as the variance per $\ln k$, so that:

$$\Delta^2(k) = \frac{k^3 P(k)}{2\pi^2}\,. \tag{116}$$

11.3.1. Nonlinear Evolution

The power spectrum gives us the evolution of the initial matter density fluctuations. However, the linear evolution breaks down at small scales, when complex structures begin to form, and overdensities can no longer treated as perturbations on a smooth background. This is the nonlinear regime of gravitational evolution. The scale above which nonlinearities cannot be ignored is approximately set by $\Delta(k_{\rm NL}) \simeq 1$, which corresponds to $k_{\rm NL} \simeq 0.2\,h\,\mathrm{Mpc}^{-1}$ in most cosmological models. The standard way to model nonlinear evolution is by using phenomenological fits based on N-body simulations.

One strategy is to build a models based on a stable clustering hypothesis [305–307], which assumes that the nonlinear collapsed objects form isolated, virialised systems that are decoupled from the expansion of the universe.

A different approach is used in the halo model [308]. Here, the density field is decomposed into individual clumps of matter with some density profile and varying mass. By using this model to calculate the number of clumps within a given volume, the galaxy halo profile can be calculated. This is the equivalent of the power spectrum for these matter halos. A functional relation between the linear power spectrum and this halo profile is then derived and calibrated using large N-body simulations. This relation is then used to calculate the nonlinear power spectrum, using a fitting formula, for instance (see, e.g., [309,310]).

Whichever approach is used, it must account for a range of small-scale physical processes, such as baryonic physics, stellar formation, galactic magnetic fields, and Dark Matter and neutrino properties, which have become more significant with the ability of future astrophysical experiments to probe the nonlinear regime with ever increasing precision (see, e.g., [311–317]).

The need for realistic models of the universe has come to the fore in recent years, due the massive improvement in the quality and volume of cosmological measurements. Most of the N-body simulations rely on a perturbative approach. They use an FLRW cosmological background (perfectly homogeneous and isotropic)), and assume that any sub-horizon inhomogeneous structure of the universe will contribute to an average expansion on horizon-sized volumes driven by the horizon-averaged density. With the next generation of cosmological probes, these may not be accurate enough to model the universe realistically. There are several ongoing efforts to build fully relativistic, nonlinear, inhomogeneous and asymmetric models using numerical methods [318–320]. This is a significant contribution to the study of the backreaction effect and the question of the expansion rate of the universe.

11.3.2. The Primordial Perturbations

In the very early universe, the tiny initial perturbations are thought to have formed a Gaussian random field whose covariance function is diagonal and nearly scale-invariant. This form, known as the Harrison-Peebles-Zel'dovich spectrum, was assumed in most cases within the Standard Model, as it corresponds very closely to the observed power spectrum in the universe.

This type of spectrum was first proposed in the 1970s by Edward Robert Harrison [321], Yakov Zel'dovich [322], and Phillip James Edwin Peebles [323], who were working independently, as the spectrum for initial density fluctuations. This hypothesis was subsequently closely borne out by observations. The defining characteristic of a Harrison-Peebles-Zel'dovich spectrum spectrum is that it describes a fractal metric, where the degree of perturbation is the same on all scales (hence the term "scale-invariant"), so that $P(k) \propto k$. If we assume scale invariance for the power spectrum on large scales, and combine this with Equation (101), this implies the following general shape for the matter power spectrum in the Einstein-de Sitter scenario:

$$P(k) \propto \begin{cases} k & \text{for} \quad k \ll k_0 \\ k^{-3} & \text{for} \quad k \gg k_0 \,. \end{cases} \tag{117}$$

The actual form of the spectrum depends in non-trivial ways on the parameters in the cosmological model, including the 'slope' of the initial power spectrum n_s, where $P(k) \propto k^{n_s}$. In a scale-invariant spectrum in the linear regime, the fiducial value of n_s is taken to be 1.

Cosmological observations were consistent with scale invariance up until the early data releases from the *Wilkinson Microwave Anisotropy Probe* (WMAP) [214,324], but showed some tension. Subsequent observations cast further doubts on scale invariance [216]. In 2013, the *Planck* probe, mapping the anisotropies in the CMB, led to an important and conclusive result: it ruled out scale invariance at over 5σ. The primordial power spectrum was found to be scale-dependent, with $n_s = 0.9603 \pm 0.0073$ [221]. This was confirmed by the 2015 data release, which found that $n_s = 0.968 \pm 0.006$ [205]. This significant result is a powerful demonstration of the importance of multiple sources of data, joint observations from different probes, and the increasing reliance on complex statistical techniques for cosmological model selection [325–328].

Thus it can be seen that the current Concordance Model, consisting of ΛCDM with an inflationary epoch in the early universe, was built in stages over the last 100 years. It is the result of a process of accumulation of evidence and testing of competing models and theories. At each step, theory provides the basis for adjustments to the model, and observations from different probes provided the evidence (Figure 4). The current model should in no way be seen as "true". It is merely the best model that fits all the data available so far. Future data may very well require an adjustment to the Concordance Model.

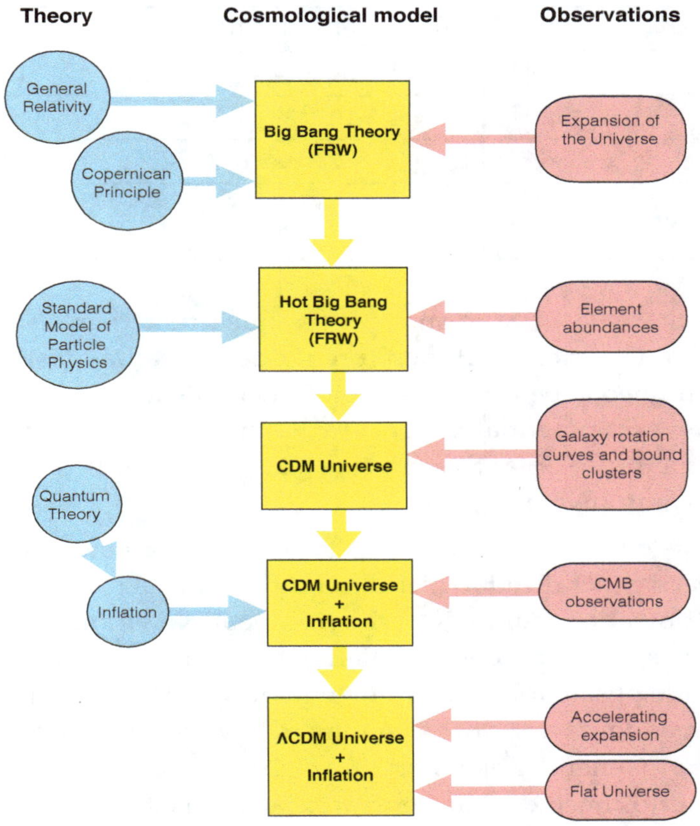

Figure 4. How the Concordance Model of Cosmology was developed. Theories and observations motivated the development of cosmological models, which were adjusted as new observations challenged the older models.

12. How Do We Test General Relativity?

The assumption of metric coupling given by Equation (5) has been tested accurately many times over the last 100 years, at scales from 10^{-4} m in laboratories [329,330], up to 10^{14} m in the Solar System [331]. These experiment test the implications of the metric coupling: the spacetime independence of non-gravitational constants, the isotropy of the coupling of all matter field to a unique metric tensor, the universality of free-fall, and gravitational redshift.

The standard way to test constraints on $S_{\rm gravity}$ and therefore the dynamics of General Relativity is by using the parameterised post-Newtonian (PPN) formalism [65]. This assumes that gravity is described by a metric over all scales. The idea is to write the most general form that $g_{\mu\nu}$ can take in the presence of matter, when considering correction of order $1/c^2$ with respect to the Newtonian limit. This method was first used in 1923 by Arthur Eddington [332]. In its simplest form, it provides us with two phenomenological parameters $\beta^{\rm PPN}$ and $\gamma^{\rm PPN}$ entering the Schwarzschild metric in isotropic coordinates:

$$g_{00} = -1 + \frac{2Gm}{rc^2} + 2\beta^{\rm PPN}\left(\frac{2Gm}{rc^2}\right)^2 \quad \text{and} \quad g_{ij} = \left(1 + 2\gamma^{\rm PPN}\frac{2Gm}{rc^2}\right)\delta_{ij}\,. \tag{118}$$

According to General Relativity, $\beta^{\rm PPN} = \gamma^{\rm PPN} = 1$. The experimental constraints on these parameters are summarised in Table 3.

Table 3. Some constraints on parameterised post-Newtonian (PPN) parameters from recent tests.

Method	Constraint	Experiment
Shift of perihelion of Mercury	$\lvert 2\gamma^{\rm PPN} - \beta^{\rm PPN} - 1\rvert < 3 \times 10^{-3}$	Data to 1990 [333]
Lunar laser ranging	$\lvert 4\beta^{\rm PPN} - \gamma^{\rm PPN} - 3\rvert = (4.4 \pm 4.5) \times 10^{-4}$	Data to 2004 [334,335]
Very long baseline interferometry	$\lvert \gamma^{\rm PPN} - 1\rvert = 4 \times 10^{-4}$	Data from 1979 to 1999 [336]
Time-delay variation	$\gamma^{\rm PPN} - 1 = (1.2 \pm 2.3) \times 10^{-5}$	Cassini spacecraft [67]
Planetary perihelion precessions	$\beta^{\rm PPN} - 1 = (-2 \pm 3) \times 10^{-5}$ $\gamma^{\rm PPN} - 1 = (4 \pm 6) \times 10^{-5}$	Solar System ephemerides to 2013 [337]

It should be noted that the PPN formalism assumes that there is no characteristic length scale for the gravitational interaction, and therefore it does not allow testing of finite-range effects. These too have been constrained to be very close their General Relativistic value of zero. The deviation of amplitude α from a Newton potential on a characteristic scale λ is typically $\alpha < 10^{-2}$ on scales ranging from a few millimetres to Solar System size [329]. This implies no deviation from GR over more than 15 orders of magnitude in length scale.

The Solar System tests constrain Eddington's two PPN parameters to a tiny region very close to 1. The formalism has been generalized to include eight additional phenomenological parameters [65] to describe any possible deviation from GR at the first post-Newtonian order. They have all been constrained to be very close to their General Relativistic values. The latest data use Solar System ephemerides, which include perihelion measurements for Mercury and the other planets, as well as lunar ephemerides. Independent teams of astronomers have estimated corrections to the standard first-order post-Newtonian General Relativistic formalism to be all statistically compatible with zero [337,338]. This leads us to conclude that General Relativity is the only theory consistent with Solar System experiments at the post-Newtonian order.

It should be noted that, so far, it is only the first-order post-Newtonian, static, Schwarzschild-like part of the spacetime metric that has been modelled and tested using Solar System dynamics. The first-order post-Newtonian gravitomagnetic or Lense-Thirring part of the spacetime metric has neither been modelled nor tested yet in the Solar System. However, it has recently been pointed out that this could be possible in the next few years by focussing on particular models (Sun and planets, planets and spacecraft, or planets and planets) [339]. This would open the field to the possibility of constraining the PPN parameters using a more complete model of General Relativistic effects.

What about larger or smaller scales? The size of the universe was around 10^{-35} m in the beginning, and the present size is around 10^{26} m. GR remains untested at these extreme scales. But one fundamental assumption of GR is that it describes gravity at all scales (i.e., the theory assumes the analytical continuity of solutions).

The evolution of the universe itself therefore provides a useful test of General Relativity. The challenge lies in testing a wide range of potentials (weak and strong field regimes) over cosmological scales. The strong field regime is tested using compact objects. However, over cosmological scales, the kind of objects that would produce gravitational potentials approaching the strong field regime simply do not exist. One solution is to observe the evolution of the universe and check whether the evolution of large-scale structure corresponds to the predictions of General Relativity.

13. Cosmological Tests

General Relativity has been submitted to 100 years of Solar System tests, which it has passed with flying colours. The discoveries of the last two decades: cosmic acceleration, the scale dependence of the primordial power spectrum, and also the open question of Dark Matter, have made it clear that GR must also be tested at astrophysical and cosmological scales.

This presents a serious challenge. We derive our knowledge of the universe from measurements of distances and times, and statistical properties like the distribution of matter. The main difficulty in extending the Solar System tests to cosmological scales is that these measurements depends strongly on the construction of cosmological models.

These models depend on four hypotheses:

(1) A theory of gravitational interactions.
(2) A description of the matter in the universe and the non-gravitational interactions such as electromagnetic emissions.
(3) A hypothesis on the symmetry.
(4) A hypothesis on the topology, or the global structure of the universe.

Some of the hypotheses are hard to verify, and some have unverifiable implications. Assuming the symmetry of our solutions means that we also assume the laws of physics are the same throughout the universe, including its unobservable part outside the cosmological horizon (delimited by a radius of around 15.7 Gpc). This is a very strong assumption, but one that is unverifiable.

Any cosmological model needs all four hypotheses. The first two are the physical theories, but their equations cannot be solved without some kind of assumption on the symmetry of the solutions, given by the third hypothesis. The fourth hypothesis is then an assumption on the global properties of these solutions.

The simplest ΛCDM Concordance Model assumes that gravity is described by General Relativity (Hypothesis 1), and that the universe contains the particles and fields of the Standard Model of particle physics [171], together with Dark Matter and a cosmological constant (Hypothesis 2). The Einstein equations require an effective stress-energy tensor averaged out on large scales, so the model requires an extra assumption on the averaging procedure. This averaging and the validity or otherwise of this assumption is the subject of research on the backreaction effect. The model assumes the Copernican Principle (Hypothesis 3), and the continuity of the solutions of Einstein equations across all spatial sections (Hypothesis 4).

The number of alternative cosmological models proposed, especially since the 1990s, is too numerous to list here. Many of them are impossible to rule out simply by fitting them to the data, because their predictions are so close to those of ΛCDM. It is easier to test the four hypotheses rather than trying to test the observables predicted by the models. As with any null test, a significant violation would indicate the need to modify one or more hypotheses.

As far as tests of GR (the theory of gravity in ΛCDM) are concerned, many experiments test more than one assumption. We list the main experiment types, and the assumptions they test, in Table 4.

Table 4. What we are actually testing. Experimental tests of General Relativity (GR) often probe more than one assumption, and isolating the effects is a challenge in itself.

Experiment	Assumption Tested
Solar System tests	Metric coupling
Quadrupolar shift of nuclear energy levels	Isotropy [340–342]
Lunar laser ranging and orbiting gyroscopes	Universality of freefall [343] and structure of metric
Space-borne clocks	Gravitational redshift [344]
Shift in perihelion of planets	Structure of metric [345]
Time invariance of physical constants	Metric coupling [346]
Detection of gravitational waves	Lorenz gauge condition and inhomogeneous wave equation

13.1. Testing the Description of Matter and Non-Gravitational Interactions

General Relativity can be tested by measuring the fundamental constants of the theory [346–349]. Any variation would require a modification of GR [350,351]. A local measurement of a fundamental constant, such as a determination of the fine-structure constant from the Oklo phenomenon [352–354], is actually a cosmological-scale measurement along the time dimension. Astrophysical probes such as 21-cm radiation [355] or the cosmic microwave background [356], can be used to test the constancy of the fine-structure constant α, or to constrain simultaneous variations of α and Newton's gravitational constant G [357]. A non-constant G would have serious implications both for Newtonian physics and for General Relativity. This motivates ongoing efforts to devise new methods to measure this quantity, and to push the limits of experimental accuracy in order to test the constancy of G [358–362].

13.2. Testing the Assumption of Symmetry

The assumption of symmetry can be tested by checking for any deviations from isotropy. This requires statistical ensembles of data, so the best observables are the CMB, and, on a smaller scale, large-scale structure [363–365].

13.3. Testing the Gravitational Interactions

The gravitational interactions have been tested many times at laboratory scales and all the way up to Solar System scales. The challenge today is to test them at cosmological scales.

Tests of General Relativity may be placed in three broad categories: laboratory, astrophysical, and cosmological. Experiments cannot span all length scales, and so they cannot test all theories. We are forced to design experiments which can test alternative theories, or the effects of GR, at particular length scales (Table 5).

Laboratory tests probe effects from sub-millimetre scales to a few hundred metres. Galileo's experiments on weights dropped from the Leaning Tower of Pisa [366] are one example of a laboratory-scale test. The torsion balance experiments of Cavendish and Eötvös [367], and their modern versions [368–371], are another.

Astrophysical tests probe gravity from Solar System scales all the way up to galactic cluster scales. They include experiments such as laser ranging off the Moon (lunar laser ranging) and several of the Earth's artificial satellites [372–375], and proposals to extend laser ranging to other planets in the Solar System (planetary laser ranging) [376–380], radar astrometry on near-Earth objects [381], observations of the precession of the perihelion as well as higher-order effects for Mercury and other planets [338,382,383], observations of spacecraft in the Solar System [337], actual or proposed measurements of the acceleration of the Pioneer [384–386] and New Horizons probes [387–389], observations of the orbits of compact objects such as neutron stars [390–392], and measurements of galaxy rotation curves [393,394]. In the last decade, observations of extrasolar planetary systems have been proposed and used in order to test GR [395–404]. All of these systems are characterised by their spherical symmetry. Many can be approximated by a test object orbiting at a distance r around a central mass M. The gravitational field of the more massive central object is then probed by observing the orbit of the test body.

Table 5. Bridging the length scales to test the cosmological model. Experimental tests of gravity and dark sector couplings, at their typical length scales. Massive gravity (MG) screening mechanisms would show up at short ranges, while smooth dark energy manifests itself at cosmological scales. We give the experimental accuracy from current and future experiments planned over the next decade. The comparison between growth and expansion history comes from combined BAO, supernova, weak lensing, redshift distortion, cosmic microwave background (CMB) lensing, and cluster data. Lensing effects and dynamical mass comparisons can be carried out over a range of scales: inside galaxies, using strong lensing and stellar velocities, and on cosmological scales, using cross-correlations. This is a route to testing screening effects in alternative theories. Laboratory and Solar System tests can also probe dark sector couplings besides short-range effects, but many of the constraints obtained depend on the cosmological model.

Test	Length Scale	Theories Probed	Current Status (and Future)
Growth vs. expansion history	100 Mpc − 1 Gpc	GR with smooth dark energy	10% accuracy (2%–4%)
Lensing vs. Dynamical mass	0.01 − 100 Mpc	Test of GR	20% accuracy (5%)
Astrophysical tests	0.01 AU − 1 Mpc	MG screening mechanisms	∼10% (Up to 10 times improvement)
Laboratory and Solar System tests	1 mm − 1 AU	PPN parameters in MG	Constraints are model dependent (Up to a tenfold improvement)

For such systems, the deviation of the metric from the Minkowski form is characterised by the magnitude of the Newtonian gravitational potential

$$\epsilon = \frac{GM}{rc^2} \,. \tag{119}$$

The strongest gravitational fields accessible to an observer occur in the limit $\epsilon \to \mathcal{O}(1)$. In this limit, the central object is a black hole and the test object is close to the event horizon.

We know that the Riemann curvature tensor is an essential quantity in General Relativity. The approximate magnitude of this tensor is expressed by the Kretschmann scalar (the fully contracted Ricci scalar) for the Schwarzchild metric:

$$\xi \propto \frac{GM}{r^3c^2} \,. \tag{120}$$

Note that this expression is more complicated for rotating objects [405]. For the purposes of our description, however, we may use this simple expression, which we call the curvature.

The third broad category is the cosmological tests. First, the measurements are made on a statistical ensemble. The masses must be treated as power spectra. Second, the gravitational field assigned to each wavenumber k is very weak. The description of the system often depends on the cosmological model. At this scale, the description must take into account the background perturbations and the expansion of the universe.

The position of the various systems on a potential-curvature parameter space is shown in Figure 5. The potential accessible to observers is bounded by the line $\epsilon \simeq 1$. There is no limit to the maximum curvature that can be observed except for the Planck limit, where $r \simeq 1.6 \times 10^{-33}$ cm. This lies many orders of magnitude above the boundaries of the figure. General Relativity is a complete theory, and does not fail below the Planck scale. To test or even probe GR below the Planck scale is a different matter, and a goal that has not yet been achieved. Many alternative theories of gravity that attempt to explain cosmic acceleration do so through modifications to GR at the Planck scale, so it is important to explore ways in which GR can be tested at these scales.

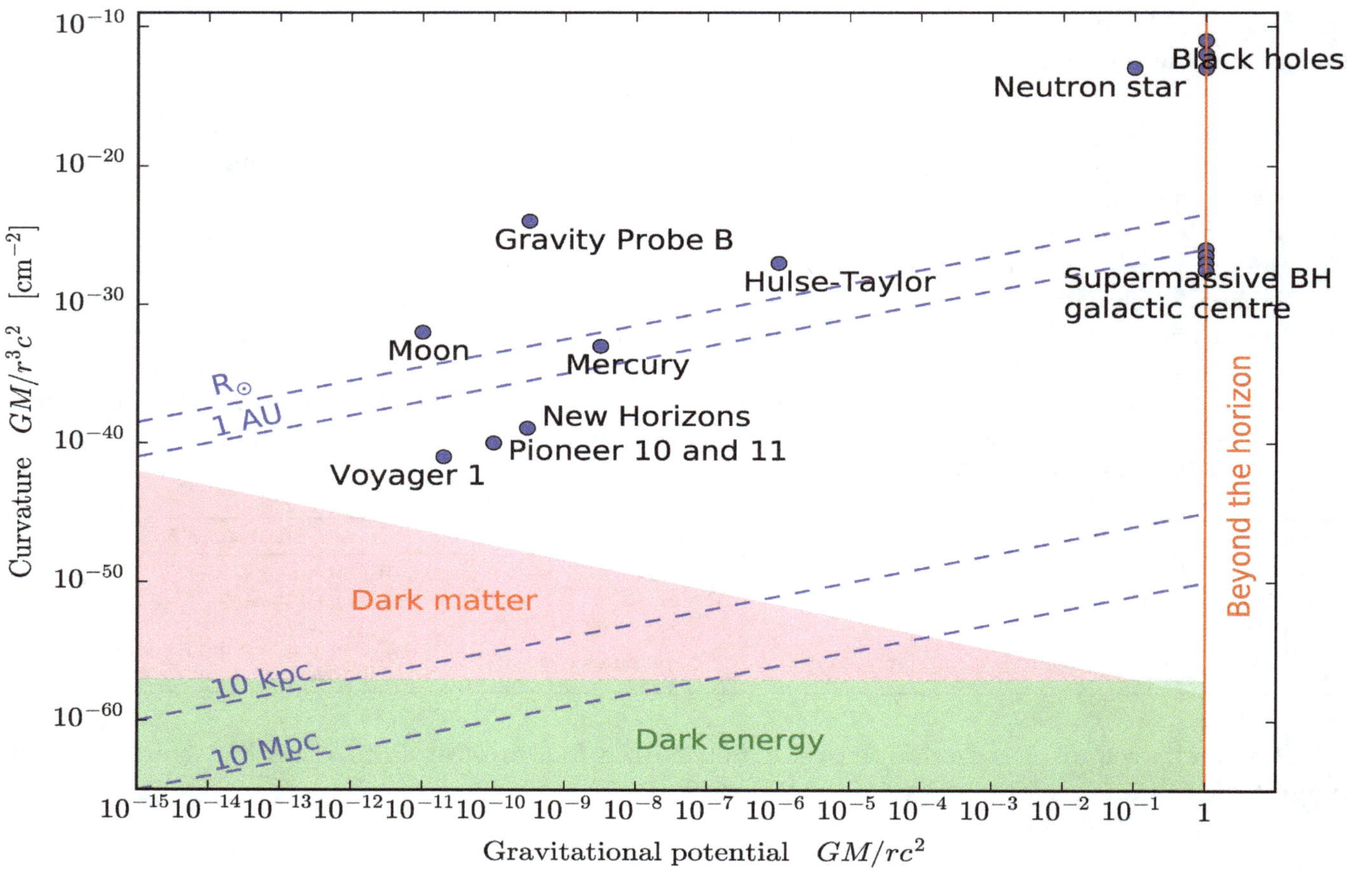

Figure 5. The parameter space for quantifying the strength of a gravitational field. The horizontal axis measures the potential. The vertical axis measures the spacetime curvature of the gravitational field at a radius r away from a central object of mass M. Regions of this parameter space with potential greater than 1 represent distances from a gravitating object that are smaller than the event horizon radius and are therefore inaccessible to observers. The red vertical line on the right-hand side of the plot marks the horizon limit. This is a schematic plot, and in no way do we show an exhaustive list of objects and systems that have been used or could be used to test GR. The region of Solar System-scale tests is broadly bounded by the Moon, Gravity probe B [406,407], Mercury, and the Pioneer and New Horizons spacecraft. We have included the Voyager spacecraft in the diagram, even though, unlike the Pioneer probe, it was never suitable for tests of GR. The famous Hulse-Taylor binary pulsar [408], although a neutron star binary, is also roughly in the region of Solar System tests. Black holes and neutron stars are in the strong field regime, and the former are at at the limit of the event horizon boundary. Adapted from [409].

What is the minimum curvature? The unperturbed FLRW metric is isotropic, and the unperturbed Kretschmann scalar is a function of time only. The curvature for the homogeneous universe drops as the universe expands. The present universe has a curvature which is just above the boundary of the region marked 'Dark energy', since dark energy is not yet completely dominant over pressureless matter. However, the curvature will approach this limit asymptotically. In this paradigm, this represents a fundamental minimum curvature scale. It is shown in the figure by the region labelled "Dark energy".

Galaxies are astrophysical probes of GR. Their innermost regions can be modelled as test particles orbiting a central supermassive black hole. Galactic rotation curves, which can only be explained by introducing Dark Matter, describe galaxy velocities up to the outermost regions. Systems below a constant acceleration scale of approximately $1.2 \times 10^{-10}\,\mathrm{ms}^{-2}$ cannot be modelled without adding a contribution to the gravitational filed in the form of Dark Matter. This constant acceleration is the diagonal boundary of the region labelled "Dark matter" in the figure.

Note that the regions of parameter space occupied by Dark Matter and dark energy overlap—there is a degeneracy between these two unknown components of the cosmological model. However, it is not impossible to distinguish between their effects, since their properties are different. In particular, Dark Matter forms clumps just like baryonic matter, while dark energy does not.

GR has not been tested in the region between curvatures of $\sim 10^{-40}$ and $\sim 10^{-50}$. This corresponds to the region between Solar System scales and cosmological scales probed by galaxy surveys and the CMB. The challenge lies in finding systems which span these scales. In theory they do exist, in the form of galaxies, clusters and superclusters. Their rotation curves transition from Schwarzchild orbits in their innermost regions, to outer regions dominated by Dark Matter. However, our observations are hampered by the fact that we are limited by the resolution of our telescopes, so we can only observe the outer regions. In addition, the untested region is situated between a region where GR is extremely well-constrained (by Solar System tests), and a region where Dark Matter and dark energy have to be invoked, and where we must take the cosmological model into account, because the effect of large-scale structure on background dynamics becomes non-negligible. It has been suggested that this backreaction effect may be at the origin of the observed cosmic acceleration.

Cosmological tests of General Relativity may be broadly classified as follows:

- Tests of the consistency between the expansion history and the growth of structure. A discrepancy in the equation of state parameter of dark energy w, inferred from the two approaches can indicate a breakdown of the GR-based smooth dark energy cosmological paradigm.
- Detailed measurements of the linear growth factor across different scales and redshifts.
- Comparison of the cosmological mass distribution inferred from different probes, especially redshift space distortions and lensing.

14. Possible Modifications of GR and Cosmological Implications

It is useful to identify the regimes in which modifications to GR may appear. This enables us to get a clearer picture of the capabilities and limitations of current and future experiments to tests these alternative theories.

14.1. Weak and Strong-Field Regimes

In order to test gravity in the strong-field regime, we need to observe compact objects with a very high density [410–412]. Black holes are good candidates. Their compactness and mass takes GM/rc^2 close to the maximum limit of unity. However, they have a serious drawback. Because of the "no hair theorem", they are not characterised by any coupling to a scalar field and therefore cannot be used to test for this effect, and to discriminate between scalar-tensor theories and GR.

Neutron stars, on the other hand, are still very compact bodies, but they can be strongly coupled to a hypothetical scalar field. This property has been used to test relativistic parameters by observing the Hulse-Taylor binary pulsar PSR B1913+16 [413], and the neutron star–white dwarf binary PSR J1141-6545 [414]. Gravitational time delay has been tested using other binary pulsars [415–419], and pulsars have also been suggested as ideal probes of the Lense-Thirring effect [420,421]. General Relativity passes these tests with flying colours.

14.2. Small and Large Distances

Distance-dependent modifications can be induced by a massive degree of freedom, which will cause a Yukawa-like coupling. General Relativity is very well constrained on the size of the Solar System, and there are several tests constraining Yukawa interactions at Solar System scales [422–424], but there are no constraints on scales larger than $10\,h\,\mathrm{Mpc}^{-1}$, at least without assuming some cosmological model. Some theories put forward to explain cosmic acceleration, such as Chameleon mechanisms [425], are essentially modifications of GR at cosmological distance scales.

14.3. Low and High Accelerations

Galaxy rotation curves and galaxy dynamics motivated the Dark Matter paradigm. The Tully-Fischer law [426] tells us that Dark Matter cannot be explained by a modification of General Relativity at a fixed distance. MOND instead explains it by modifying gravity at low accelerations below the typical acceleration $a_0 \sim 10^{-8}\,\mathrm{cm}\cdot\mathrm{s}^{-2}$.

14.4. Low and High Curvature

Curvature R is important in distinguishing possible extensions of the Einstein-Hilbert action. A curvature-dependent may become important even if the potential Φ remains small. In the Solar System, the curvature $R_\odot \sim 4\times 10^{-28}\,\mathrm{cm}^{-2}$. The curvature of the homogeneous universe according to the Friedmann equation is

$$R_{\mathrm{FLRW}}(z) = 3H_0^2\left(\Omega_{\mathrm{m}}(1+z)^3 + 4\Omega_\Lambda\right), \tag{121}$$

from which we can see that the curvature of the universe evolved with time, from $\sim 10^{-33}\,\mathrm{cm}^{-2}$ at the time of nucleosynthesis, to $\sim 10^{-56}\,\mathrm{cm}^{-2}$ at $z = 1$. The curvature scale associated with the cosmological constant is $R_\Lambda = (1/6)\Lambda$, so the phenomenology of the cosmological constant occurs in low curvature regime

$$R < R_\Lambda \sim 1.2\times 10^{-30} R_\odot\,. \tag{122}$$

This is of particular interest to paradigms which seek to explain cosmic acceleration through the backreaction effect. For cosmological-scale perturbations, we are always in the weak field regime. However, the curvature perturbation associated with large-scale structure becomes of the order of the background curvature at redshift $z \sim 0$, even if we are still in the weak field limit. This means that the effect of large-scale structure on the background dynamics may be non-negligible [427].

In summary, in order to explain the dark energy or Dark Matter problem by modifying General Relativity, the modifications have to be either at large scales (typically Hubble scales), low accelerations (typically below a_0, or small curvatures (typically R_Λ). The regions corresponding to dark energy and Dark Matter curvature and potentials are shown in Figure 5.

14.5. Cosmological Probes

The idea of testing General Relativity using large-scale structure was first proposed in 2001 [428]. This relies on the ingenious idea that if gravity is well-described by General Relativity, and the universe well-described by ΛCDM, then on sub-Hubble scales, and considering only scalar perturbations, the spacetime metric can be written as

$$\mathrm{d}s^2 = -(1+2\Phi)\,\mathrm{d}t^2 + (1-2\Psi)a^2(t)\gamma_{ij}\,\mathrm{d}x^i\,\mathrm{d}x^j \tag{123}$$

where Φ and Ψ are the two potentials, and γ_{ij} is the metric of the spatial section [429–431]. The Einstein equations reduce to the Poisson equation

$$\triangle\Psi = 4\pi G\rho_{\mathrm{matter}} a^2 \delta_{\mathrm{matter}} \tag{124}$$

and

$$\Psi - \Phi = 0\,, \tag{125}$$

since the matter anisotropic stress is negligible. The spectrum of the two gravitational potentials has to be proportional to the matter power spectrum. The scale dependence of the gravitational potential P_Φ and of the matter distribution $P(k)$ are related by:

$$k^4 P_\Phi(k,a) = \frac{9}{4}\Omega_{\mathrm{m}} H^2 a^{-2} P(k,a)\,. \tag{126}$$

If the Poisson equation is modified by some modification of gravity, the matter power spectrum changes shape according to the cosmological model assumed, so the above relation is not model-independent. However, the fact that the two spectra differ is independent of the cosmological model. It therefore provides a test of the underlying gravitational theory. Such a test can be carried out by comparing weak lensing data to galaxy surveys.

Various similar approaches have been proposed [432]. This approach allows us to test various classes of alternative theories by means of large-scale structure. Some of these theories include Dvali-Gabadadze-Porrati models [134,433,434], quintessence [435–439], and scalar-tensor theories [440–443]. The difficulty lies in finding a parameterisation of the perturbation equations which is consistent with the one used for the background evolution, since both assume the same theory for gravitational interactions.

The cosmic microwave background is another potential testing ground for General Relativity. The amplitude and position of the peaks of the cosmic microwave background allows us to probe the potential wells present during the recombination era by extracting information on the primordial power spectrum created during inflation. *Planck* data, alone or in combination with weak lensing has been used to test GR [444,445] and modified gravity [446–448].

Despite the fact that GR is extremely well tested on laboratory and Solar System scales, cosmology provides plenty of scope for alternatives to General Relativity. Let us consider cosmic acceleration, which is now a confirmed observation. It requires an explanation. The cosmological constant paradigm rests on three assumptions: that the observations are correct, that GR is the correct theory of gravity, and that our FLRW model of the universe is correct.

More than two decades of precision measurements have removed any doubt that the observed acceleration may be due to incorrect modelling of experimental errors. So in order to do away with the cosmological constant, we must seek possible answers in the other two assumptions.

Backreaction is an "alternative" to the alternative theories, or to the Dark Matter paradigm. It is firmly within the General Relativity paradigm, and seeks to explain cosmic acceleration by modelling the universe and the structures within it, and therefore its expansion history, in more detail than is the case with the FLRW model. It keeps the first and last assumption, but does away with the assumption of an FLRW universe. However, is it enough to explain the observed acceleration?

If GR is assumed to be the correct theory together with the other two assumptions, cosmological measurements are usually interpreted as providing evidence for Dark Matter and a nonzero cosmological constant or dark energy. This poses conceptual problems. Why is the observed value of the cosmological constant so small in Planck units? It also poses a coincidence problem. Why is the energy density of the cosmological constant so close to the present matter density? No dynamical solution of the cosmological constant problem is possible within GR—the cosmological constant is not the attractor of some dynamical function.

This opens the field to possible modifications of GR. Should GR be modified at low and high energies? This is a serious challenge for theorists. Einstein's theory is the unique interacting theory of a Lorentz-invariant massless spin-2 particle. New physics in the gravitational sector must introduce additional degrees of freedom. These additional degrees of freedom must modify the theory at low or high energies, or both, while being consistent with GR in the intermediate-energy regime, that is, at length scales $1\,\mu \leqslant \ell \leqslant 10^{11}$ m, where the theory is extremely well tested.

Figure 6 illustrates some of the difficulties in testing the completeness of General Relativity. There is an obvious "scale desert" between Solar System scales, and cosmological scales. At the other end of the scale, there is a gap which is often overlooked: between sub-millimetres scales, which is the current limit where GR has been tested, and the Planck scale. A modification of gravity at very small scales would be apparent in this regime. Even ignoring quantum effects, it is a serious challenge to test gravity in a regime in which vastly stronger forces come into play.

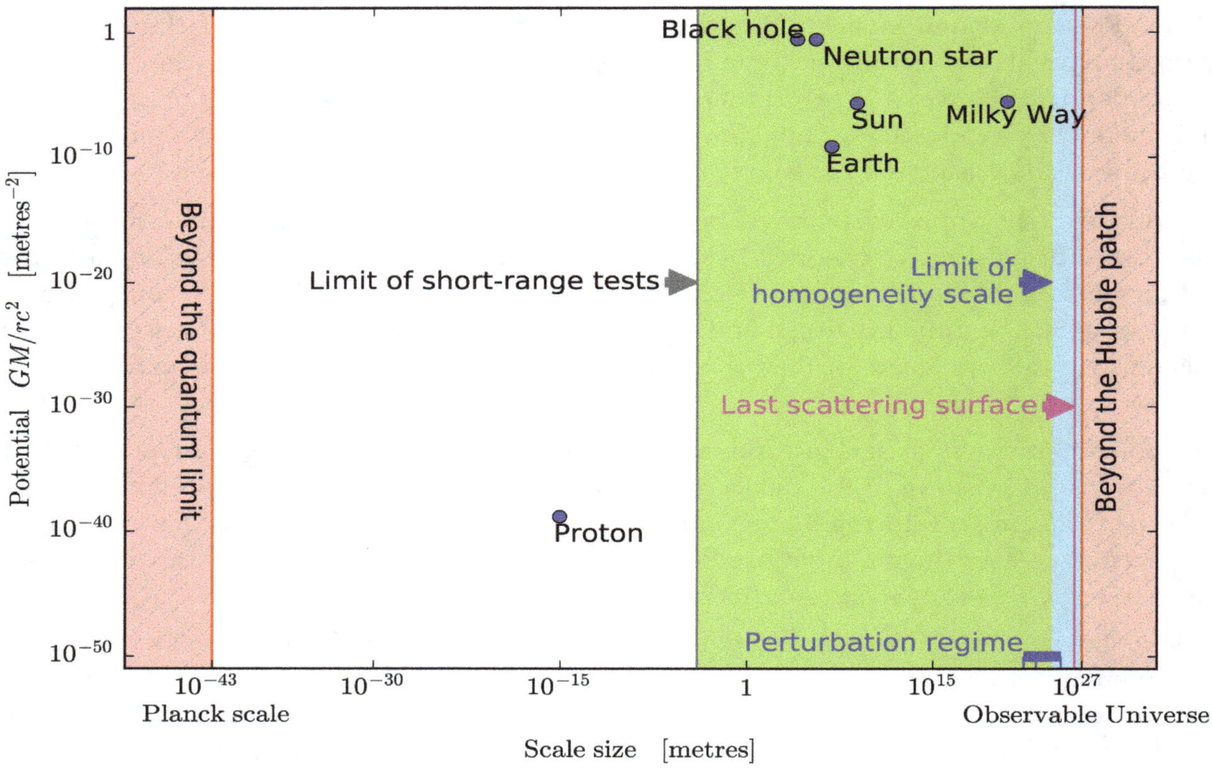

Figure 6. The parameter space for experiments. The horizontal axis is the typical length scale of the object in question. The vertical axis measures the gravitational potential. The red vertical line on the left-hand side marks the Planck scale. The vertical line on right-hand side of the plot marks radius of the observable universe, or Hubble radius. Experimental verification of GR is impossible beyond these limits. The radius of the surface of last scattering is only slightly smaller than our Hubble radius. Assuming GR implies assumptions far beyond the range that has been experimentally tests. For instance, if we define the Planck mass as $m_{\text{Planck}} = \sqrt{\hbar c/G}$, we are assuming that the gravitational constant remains constant down to the Planck length. This extrapolates the inverse square law over a scale of more than 10^{30} from what ha s been tested. The green region is where Solar System tests have been carried out. Beyond $\sim$100 Mpc, assumption of a homogeneous and isotropic metric becomes accurate enough to use in physical models. The blue region shows length scales at which the FLRW metric is valid. By way of comparison to this parameter space, the nonlinear regime for perturbation theory, which gives use the matter power spectrum, covers $\sim 10^{22}$ to $\sim 10^{23}$ m, the linear, quasi-static regime covers $\sim 10^{23}$ to $\sim 10^{25}$ m, and beyond that is the superhorizon regime. These length scales ranging from 1 Mpc to above 1 Gpc fit on a tiny part of the horizontal axis above, shown by the thick blue horizontal line.

15. The Nature of Dark Energy and the Implications for General Relativity

The Concordance Model of cosmology assumes that General Relativity is correct, an assumption which is justified by the tests which GR has undergone. Within this model, ~95% of the content of the universe is unaccounted for. Dark matter, which makes up around 25% of the mass-energy of the universe is a matter-like component which is cold (sub-relativistic) and weakly interacting. The discrepancy between the observed acceleration of the expansion of the universe and the predictions of GR leads to the conclusion that there must be a cosmological component with a negative equation of state parameter making up around 70% of the mass-energy content of the universe: dark energy.

But the dark energy paradigm does not fix the nature of this component. There exist many theories which attempt to explain its nature. In particular, one can ask whether the basic assumptions of the Concordance Model—homogeneity and isotropy—are correct.

If the universe is not homogeneous and isotropic, the FLRW equations are no longer valid. Over the last decade, one line of research has attempted to explain the accelerated expansion by exploring the implications of an inhomogeneous universe on a General Relativistic cosmology. This effect is the backreaction. This approach does away with dark energy.

There are various ongoing investigations on the effect of the backreaction due to an inhomogeneous universe [269,427,449–455], with different lines of research offering different interpretations of the Buchert equations, where the Friedmann equations are supplemented by an additional backreaction term [456]. Whether one can explain all of the observed expansion history of the universe as a consequence of the growth of inhomogeneities without invoking some additional fluid component is the subject of ongoing debate [452,457]. The backreaction, even if it turns out to be incapable of replacing the dark energy paradigm, is still a subject worth investigating, if anything as a correction to the homogeneity assumption, which obviously breaks down at small scales.

If we assume that the Copernican Principle holds, then the universe is well described by a Friedmann-Robertson-Walker spacetime. The dynamics of the background expansion are determined by the content of the universe: the list of fluids (perfect fluids, due to the assumption of the Copernican Principle), with their equations of state. Within the dark energy paradigm, we can distinguish two main strategies for formulating hypotheses:

(1) there is some new kind of component in the universe, or
(2) there is some new property of gravity.

Let us first recall that General Relativity rests on two assumptions: the gravitational interaction is described by a massless spin-2 field, and matter is minimally coupled to the metric, which implies the weak equivalence principle. The Einstein-Hilbert action described by Equation (6) implies that

$$S_{\text{gravity}} = \frac{c^3}{16\pi G}\int R\sqrt{-g}\,\mathrm{d}^4x + S_{\text{matter}}[\text{matter}; g_{\mu\nu}]\,, \tag{127}$$

where R is the Ricci scalar of the metric tensor $g_{\mu\nu}$, and $S_{\text{matter}}[\text{matter}; g_{\mu\nu}]$ is the action of the matter fields. If we only consider field theories, this gives us a useful classification scheme for the different theories that seek to explain the nature of dark energy.

The first strategy listed above assumes that gravitation is described by General Relativity, and introduces new forms of gravitating components beyond the Standard Model of particle physics. This adds a new term $S_{\text{DE}}[\psi; g_{\mu\nu}]$ to the action in Equation (127) while keeping the Einstein-Hilbert action of all the standard and Dark Matter unchanged.

The second strategy modifies gravity, and therefore extends the action, either by modifying the Einstein-Hilbert action of the coupling of matter. These theories also involve new forms of matter.

A cosmological constant Λ is the simplest modification which can be made to gravity, and it is equivalent to dark energy with a constant equation of state. To explain the observed acceleration, the new form of matter must have an equation of state $w < -1/3$. Dark energy can also be attributed to the energy of the vacuum, although the energy predicted by the Standard Model of particle physics is either 0 (using super-symmetry), or 10^{120} orders of magnitude larger than the observed cosmological value [458]. There are ongoing attempts to solve this 'fine-tuning problem' using string theory [459–461], causal sets [462,463], or by using anthropic arguments [464,465] .

The other approach is to attribute dark energy to a scalar field whose potential has evolved in some way that it currently exerts a negative pressure. Such fields, in theories within the framework of GR, are termed Quintessence. Their distinguishing feature is that they allow the equation of state of dark energy to evolve. Alternatives to Quintessence within the same approach include K-essence, Phantom Fields, or the Chaplygin Gas [466–469].

Another strategy is to depart from General Relativity and modify the laws of gravity and posit dark energy as the manifestation of an effect arising from extra dimensions, or higher-order corrections. Within this category, the more successful theories have been of two types. Dvali-Gabadadze-Porrati (DGP) dark energy considers the universe as a 4D brane within a 5D Minkowskian bulk [134]. The weakness of gravity relative to the other forces is explained by gravity "leaking" into the higher dimensions as it acts through the bulk [470], whereas the other forces act within the brane. The other class of theories is $f(R)$ gravity, where the Ricci scalar R in the Lagrangian is replaced by some function $f(R)$. Such theories correspond to scalar-tensor gravity with vanishing Brans-Dicke parameter [471].

We can therefore identify four main classes of theories, as shown in Figure 7. Classes 1 and 2 assume GR and introduce new forms of gravitating matter, the difference being that in class 2, the distance-duality relationship may be violated due to mechanisms such as photon decay. Classes 3 and 4 modify gravity in some way. In class 3, a new field introduces a long-range force so that gravity is no longer described by a massless spin-2 graviton. In this class, there may be a variation of the fundamental constants. Class 4 includes theories in which there may exist an infinite number of new degrees of freedom, such as brane models or multidimensional models. These models predict a violation of the Poisson equation on large scales. The constraints which may be placed on these theories, and the experiments to test them, correspond to the regimes in which gravity is modified, as summarised in Section 14.

An exhaustive list of dark energy or cosmic acceleration theories is beyond the scope of this review. We simply note that most of them give different predictions for the equation of state of dark energy, its evolution, or the expansion history of the universe . To distinguish between these proposals we need to track the evolution of these parameters throughout the history of the universe.

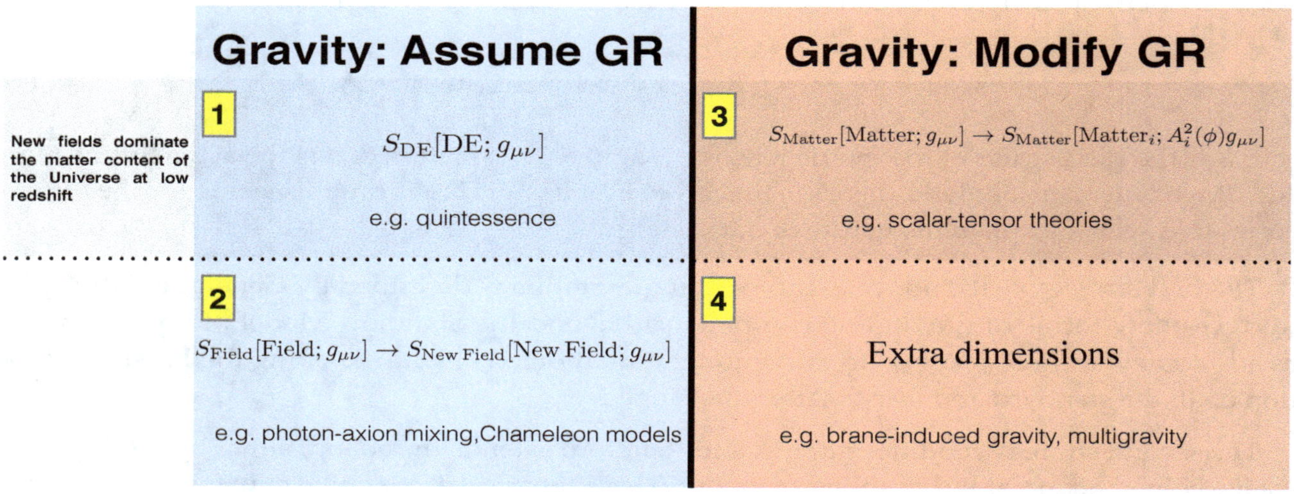

Figure 7. The four main classes of dark energy theories, within the two broad strategies, classified as modifications of the General Relativistic action. Classes 1 and 2 assume the gravitational metric coupling of GR, whereas classes 3 and 4 modify this metric coupling, and are therefore modifications of gravity. In the upper line of classes (1 and 3), new fields dominate the matter content of the recent universe. Adapted from [431].

16. The Current Status of General Relativity

General Relativity has been subjected to a multitude of tests in its 100 years of existence. As of 2016, the main predictions of GR have been tested and confirmed. Whereas it is sufficient for most purposes at ordinary accelerations and energy scales to use Newtonian calculations, General Relativity has found its way into daily life, in the Global Positioning System and geodesy. The postulates of General Relativity have been confirmed with ever-increasing accuracy. Deviations from the Einstein Equivalence Principle are now constrained to below $\sim 10^{-14}$, deviations from Local Lorentz Invariance down to $\sim 10^{-20}$, and deviations from Local Position Invariance down to $\sim 10^{-6}$.

General Relativity has been probed down to scales of $\sim 10^{-6}$ metres in laboratories, and up to 1000 AU in space-based experiments and observations. Astrophysical and cosmological observations have probed GR at scales from 1 Mpc up to gigaparsec scales. In the Solar System, the dynamics of GR have been tested with radar and laser-ranging. We can track the ephemerides of the planets in the Solar System, right up to the minor outlying planets (such as Sedna, with an aphelion just short of 1000 AU). The farthest objects whose trajectory has been followed from the moment they were "thrown" are now outside the Solar System. They are the Pioneer 10 and 11 spacecraft, and are now around 70 AU distant, and Voyager 1, which is at a distance of 135 AU, making it the only object to have reached interstellar space. Communication with the Pioneer spacecraft ceased in 2003. The Pioneer spacecraft exhibited an anomalous constant acceleration towards the Sun which could not be explained using GR. This prompted a reexamination of all the recorded data. It is now generally accepted that the anomaly is caused by thermal radiation, and that once this is accounted for, there is no remaining anomalous acceleration [472].

Gravitational waves, among the last untested predictions of GR, were first detected in late 2015, ushering a whole new era of observational astrophysics, in which the strong field regime can be probed and where the predictions of GR can be tested.

The evidence for General Relativity is extremely strong. The theory has passed all tests in the weak-field limit at Solar System scales, including tests of the assumptions (the Equivalence Principle) and the predictions specific to GR (frame-dragging, gravitational time dilation), and in the strong field with the observation of a binary black hole merger and the resulting gravitational waves. We give a summary of the experimental milestones in Table 6. However, there are still issues that allow room for speculation.

The first is the question of the completeness of GR. Is it valid at all scales? There is a scale gap in our tests of GR between laboratory, Solar System and galactic scales, and cosmological ones. This gives rise to a multitude of domains of validity for different alternative theories, whereas ideally we should seek a universal theory that can explain phenomena at all scales.

The second issue is the accuracy of our approximations. General Relativity may be conceptually simple in that it is based on a minimum number of postulates. But the resulting field equations, when applied to real physical systems, can be very hard to solve. We get around this by making approximations such as spherical symmetry, or homogeneous and isotropic perfect fluids, which allows us to obtain analytical solutions. However, the accuracy of these approximations may not always be good enough. This is evident in the case of cosmological perturbations and large-scale structure. At which scales is it valid to use a Friedmann-Lemaître-Robertson-Walker metric? Can the backreaction explain some or all of the observed cosmic acceleration? For smaller systems such as aspherical collapsing bodies, we still need accurate models in order to match theory and observation. Such questions have spurred the development of numerical methods in General Relativity. Physicists now have the necessary computing power to go far beyond simple first-order approximations.

The remaining open questions are of a cosmological nature: Dark Matter and dark energy, and inflation. Dark matter and dark energy account for around 25% and 70% of the mass-energy content of the universe, respectively. They are not a problem for the theory of gravity itself, but it does mean that we do not know the nature of 95% of the content of the universe. Inflation solves a number of cosmological problems, but whatever theory we choose to explain inflation, we still need to introduce new physics beyond GR.

Table 6. The major predictions of GR, and their experimental status.

Effect	Milestones	Current, Future and Proposed Experiments	Status
Mass equivalence	Galileo [366,473] Eötvös [367]	Eöt-Wash Group [371] Lunar Laser Ranging [343] MICROSCOPE[474] STEP [475] Galileo Galilei satellite [476]	Confirmed
Gravitational time dilation [478]	Eddington solar eclipse [4] Pound-Rebka experiment [479] Space-borne hydrogen masers [481]	Quantum interference of atoms [477] ACES [480] Galileo 5 and 6 satellites[344] Einstein Gravity Explorer [482]	Confirmed
Precession of orbits	Orbit of Mercury (Einstein [3])	Binary pulsar observations [483–485] Solar System and extrasolar planets [337,402]	Confirmed
De Sitter precession	Gravity Probe B [486] Lunar laser ranging [488–490] Binary pulsars[493,494]	Binary pulsars [487] Improved lunar laser ranging [491,492]	Confirmed
Lense-Thirring precession	Gravity Probe B [486] LARES [420,421,496]	LARASE [495] Laboratory tests [497] Solar System bodies [339] Binary pulsars [420,421] Black holes [478,496,498,499]	Confirmed
Gravitational waves	LIGO [80]	Advanced LIGO [500] eLISA https://www.elisascience.org	Recently confirmed in two events [80,501]
Strong field effects	PSR J0348 + 0432 [483]	Black holes [411,502] Binary pulsars [117]	Recently confirmed
Orbital precession due to oblateness of central body		Low-orbit satellites [503–506] Stars orbiting black holes [506] Juno spacecraft around Jupiter [507]	Not yet observed

17. Future Developments

The current Concordance Model of cosmology was built in successive (and sometimes concurrent) steps. General Relativity applied to a spacetime under the Copernican Principle, filled with pressureless matter, produced the Einstein-de Sitter model. Motivated by the observed Hubble expansion, it resulted in the Big Bang model. Clues from element abundances, baryon assymmetry, and knowledge of nucleosynthesis from the Standard Model of particle physics meant that the universe had to have a thermal history, which resulted in the Hot Big Bang Model. When evidence for missing mass became incontrovertible, cold Dark Matter had to be added to the inventory of cosmic components. This model worked well, but not well enough. It could not explain the observed homogeneity of the universe across regions which were causally disconnected, nor its flatness. So Inflation was introduced. Observations of an accelerated cosmic expansion motivated a search for explanations within the then current paradigm, which resulted in various hypotheses: a curved geometry, supermassive neutrinos, or perhaps a particular cosmological topology. In the end, the paradigm had to be shifted yet again with the introduction of dark energy.

The Concordance Model can explain the observations with just six parameters: the physical baryon density parameter $\Omega_b h^2$, where h is the Hubble parameter, the physical Dark Matter density parameter $\Omega_c h^2$, the age of the universe t_0, the scalar spectral index n_s, the curvature fluctuation amplitude Δ^2_R, and the reionisation optical depth τ. That such a degree of fit is offered by such a simple model is remarkable.

The success of the Concordance Model has been its ability to include physical effects at extremely different scales, from primordial nucleosynthesis to large-scale structure evolution, in one coherent theory. However, this does not allow us to state that the ΛCDM model is correct. It merely implies that deviations from ΛCDM are too small compared to the current observational uncertainties to be inferred from cosmological data alone. This leaves room for some very fundamental open questions, which we have described in this review.

The science of cosmology finds itself at a critical point where it has to make sense of the vast quantity of data that has become available. Different probes have allowed us to piece together interlocking information which, so far, confirms the Concordance Model. The cosmic microwave background has provided conclusive evidence of a flat geometry, super-horizon features, the correct harmonic peaks, adiabatic fluctuations, Gaussian random fields, and most recently, a departure from scale invariance. We have not yet observed primordial inflationary gravitational waves. Large-scale structure observations, which probe the recent universe, provide firm evidence in favour of the Concordance Model's explanation of the evolution of density perturbations and the growth of structure, and provide a bridge between the effects of long-range gravitational interactions and shorter-range forces.

The recent detection of gravitational waves in two events (possibly three), one hundred years after they were predicted by Einstein, directly validates General Relativity in several ways. It shows that GR is correct in the strong-field regime, that black holes really exist, and black hole binaries too, and it proves that gravitational waves are a real physical phenomenon, and not just a mathematical artefact of GR.

We are fast approaching the point where cosmological observations will be limited only by cosmic variance, i.e., no more data will be available from our Hubble patch [508]. What does that imply for the development of new theories? How do we test the predictivity of these theories without new data?

We will have to look for new effects in old data, and for new correlations in future data. Large-scale structure is extremely useful in testing General Relativity and cosmological models, but the future may bring other observational windows. In particular, the next fifty to one hundred years may see the development of a gravitational wave astronomy. This is likely to be an even more significant development than even CMB astronomy. The observation of primordial gravitational waves would provide vital information on the inflationary epoch [509,510]. B-mode polarisation of the CMB offers an indirect pathway to the observation of this gravitational wave background [511,512]. The second major development could be the observation of a cosmic neutrino background [513], which is the result of neutrino decoupling in the lepton era. This would push the observations along our past light cone even further back in redshift, providing information on the universe before recombination and the CMB.

17.1. Plausible Conclusions from Incomplete Information

The statistical questions facing cosmologists pose some particular problems. We observe a finite region of our universe, which is itself a single realisation of the cosmological theory. We can only observe whatever is on or inside our past light cone, as shown in Figure 1. Not only are we limited by cosmic variance, we also have just a single data point for the cosmological model.

There exist several alternatives to GR and to ΛCDM that have not been ruled out by experiment. Constructing viable physical models is not just a question of fitting the model to the data. It is a question of model selection, which requires robust statistical techniques that allow us to make sensible decisions using our incomplete information. Bayesian inference provides a quantitative framework for plausible conclusions [514,515]. We can identify three levels of Bayesian inference.

(1) Parameter inference (estimation). We assume that a model M is true, and we select a prior for the parameters $\boldsymbol{\theta}$, or the $\mathrm{Prob}(\boldsymbol{\theta}|M)$.
(2) Model comparison. There are several possible models M_i. We find the relative plausibility of each in the light of the data D, that is we calculate the ratio $\mathrm{Prob}(D|M_i)/\mathrm{Prob}(D|M_0)$.
(3) Model averaging. There is no clear evidence for a best model. We find the inference on the parameters which accounts for the model uncertainty.

At the first level of Bayesian inference, we can estimate the allowed parameter values of the theory. If we assume General Relativity as our theory, we still need to fix the values of the various constants. This is the rationale behind ongoing efforts to measure quantities like Newton's constant G ever more accurately, despite the fact that the parameter has been around for three centuries. What are the energy densities of the various components of a ΛCDM universe? If we include a dark energy equation of state parameter w, what value does it take?

Next, we can ask which parameters we should include in the theory. Should we include a cosmological constant in General Relativity? Or should we include dynamical dark energy parameter? Although current data are consistent with the six-parameter ΛCDM model based on GR, there are more than twenty candidate parameters which might be required by future data (see [516]). We cannot simply include all possible parameters to fit the data, since each one will give rise to degeneracies that weaken constraints on other parameters, including the ΛCDM parameter set (e.g., [517–519]). The landscape of alternative cosmological models is even larger if we relax our assumptions on the theory of gravity [520].

The goal in data analysis is usually to decide which parameters need to be included in order to explain the data. For physicists, those extra parameters must be physically motivated. That is, we need to know the physical effects to which our data are sensitive, so that we can relate these effects to physics. At the current state of knowledge, we have to acknowledge the possibility of more than one model. We therefore require a consistent method to discard or include parameters. This is the second level of Bayesian inference—model selection.

Bayesian model selection penalises models which introduce wasted parameter space. We can always construct a theory that fits the data perfectly, even better than GR, but we would need to introduce extra free parameters (e.g., extra fields or couplings between matter and the metric). This is the mathematical equivalent of Occam's razor. We seek a balance between goodness of fit (the degree of complexity) and predictive power (consistency with prior knowledge). General Relativity fits all the data with the minimum number of parameters. In cosmology, ΛCDM is the best model because it only involves one new parameter and no new fields.

The problem of model selection in relation to GR is as old as the theory itself. But only recently have cosmologists have started to use Bayesian methods for cosmological model selection (e.g.,

[521–524]), when the astrophysical data began to have the necessary statistical power to enable model testing. Bayesian techniques are starting to be applied to General Relativity itself [525]. With the next generation of astrophysical probes in the pipeline, model selection is likely to grow in importance [514].

Bayesian model selection cannot be completely free of assumptions. In cosmology, there is some model structure which depends on a number of unverifiable hypotheses about the nature of the universe. The Copernican Principle is one such hypothesis [9,526,527].

The third level of Bayesian inference is model averaging. In the current scenario, there is firm evidence for General Relativity as the best model for gravitational interactions. However, it is still useful to quantify our degree of certainty (or doubt), for one simple reason: we do not have the final list of alternative theories. In other words, when we choose GR against any number of alternative theories, we have no knowledge about other alternatives outside that list (such as theories yet to be developed, for example). At best, we know which alternatives are ruled out by the data. This level of Bayesian inference is the application of a principle that has been called 'Cromwell's Rule' [528]: even if all the data show our theory to be correct, we should allow a non-zero probability, even if tiny, that the theory is false.

The utility of alternative theories becomes evident when we apply Bayesian model selection. For science to advance by falsification, it is not enough to claim that the present theory is false. We need to know which alternative theory is favoured instead. Newtonian gravity would have likely have survived the 1919 eclipse if General Relativity had not been formulated.

17.2. Experimental Progress

There has been rapid progress in constraining cosmological parameters and models over the last two decades, with a multitude of experiments observing the CMB, large-scale structure, galaxies and supernovae. We will just provide a summary of the most recent data sets.

The first are anisotropies in the CMB, where the main statistic is the angular power spectrum of fluctuations C_ℓ, and polarisation of the CMB. The most recent and current are: WMAP (http://map.gsfc.nasa.gov), *Planck* (http://www.cosmos.esa.int/web/planck), Atacama Cosmology Telescope (ACT) (http://act.princeton.edu), South Pole Telescope (SPT) (https://pole.uchicago.edu), Atacama Cosmology Telescope polarisation-sensitive receiver (ACTPol) [529], SPTPol, Spider (http://spider.princeton.edu), Polarbear (http://bolo.berkeley.edu/polarbear), Background Imaging of Cosmic Extragalactic Polarization (BICEP2) (http://bicepkeck.org,https://www.cfa.harvard.edu/CMB/bicep2), Keck Array (http://bicepkeck.org,https://www.cfa.harvard.edu/CMB/keckarray).

The second source of data are surveys cataloguing the angular positions and redshifts of individual galaxies, leading to the power spectrum of fluctuations $P(k,z)$, or the two-point correlation function $\xi(r)$. The recent and current experiments are: Baryon Oscillation Spectroscopic Survey (BOSS) using Sloan Digital Sky Survey (SDSS) data (http://www.sdss3.org/surveys/boss.php), Dark Energy Survey (DES) (https://www.darkenergysurvey.org), Weave (http://www.ing.iac.es/weave/science.html), Hobby-Eberly Telescope Dark Energy Experiment (HETDEX) (http://www.hetdex.org), Extended Baryon Oscillation Spectroscopic Survey (eBOSS) (https://www.sdss3.org/future/eboss.php), Mid-Scale Dark Energy Spectroscopic Instrument (MS-DESI) (https://www.skatelescope.org), Canadian Hydrogen Intensity Mapping Experiment (CHIME) (http://chime.phas.ubc.ca), Baobab, MeerKAT (http://www.ska.ac.za/science-engineering/meerkat), and ASKAP (http://www.atnf.csiro.au/projects/askap/index.html).

The third source of data are weak lensing, which use the fact that images of distant galaxies are distorted and correlated by intervening gravitational potential wells to produce statistics such as the convergence power spectrum C_ℓ^κ [530,531]. Some current experiments are: Dark Energy Survey (DES) (https://www.darkenergysurvey.org), Red Cluster Sequence Lensing Survey (RCSLens) (http://www.rcslens.org), Canada-France Hawaii Telescope Lensing Survey (CFHTLenS) (http://www.cfhtlens.org), New Instrument of Kids Arrays (NIKA2) (http://ipag.osug.fr/nika2), and Hyper Suprime-Cam (HSC) (http://hsc.mtk.nao.ac.jp/ssp).

The final source of data are catalogues of peculiar velocities. By measuring redshifts and radial distances of galaxies and clusters it is possible to reconstruct a radially projected map of large-scale motions. Progress in this field will come from all three data sets above.

The main science goal of the next generation of cosmological probes is to test the Concordance Model of cosmology. Some major experiments will be operational in the next decade.

Planck, decommissioned in 2013, marked a major milestone in CMB experiments. Its proposed successor is the ground-based programme CMB-S4 [532], which should reach sensitivities below $10^{-3}\mu$ K whose main aim is to achieve higher resolutions, probe larger scales, and measure new observables such as polarisation.

There are various future experiments to map the mass-energy content of the universe (including baryons and Dark Matter), either in the planning phase, or close to completion. The Large Synoptic Space Telescope (LSST) (https://www.lsst.org), which should achieve first light in 2019, is a ground-based telescope which will map the entire sky.

The *Euclid* space telescope (http://www.euclid-ec.org), due for launch in 2020, will map galaxies and large-scale structure over the whole sky at visible and near-infrared wavelengths, providing a catalogue of 12 billion sources at 50 million . *Euclid* will probe the recent universe, when galaxies have formed and dark energy starts to dominate. Its main scientific objective is to understand the origin of the accelerated expansion of the universe by probing the nature of dark energy using weak-lensing observables (which include cosmic shear, higher-order distortions, and cosmic magnification), and galaxy-clustering observations [225].

The Square Kilometre Array (SKA) (https://www.skatelescope.org), which will begin operations in 2020, is a multi-wavelength radio telescope, built across multiple sites to achieve the largest collecting area ever [70,533]. It should provide the highest resolution images of the radio sky, thus providing maps of large-scale structure, and observing pulsars which should provide direct tests of General Relativity. The SKA will observe the epoch between the emission of the CMB and the formation of the first galaxies. Neutral hydrogen surveys (or 21 cm intensity mapping) [534], offer yet another promising probe, as do galaxy redshift surveys [535].

Joint observables will be key in the next generation of experiments. These include the Sachs-Wolfe effect, and the Sunyaev-Zel'dovich effect [536]. We will also need to extract more observables from the CMB. We need information on in order to constrain inflationary models.

As experiments probe larger scales at better resolutions (low ℓ and high ℓ), the data analysis and the cosmological tests (dark energy, Dark Matter, the properties of cosmic neutrinos) will require accurate calculations on the growth of large-scale structure, which can only be achieved using N-body simulations [537–544].

At the other end of the length scale, there is particle physics. The Very Large Hadron Collider, the successor to the Large Hadron Collider (https://home.cern/topics/large-hadron-collider) (which can achieve energies around 14 TeV), is still in its conceptual phase. It could, if built, probe energies around 100 TeV, allowing it to test physics beyond the Standard Model, possibly including Supersymmetry and Grand Unified Theories, and thus provide clues on the nature of Dark Matter and dark energy (see, e.g., [545]).

Tremendous progress is also being made in Milky Way astrophysics. With the launch on 2013 of the Gaia mission (http://www.cosmos.esa.int/web/gaia), we will soon have improved data on Solar System ephemerides, and on the orbits and tidal streams in the Milky Way. This will allow precise tests of GR at Galactic scales. In particular, tidal streams provide an opportunity to probe GR at late cosmological times and to close the gap between astronomical and cosmological scales.

Now that gravitational waves have been detected, gravitational wave physics is set to become an established branch of astrophysics. The Laser Interferometer Gravitational-Wave Observatory (LIGO) (https://www.ligo.caltech.edu), Advanced LIGO (aLigo) (https://www.advancedligo.mit.edu) and Virgo (http://www.virgo-gw.eu) will subject GR to a battery of test in the strong regime. Also targeting the regime of strong curvatures and potentials, the planned Event Horizon Telescope (http://www.eventhorizontelescope.org) is a network of millimetre and sub-millimetres telescopes being used for very-long baseline interferometry to directly image supermassive black holes in galactic centres.

17.3. Theoretical and Computational Progress

On the theoretical front, there are four main lines of development in theories of gravity. They are all motivated by current open questions in physics. The ongoing attempt to find a Grand Unified Theory continues to motivate the development of string theory [546] as a framework for gravity. Aside from the theoretical difficulties of a mathematically complex theory, the challenge for string theory is to produce physical predictions which can be experimentally tested. The second approach is brane theories or supergravity, in which spacetime has more than four dimensions. The resulting field theory combines Supersymmetry (from particle physics) and General Relativity [547]. The third approach is quantum gravity [548], in which spacetime, as a dynamical field, is a quantum object. This implies a violation of Lorentz Invariance near the Planck scale, which in turn means that some particle decays forbidden by Special Relativity are allowed, and possibly charge-parity-time violations too. This motivates the search for signatures of quantum gravity in particle physics experiments.

The final approach, which is closer to the theoretical framework of the Concordance Model of cosmology, is a phenomenological use of General Relativity. This includes the various alternative theories that seek to explain cosmic acceleration and the missing mass, and also the various theories which provide dark energy and Dark Matter candidates.

The early alternatives to General Relativity were motivated by theoretical considerations. The current alternatives are mainly motivated by the open questions in cosmology. Cosmology is now in the age of Big Data. In the last decade, the data finally caught up with the theory, and we are now in a position to test many of the current alternative theories using statistical techniques.

The principles that underpin statistical techniques are simple enough. Under minimal assumptions about signal and noise, it is simply a question of maximising the Gaussian likelihoods. However, in practice they are extremely complex. The foreground contamination, the signal, the possibly non-Gaussian noise and the systematics all have to be modelled. This requires data simulation, so the number of maps that is generated is far larger than the number that is actually observed by the instrument (see, e.g., [549–554]).

A low signal-to-noise ratio, and higher-resolution observations of fainter signals over a larger frequency range, result in massively big data sets. For statistical probes such as the CMB and large-scale structure, these data sets have to be analysed as a whole, in order to correlate data. In addition to the volume of data collected, this generates a huge amount of computational data, and requires the appropriate computing power to carry out multiple complex calculations. We are witnessing a Moore's Law in cosmology, where the data volume of experiments increases by a factor of around 1000 every ten years, as shown in Figure 8.

In addition to the data volume challenge, we have an algorithmic challenge. The number of computations for cosmological data analysis depends on four main factors: the number of observations N_t, the number of pixels N_p, the number of multipoles (a function of the resolution and frequency range) N_ℓ, and the number of iterations N_i. The first three quantities determine the data volume. The data simulation scales at least as $\mathcal{O}(N_\mathrm{t})$. The map-making scales as $\mathcal{O}(N_\mathrm{i} N_\mathrm{t} \log N_\mathrm{t})$. The maximum-likelihood power spectrum estimation scales as $\mathcal{O}(N_\mathrm{i} N_\mathrm{l} N_\mathrm{p}^3)$.

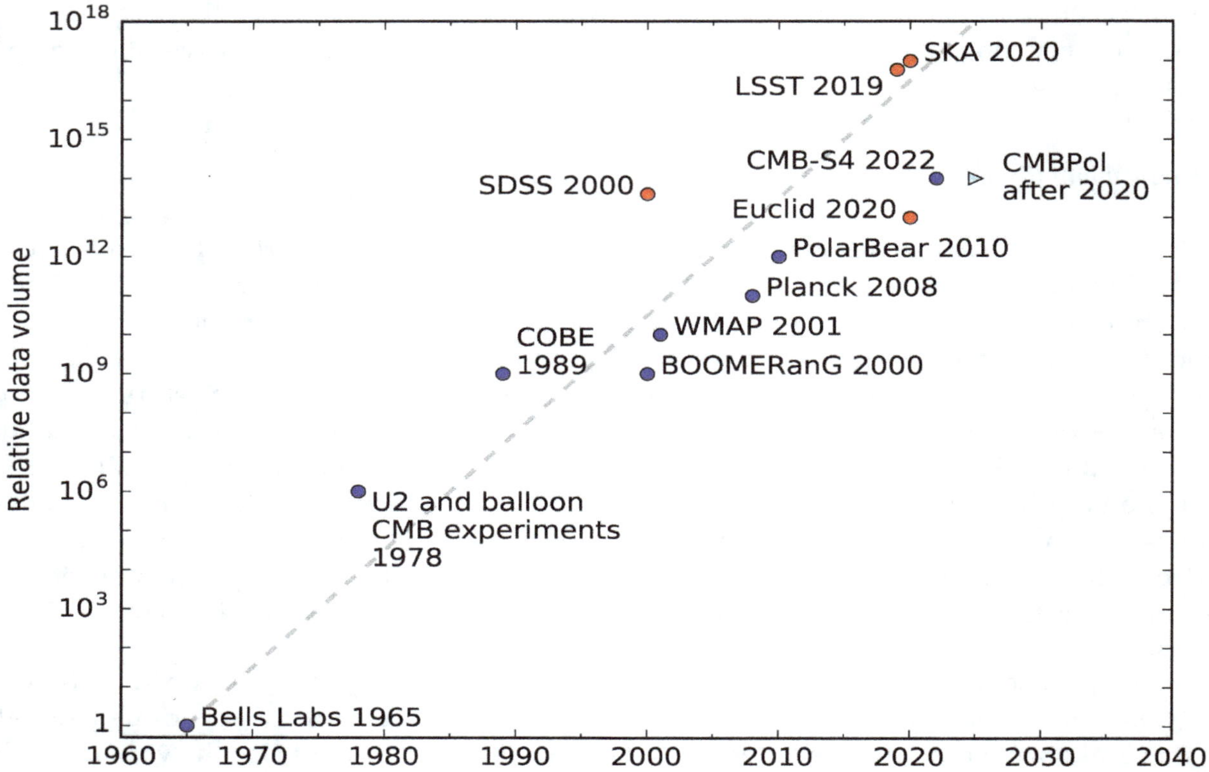

Figure 8. The growing data volume of experiments. The data volume of each experiment is shown as an order-of-magnitude multiple of the data volume of the 1965 Bells Labs experiment which detected the CMB. The labels show the year of 'first light' for each experiment. CMB surveys are marked by blue dots, while red dots show large-scale structure surveys. A first light date for CMBPol (light blue triangle) has not yet been fixed. Note that the vertical axis is logarithmic: the date volume increases about a thousandfold every ten years (grey dashed line). Note too that we plot CMB, large-scale structure, space and ground-based probes on the same graph. Ground-based probes will always tend to have a larger data volume than space-borne probes, due to the bandwidth limit on data transmission from spacecraft. Longer-running experiments will also have a larger data volume.

Planck marked a milestone in data science. It was the first CMB experiment in which the whole data treatment process was parallelised, and where Monte Carlo methods were used in order to cut down on the number of data realisation iterations that were carried out.

There are two main considerations in data analysis: the amount of data, and the complexity of the theory. Future experiments will require sophisticated techniques, and considerable computing power to process the vast amounts of data. Gaia is already collecting data [555], while Euclid, the SKA, Enhanced LIGO, and CMB-S4 will soon be operational. Farther ahead, the Evolved Laser Interferometer Space Antenna (eLISA) (https://www.elisascience.org), due for launch in 2034, will also generate huge volumes of data, and will require the necessary computing power to test GR directly, by comparing the data to simulated black hole collapse, inspiralling binaries.

What direction will experimental cosmology take over the coming decade? Extracting fainter signals, such as CMB polarisation, or going to very high resolutions requires larger data volumes to provide a higher signal-to-noise ratio, and it requires more complex models to control fainter systematic effects. Even the Solar System tests of General Relativity will depend upon vastly greater data volumes and computational complexity to get the full relativistic ephemerides, given the ever-increasing number of objects being tracked in the Solar System and the significantly greater precision of the data for each object.

This enables us to make a sensible prediction on future developments. The science we are able to extract from present and future data sets will be determined by the limits on our computational capability, and our ability and willingness to exploit it.

17.4. Conclusions

GR may well survive for another 100 years. After all, Newtonian gravity was around for 200 years. GR has just reached its peak, when data and computing power have caught up with the theory. We are at a pivotal moment in the history of GR. We are on the point of confirming beyond reasonable doubt all its predictions throughout its entire domain of validity.

We have seen how modern cosmology is faced with big questions which touch the very foundations of physics. What is this form of matter which interacts only with gravity and apparently with nothing else? Why is the expansion of the universe accelerating? What caused the universe to undergo a period of rapid expansion soon after the Big Bang? These questions, motivated by cosmological observations, lead to questions about fundamental physics. Are there forces and interactions besides the four we know of, that is, gravity, electromagnetism, and the strong and weak nuclear forces? Are there particles beyond the Standard Model? What determines the value of the fundamental constants of nature? What is the real structure of spacetime? Are there extra dimensions?

Science needs data, so each of these questions must be addressed through careful experiment. The challenge of modern experimental physics is to probe nature at extreme distances and energies, well outside the capabilities of the instruments that were available to Einstein. It has certainly come a long way, as shown by the detection of gravitational waves in 2015, a feat which was thought to be impossible by many of Einstein's contemporaries.

General Relativity is not the final theory of gravity, for there is no such thing. As General Relativity turns 100, we would do well to celebrate it with a healthy does of scientific scepticism. Long live General Relativity, and a big welcome to its eventual replacement, whether in our lifetime or not.

Acknowledgments: George F. Smoot acknowledges support through his Chaire d'Excellence Université Sorbonne Paris Cité and the financial support of the UnivEarthS Labex programme at Université Sorbonne Paris Cité (ANR-10-LABX-0023 and ANR-11-IDEX-0005-02). I.D. acknowledges that the research work disclosed in this publication is partially funded by the REACH HIGH Scholars Programme—Post-Doctoral Grants. The grant is part-financed by the European Union, Operational Programme II—Cohesion Policy 2014–2020.

Author Contributions: All the authors conceived the idea and contributed equally.

Abbreviations

The following abbreviations are used in this work:

AU	Astronomical Unit
CDM	Cold Dark Matter
CMB	Cosmic Microwave Background
EEP	Einstein Equivalence Principle
ECKS	Einstein-Cartan-Kibble-Sciama
eV	electronvolt
FLRW	Friedmann-Lemaître-Robertson-Walker
GR	General Relativity
Gy	Gigayear (10^9 years)
ΛCDM	Λ Cold Dark Matter
LLI	Local Lorentz invariance
LLR	Lunar laser ranging
LPI	Local position invariance
Mpc	Megaparsec
PPN	Parameterised post-Newtonian
SR	Special Relativity
TeV	teraelectronvolt (10^{12} electronvolts)

References

1. Feynman, R.; Leighton, R.B.; Sands, M.L. *The Feynman Lectures on Physics*; Addison-Wesley: Redwood City, CA, USA, 1989; Volume 2.
2. Gauthier-Lafaye, F.; Holliger, P.; Blanc, P.L. Natural fission reactors in the Franceville basin, Gabon: A review of the conditions and results of a "critical event" in a geologic system. *Geochim. Cosmochim. Acta* **1996**, *60*, 4831–4852.
3. Einstein, A. Erklärung der Perihelbewegung des Merkur aus der allgemeinen Relativitätstheorie. *Sitzungsber. Preuss. Akad. Wiss.* **1915**, *1915*, 831–839. (In German)
4. Dyson, F.W.; Eddington, A.S.; Davidson, C. A Determination of the deflection of light by the sun's gravitational field, from observations made at the total eclipse of 29 May 1919. *Philos. Trans. R. Soc. Lond. Ser. A* **1920**, *220*, 291–333.
5. Will, C.M. The 1919 measurement of the deflection of light. *Class. Quantum Gravity* **2015**, *32*, 124001.
6. Coles, P. Einstein, Eddington and the 1919 Eclipse. In *Historical Development of Modern Cosmology*; Martínez, V.J., Trimble, V., Pons-Bordería, M.J., Eds.; Astronomical Society of the Pacific: San Francisco, CA, USA, 2001; Volume 252, p. 21.
7. Popper, K.R. *Logik der Forschung*; Mohr Siebeck: Tübingen, Germany, 1934.
8. Lahav, O.; Massimi, M. Dark energy, paradigm shifts, and the role of evidence. *Astron. Geophys.* **2014**, *55*, 3.12–3.15.
9. Ellis, G.F.R. Cosmology and verifiability. *Q. J. R. Astron. Soc.* **1975**, *16*, 245–264.
10. Crombie, A.C. Quantification in medieval physics. *Isis* **1961**, *52*, 143–160.
11. Bennett, J. Early modern mathematical instruments. *Isis* **2011**, *102*, 697–705.
12. Franklin, A. Principle of inertia in the middle ages. *Am. J. Phys.* **1976**, *44*, 529–543.
13. Sorabji, R. *Matter, Space and Motion: Theories in Antiquity and Their Sequel*; Duckworth: London, UK, 1988.
14. Cushing, J.T. *Philosophical Concepts in Physics*; Cambridge University Press: Cambridge, UK, 1998.
15. Grant, E. Scientific Thought in Fourteenth-Century Paris: Jean Buridan and Nicole Oresme. *Ann. N. Y. Acad. Sci.* **1978**, *314*, 105–126.
16. Babb, J. Mathematical concepts and proofs from nicole oresme: Using the History of calculus to teach mathematics. *Sci. Educ.* **2005**, *14*, 443–456.
17. Transtrum, M.K.; Machta, B.B.; Brown, K.S.; Daniels, B.C.; Myers, C.R.; Sethna, J.P. Perspective: Sloppiness and emergent theories in physics, biology, and beyond. *J. Chem. Phys.* **2015**, *143*, 010901.
18. Galilei, G. *Sidereus Nuncius Magna, Longeque Admirabilia Spectacula Pandens*; Tommaso Baglioni: Venice, Italy, 1610.
19. Perryman, M. The history of astrometry. *Eur. Phys. J. H* **2012**, *37*, 745–792.
20. Newton, I. *Philosophiae Naturalis Principia Mathematica*, 1st ed.; Joseph Streater: London, UK, 1687.
21. De Maupertuis, P.L.M. Les Loix du mouvement et du repos déduites d'un principe metaphysique. In *Histoire de l'Académie Royale des Sciences et des Belles Lettres*; Académie Royale des Sciences et des Belles Lettres de Berlin: Berlin, Germany, 1746.
22. Euler, L. Réfléxions sur quelques loix générales de la nature qui s'observent dans les effets des forces quelconques. *Acad. R. Sci. Berl.* **1750**, *4*, 189–218.
23. Lagrange, L. *Méchanique Analytique*; Chez La Veuve Desaint: Paris, France, 1788.
24. Hamilton, W.R. On a general method in dynamics. *Philos. Trans. R. Soc. Lond.* **1834**, *124*, 247–308.
25. Hamilton, W.R. Second essay on a general method in dynamics. *Philos. Trans. R. Soc. Lond.* **1835**, *125*, 95–144.
26. Galle, J.G. Account of the discovery of Le Verrier's planet Neptune, at Berlin, Sept. 23, 1846. *Mon. Not. R. Astron. Soc.* **1846**, *7*, 153.
27. Danjon, A. Le centenaire de la découverte de Neptune. *Ciel Terre* **1946**, *62*, 369.
28. Lexell, A.J. Recherches sur la nouvelle planete, decouverte par M. Herschel & nominee Georgium Sidus. *Acta Acad. Sci. Imp. Petropolitanae* **1783**, *1*, 303–329.
29. Sheehan, W. News from Front (of the Solar System): The problem with Mercury, the Vulcan hypothesis, and General Relativity's first astronomical triumph. In Proceedings of the 227th AAS Meeting, Kissimmee, FL, USA, 4–8 January 2016; Volume 227.
30. Poincaré, H. Sur le problème des trois corps et les équations de la dynamique. *Acta Math.* **1890**, *13*, 1–270.

31. Poincaré, H. *Thorie Mathématique de la Lumiére*; Carré & C. Naud: Paris, France, 1889.
32. Maxwell, J.C. A dynamical theory of the electromagnetic field. *Philos. Trans. R. Soc. Lond.* **1865**, *155*, 489–512.
33. Einstein, A. Zur Elektrodynamik bewegter Körper. *Ann. Phys. (Germany)* **1905**, *17*, 891–921.
34. Einstein, A. Über den Einfluss der Schwerkraft auf die Ausbreitung des Lichtes. *Ann. Phys. (Germany)* **1911**, *35*, 898–908.
35. Einstein, A. Lichtgeschwindigkeit und Statik des Gravitationsfeldes. *Ann. Phys. (Germany)* **1912**, *38*, 355–369.
36. Einstein, A. Theorie des statischen Gravitationsfeldes. *Ann. Phys. (Germany)* **1912**, *38*, 443–458.
37. Einstein, A.; Grossmann, M. Entwurf einer verallgemeinerten Relativitätstheorie und eine Theorie der Gravitation. I. Physikalischer Teil von A. Einstein II. Mathematischer Teil von M. Grossmann. *Z. Math. Phys.* **1913**, *62*, 225–244 and 245–261.
38. Le Verrier, U. Lettre de M. Le Verrier à M. Faye sur la théorie de Mercure et sur le mouvement du périhélie de cette planète. *C. R. Hebd. Séances L'Acad. Sci.* **1859**, *49*, 379–383.
39. Will, C.M. The confrontation between general relativity and experiment. *Living Rev. Relativ.* **2014**, *17*, 4.
40. Fierz, M.; Pauli, W. Relativistic wave equations for particles of arbitrary spin in an electromagnetic field. *Proc. R. Soc. Lond.* **1939**, *173*, 211–232.
41. Schild, A. Time. *Tex. Q.* **1960**, *3*, 42–62.
42. Schild, A. Gravitational Theories of the Whitehead type and the principle of equivalence. In *Evidence for Gravitational Theories*; Møller, C., Ed.; Academic Press: New York, NY, USA, 1962.
43. Schild, A. Lectures on General Relativity Theory. In *Relativity Theory and Astrophysics: I, Relativity and Cosmology: II, Galactic Structure: III, Stellar Structure*; Ehlers, J., Ed.; American Mathematical Society: Providence, RI, USA, 1967.
44. Riemann, B. *Bernhard Riemann's Gesammelte mathematische Werke und wissenschaftlicher Nachlass*, 1st ed; Weber, H., Ed.; Teubner: Leipzig, Germany, 1876.
45. Riemann, B. Ueber die Hypothesen, welche der Geometrie zu Grunde liegen. *Abh. Kön. Ges. Wiss. Gött.* **1868**, *13*, 133–150.
46. Einstein, A. Die Feldgleichungen der Gravitation. *Sitzungsber. Preuss. Akad. Wiss.* **1915**, *1915*, 844–847. (In German)
47. Einstein, A. Zur Allgemeinen Relativitätstheorie. *Sitzungsber. Preuss. Akad. Wiss.* **1915**, *1915*, 799–801. (In German)
48. Ni, W.T. Theoretical Frameworks for Testing Relativistic Gravity. IV. A Compendium of Metric Theories of Gravity and Their Post-Newtonian Limits. *Astrophys. J.* **1972**, *176*, 769–796.
49. Thorne, K.S.; Ni, W.T.; Will, C.M. Theoretical frameworks for testing relativistic gravity: A review. In Proceedings of the Conference on Experimental Tests of Gravitational Theories, Pasadena, CA, USA, 11–13 November 1970; pp. 10–31.
50. Lovelock, D. The uniqueness of the Einstein field equations in a four-dimensional space. *Arch. Ration. Mech. Anal.* **1969**, *33*, 54–70.
51. Lovelock, D. The Einstein Tensor and Its Generalizations. *J. Math. Phys.* **1971**, *12*, 498–501.
52. Lovelock, D. The Four-Dimensionality of Space and the Einstein Tensor. *J. Math. Phys.* **1972**, *13*, 874–876.
53. Cartan, E. Sur les équations de la gravitation d'Einstein. *J. Math. Pures Appl.* **1922**, *1*, 141–204.
54. Anderson, J.L. *Principles of Relativity Physics*; Academic Press: New York, NY, USA, 1967.
55. Norton, J.D. General covariance, gauge theories and the Kretschmann objection. In *Symmetries in Physics: Philosophical Reflections*; Brading, K.A., Castellani, E., Eds.; Cambridge University Press: Cambridge, UK, 2003.
56. Kretschmann, E. Über den physikalischen Sinn der Relativitätspostulate, A. Einsteins neue und seine ursprungliche Relativitätstheorie. *Ann. Phys. (Germany)* **1917**, *53*, 575–614.
57. Nordstrøm, G. Zur Theorie der Gravitation vom Standpunkt des Relativitäsprinzip. *Ann. Phys. (Germany)* **1913**, *42*, 533–554.
58. Einstein, A.; Fokker, A.D. The North current gravitation theory from the viewpoint of absolute differential calculus. *Ann. Phys. (Germany)* **1914**, *44*, 321–328.
59. Whitehead, A.N. *The Principle of Relativity*; Cambridge University Press: Cambridge, UK, 1922.
60. Will, C.M. Relativistic gravity in the solar system, II: Anisotropy in the Newtonian gravitational constant. *Astrophys. J.* **1971**, *169*, 141–156.

61. Birkhoff, G.D. Matter, electricity and gravitation in flat spacetime. *Proc. Natl. Acad. Sci. USA* **1943**, *29*, 231–239.
62. Jordan, P. Zum gegenwärtigen Stand der Diracschen kosmologischen Hypothesen. *Z. Phys.* **1959**, *157*, 112–121.
63. Brans, C.; Dicke, R.H. Mach's principle and a relativistic theory of gravitation. *Phys. Rev.* **1961**, *124*, 925–935.
64. Dicke, R.H. Mach's principle and invariance under transformation of units. *Phys. Rev.* **1962**, *125*, 2163–2167.
65. Will, C.M. *Theory and Experiment in Gravitational Physics*, 2nd ed.; Cambridge University Press: Cambridge, UK, 1993.
66. Psaltis, D. Constraining Brans-Dicke Gravity with Accreting Millisecond Pulsars in Ultracompact Binaries. *Astrophys. J.* **2008**, *688*, 1282–1287.
67. Bertotti, B.; Iess, L.; Tortora, P. A test of general relativity using radio links with the Cassini spacecraft. *Nature* **2003**, *425*, 374–376.
68. Avilez, A.; Skordis, C. Cosmological Constraints on Brans-Dicke Theory. *Phys. Rev. Lett.* **2014**, *113*, 011101.
69. Ooba, J.; Ichiki, K.; Chiba, T.; Sugiyama, N. Planck constraints on scalar-tensor cosmology and the variation of the gravitational constant. *Phys. Rev. D* **2016**, *93*, 122002.
70. Bull, P. Extending Cosmological Tests of General Relativity with the Square Kilometre Array. *Astrophys. J.* **2016**, *817*, 26.
71. Damour, T.; Nordtvedt, K. General relativity as a cosmological attractor of tensor-scalar theories. *Phys. Rev. Lett.* **1993**, *70*, 2217–2219.
72. Damour, T.; Nordtvedt, K. Tensor-scalar cosmological models and their relaxation toward general relativity. *Phys. Rev. D* **1993**, *48*, 3436–3450.
73. Kustaanheimo, P. Route dependence of the gravitational redshift. *Phys. Lett.* **1966**, *23*, 75–77.
74. Kustaanheimo, P.E.; Nuotio, V.S. *Relativistic Theories of Gravitation I: The One-Body Problem*; University of Helsinki: Helsinki, Finland, 1967; unpublished.
75. Einstein, A. Näherungsweise Integration der Feldgleichungen der Gravitation. *Sitzungsber. Preuss. Akad. Wiss.* **1916**, *1916 Pt 1*, 688–696. (In German)
76. Kumar, D.; Soni, V. Single particle Schrödinger equation with gravitational self-interaction. *Phys. Lett. A* **2000**, *271*, 157–166.
77. Deser, S. Gravity from self-interaction redux. *Gen. Relativ. Gravit.* **2010**, *42*, 641–646.
78. Hertz, H. *Untersuchungen über die Ausbreitung der Elektrischen Kraft*; J.A. Barth: Leipzig, Germany, 1894.
79. Einstein, A. Über einen die Erzeugung und Verwandlung des Lichtes betreffenden heuristischen Gesichtspunkt. *Ann. Phys. (Germany)* **1905**, *17*, 132–148.
80. LIGO Collaboration. Observation of Gravitational Waves from a Binary Black Hole Merger. *Phys. Rev. Lett.* **2016**, *116*, 061102.
81. Yang, C.L.; Mills, R.L. Conservation of isotopic spin and isotopic gauge invariance. *Phys. Rev.* **1954**, *96*, 191–195.
82. Yang, C.L.; Mills, R.L. Isotopic Spin Conservation and a Generalized Gauge Invariance. *Phys. Rev.* **1954**, *95*, 631.
83. Weyl, H. Gravitation und Elektrizität. In *Das Relativitätsprinzip*; J. A. Barth: Leipzig, Germany, 1918.
84. Weyl, H. *Raum-Zeit-Materie*; Springer: Berlin, Germany, 1918.
85. Christoffel, E.B. Über die Transformation der homogenen Differentialausdrücke zweiten Grades. *J. Reine Angew. Math.* **1869**, *70*, 46–70.
86. Goenner, H.F.M. On the History of Unified Field Theories. *Living Rev. Relativ.* **2004**, *7*, 2.
87. Smiciklas, M.; Brown, J.M.; Cheuk, L.W.; Smullin, S.J.; Romalis, M.V. New test of local lorentz invariance using a ^{21}Ne-Rb-**K** comagnetometer. *Phys. Rev. Lett.* **2011**, *107*, 171604.
88. Peck, S.K.; Kim, D.K.; Stein, D.; Orbaker, D.; Foss, A.; Hummon, M.T.; Hunter, L.R. Limits on local Lorentz invariance in mercury and cesium. *Phys. Rev. A* **2012**, *86*, 012109.
89. Hohensee, M.A.; Leefer, N.; Budker, D.; Harabati, C.; Dzuba, V.A.; Flambaum, V.V. Limits on violations of lorentz symmetry and the einstein equivalence principle using radio-frequency spectroscopy of atomic dysprosium. *Phys. Rev. Lett.* **2013**, *111*, 050401.
90. Utiyama, R. Invariant theoretical interpretation of interaction. *Phys. Rev.* **1956**, *101*, 1597–1607.
91. Kibble, T.W.B. Lorentz invariance and the gravitational field. *J. Math. Phys.* **1961**, *2*, 212–221.

92. Carmeli, M.; Malin, S. Reformulation of general relativity as a gauge theory. *Ann. Phys. (Germany)* **1977**, *103*, 208–232.
93. Ivanenko, D.; Sardanashvily, G. The gauge treatment of gravity. *Phys. Rep.* **1983**, *94*, 1–45.
94. Sardanashvily, G.; Zakharov, O. *Gauge Gravitation Theory*; World Scientific: Singapore, 1992.
95. Jackson, J.D.; Okun, L.B. Historical roots of gauge invariance. *Rev. Mod. Phys.* **2001**, *73*, 663–680.
96. Sardanashvily, G. Classical gauge theory of gravity. *Theor. Math. Phys.* **2002**, *132*, 1163–1171.
97. Cartan, É. Sur une généralisation de la notion de courboure de Riemann et les espaces à torsion. *C. R. Acad. Sci. (Paris)* **1922**, *174*, 593–595.
98. Cartan, É. Sur les variétés à connexion affine et la théorie de la relativité généralisée (première partie). *Ann. Sci. École Norm. Super.* **1923**, *40*, 325–412.
99. Cartan, É. Sur les variétés à connexion affine et la théorie de la relativité généralisée (suite). *Ann. Sci. École Norm. Super.* **1924**, *41*, 1–25.
100. Cartan, É. Sur les variétés à connexion affine et la théorie de la relativité généralisée (deuxième partie). *Ann. Sci. École Norm. Super.* **1925**, *42*, 17–88.
101. Weyl, H. A remark on the coupling of gravitation and electron. *Phys. Rev.* **1950**, *77*, 699–701.
102. Sciama, D. The physical structure of general relativity. *Rev. Mod. Phys.* **1964**, *36*, 1103.
103. Weyssenhoff, J.; Raabe, A. Relativistic Dynamics of spin-fluids and spin-particles. *Acta Phys. Pol.* **1947**, *9*, 7–18.
104. Costa de Beauregard, O. Translational inertial spin effect. *Phys. Rev.* **1963**, *129*, 466–471.
105. Cosserat, E.; Cosserat, F. *Théorie des Corps déformables*; Hermann: Paris, France, 1909.
106. Puetzfeld, D.; Obukhov, Y.N. Prospects of detecting spacetime torsion. *Int. J. Mod. Phys. D* **2014**, *23*, 1442004.
107. Hehl, F.W.; von der Heyde, P.; Kerlick, G.D.; Nester, J.M. General relativity with spin and torsion: Foundations and prospects. *Rev. Mod. Phys.* **1976**, *48*, 393–416.
108. Lasenby, A.N.; Doran, C.J.; Gull, S.F. Cosmological consequences of a flat-space theory of gravity. In *Clifford Algebras and Their Applications in Mathematical Physics*; Brackx, F., Delanghe, R., Serras, H., Eds.; Kluwer: Dordrecht, The Netherlands, 1993; p. 387.
109. Lasenby, A.N.; Doran, C.J.; Gull, S.F. Grassmann calculus, pseudoclassical mechanics and geometric algebra. *J. Math. Phys.* **1993**, *34*, 3683–3712.
110. Lasenby, A.N.; Doran, C.J.L.; Gull, S.F. A multivector derivative approach to Lagrangian field theory. *Found. Phys.* **1993**, *23*, 1295–1327.
111. Lasenby, A.N.; Doran, C.J.L.; Gull, S.F. 2-spinors, twistors and supersymmetry in the spacetime algebra. In *Spinors, Twistors, Clifford Algebras and Quantum Deformations*; Oziewicz, Z., Jancewicz, B., Borowiec, A., Eds.; Kluwer: Dordrecht, The Netherlands, 1993; p. 233.
112. Lasenby, A.N.; Doran, C.J.L.; Gull, S.F. Astrophysical and cosmological consequences of a gauge theory of gravity. In *Current Topics in Astrofundamental Physics*; Sánchez, N., Zichichi, A., Eds.; World Scientific: Singapore, 1995; p. 359.
113. Lasenby, A.N.; Doran, C.J.; Gull, S.F. Gravity, gauge theories and geometric algebra. *Philos. Trans. R. Soc. Lond.* **1998**, *A356*, 487–582.
114. Carmeli, M.; Leibowitz, E.; Nissani, N. *Gravitation: $SL(2,\mathbb{C})$ Gauge Theory and Conservation Laws*; World Scientific: Singapore, 1990.
115. Mukunda, N. An elementary introduction to the gauge theory approach to gravity. In *Gravitation, Gauge Theories and the Early Universe*; Iyer, B.R., Mukunda, N., Vishvershwara, C.V., Eds.; Kluwer: Dordrecht, The Netherlands, 1989.
116. Misner, C.W.; Thorne, K.S.; Wheeler, J.A. *Gravitation*; Freeman: San Francisco, CA, USA, 1973.
117. Berti, E.; Barausse, E.; Cardoso, V.; Gualtieri, L.; Pani, P.; Sperhake, U.; Stein, L.C.; Wex, N.; Yagi, K.; Baker, T.; et al. Testing general relativity with present and future astrophysical observations. *Class. Quantum Gravity* **2015**, *32*, 243001.
118. Poincaré, H. L'état actuel et l'avenir de la physique mathématique. *Bull. Sci. Math.* **1904**, *2*, 302–324.
119. Kaluza, T. *Zum Unitätsproblem in der Physik*; Sitzungsberichte der Königlich Preußischen Akademie der Wissenschaften (Berlin): Berlin, Germany, 1921; pp. 966–972.
120. Klein, O. Quantentheorie und fünfdimensionale Relativitätstheorie. *Z. Phys.* **1926**, *37*, 895–906.

121. Milne, E.A. *Kinematic Relativity; A Sequel to Relativity, Gravitation and World Structure*; Clarendon Press: Oxford, UK, 1948.
122. Thiry, Y. Sur la régularité des champs gravitationels et électromagnétiques dans les théories unitaires. *C. R. Hebd. Séances Acad. Sci.* **1948**, *226*, 1881–1882.
123. Belinfante, F.; Swihart, J. Phenomenological linear theory of gravitation: Part I. Classical mechanics. *Ann. Phys. (Germany)* **1957**, *1*, 168–195
124. Will, C.M.; Nordtvedt, K., Jr. Conservation laws and preferred frames in relativistic gravity. I. Preferred-frame theories and an extended PPN formalism. *Astrophys. J.* **1972**, *177*, 757–774.
125. Barker, B.M. General scalar-tensor theory of gravity with constant G. *Astrophys. J.* **1978**, *219*, 5–11.
126. Rosen, N. A bi-metric theory of gravitation. *Gen. Relativ. Gravitat.* **1973**, *4*, 435–447.
127. Rosen, N. A bi-metric theory of gravitation. II. *Gen. Relativ. Gravitat.* **1975**, *6*, 259–268.
128. Rastall, P. A theory of gravity. *Can. J. Phys.* **1976**, *54*, 66–75.
129. Buchdahl, H.A. Non-linear lagrangians and cosmological theory. *Mon. Not. R. Astron. Soc.* **1970**, *150*, 1–8.
130. Starobinsky, A. A new type of isotropic cosmological models without singularity. *Phys. Lett. B* **1980**, *91*, 99–102.
131. Milgrom, M. A modification of the Newtonian dynamics as a possible alternative to the hidden mass hypothesis. *Astrophys. J.* **1983**, *270*, 365–370.
132. Milgrom, M. A modification of the newtonian dynamics—Implications for galaxies. *Astrophys. J.* **1983**, *270*, 371–389.
133. Milgrom, M. A modification of the newtonian dynamics—Implications for galaxy systems. *Astrophys. J.* **1983**, *270*, 384–415.
134. Dvali, G.; Gabadadze, G.; Porrati, M. 4D gravity on a brane in 5D Minkowski space. *Phys. Lett. B* **2000**, *485*, 208–214.
135. Bekenstein, J.D. Relativistic gravitation theory for the modified Newtonian dynamics paradigm. *Phys. Rev. D* **2004**, *70*, 083509.
136. Seifert, M.D. Stability of spherically symmetric solutions in modified theories of gravity. *Phys. Rev. D* **2007**, *76*, 064002.
137. Reyes, R.; Mandelbaum, R.; Seljak, U.; Baldauf, T.; Gunn, J.E.; Lombriser, L.; Smith, R.E. Confirmation of general relativity on large scales from weak lensing and galaxy velocities. *Nature* **2010**, *464*, 256–258.
138. Einstein, A. Kosmologische Betrachtungen zur allgemeinen Relativitätstheorie. *Sitzungsber. Preuss. Akad. Wiss.* **1917**, *1917*, 142–152.
139. Friedmann, A. Über die Krümmung des Raumes. *Z. Phys.* **1922**, *10*, 377–386.
140. Friedmann, A. Über die Möglichkeit einer Welt mit konstanter negativer Krümmung des Raumes. *Z. Phys.* **1924**, *21*, 326–332.
141. Hubble, E. A relation between distance and radial velocity among extra-galactic nebulae. *Proc. Natl. Acad. Sci. USA* **1929**, *15*, 168–173.
142. Lemaître, G. Un univers homogène de masse constante et de rayon croissant rendant compte de la vitesse radiale des nébuleuses extra-galactiques. *Ann. Soc. Sci. Brux.* **1927**, *A47*, 49–59.
143. Robertson, H.P. Kinematics and world structure. *Astrophys. J.* **1935**, *82*, 248–301.
144. Robertson, H.P. Kinematics and world structure II. *Astrophys. J.* **1936**, *83*, 187–201.
145. Robertson, H.P. Kinematics and world structure III. *Astrophys. J.* **1936**, *83*, 257–271.
146. Walker, A.G. On Milne's theory of world-structure. *Proc. Lond. Math. Soc.* **1937**, *42*, 90–127.
147. Ellis, G.F.R.; Stoeger, W. Horizons in inflationary universes. *Class. Quantum Gravity* **1988**, *5*, 207–220.
148. Einstein, A.; de Sitter, W. On the relation between the expansion and the mean density of the universe. *Proc. Natl. Acad. Sci. USA* **1932**, *18*, 213–214.
149. Adams, F.C.; Laughlin, G. A dying universe: The long-term fate and evolution of astrophysical objects. *Rev. Mod. Phys.* **1997**, *69*, 337–372.
150. Caldwell, R.R.; Kamionkowski, M.; Weinberg, N.N. Phantom energy: Dark energy with $w < -1$ causes a cosmic doomsday. *Phys. Rev. Lett.* **2003**, *91*, 071301.
151. Planck Collaboration. Planck 2015 Results. XIII. Cosmological Parameters. *Astron. Astrophys.* **2016**, *594*, A13.
152. Alpher, R.A.; Bethe, H.; Gamow, G. The origin of chemical elements. *Phys. Rev.* **1948**, *73*, 803–804.
153. Alpher, R.A.; Herman, R. Evolution of the universe. *Nature* **1948**, *162*, 774–775.

154. Gamow, G. The evolution of the universe. *Nature* **1948**, *162*, 680–682.
155. Gamow, G. The origin of elements and the separation of galaxies. *Phys. Rev.* **1948**, *74*, 505–506.
156. Penzias, A.A.; Wilson, R.W. A measurement of excess antenna temperature at 4080 Mc/s. *Astrophys. J.* **1965**, *142*, 419–421.
157. Christenson, J.H.; Cronin, J.W.; Fitch, V.L.; Turlay, R. Evidence for the 2π decay of the K_2^0 meson. *Phys. Rev. Lett.* **1964**, *13*, 138–140.
158. Sakharov, A.D. Violation of CP invariance, c asymmetry, and baryon asymmetry of the universe. *JETP Lett.* **1967**, *5*, 24–27.
159. Kuzmin, V.A. CP-noninvariance and baryon asymmetry of the universe. *J. Exp. Theor. Phys. Lett.* **1970**, *12*, 228–230.
160. Khlopov, M. Cosmological reflection of particle symmetry. *Symmetry* **2016**, *8*, 81.
161. Fixsen, D.J. The temperature of the cosmic microwave background. *Astrophys. J.* **2009**, *707*, 916–920.
162. Dunkley, J.; Komatsu, E.; Nolta, M.R.; Spergel, D.N.; Larson, D.; Hinshaw, G.; Page, L.; Bennett, C.L.; Gold, B.; Jarosik, N.; et al. Five-year Wilkinson Microwave Anisotropy Probe observations: Likelihoods and parameters from the WMAP data. *Astrophys. J. Suppl.* **2009**, *180*, 306–329.
163. Hannestad, S. Primordial neutrinos. *Annu. Rev. Nucl. Part. Sci.* **2006**, *56*, 137–161.
164. Lesgourgues, J.; Pastor, S. Massive neutrinos and cosmology. *Phys. Rep.* **2006**, *429*, 307–379.
165. Lesgourgues, J.; Pastor, S. Neutrino cosmology and Planck. *New J. Phys.* **2014**, *16*, 065002.
166. Elgarøy, Ø.; Lahav, O. Neutrino masses from cosmological probes. *New J. Phys.* **2005**, *7*, 61.
167. Abazajian, K.; Fuller, G.M.; Patel, M. Sterile neutrino hot, warm, and cold Dark Matter. *Phys. Rev. D* **2001**, *64*, 023501.
168. Boyarsky, A.; Ruchayskiy, O.; Shaposhnikov, M. The role of sterile neutrinos in cosmology and astrophysics. *Annu. Rev. Nucl. Part. Sci.* **2009**, *59*, 191–214.
169. Mangano, G.; Miele, G.; Pastor, S.; Peloso, M. A precision calculation of the effective number of cosmological neutrinos. *Phys. Lett. B* **2002**, *534*, 8–16.
170. Feeney, S.M.; Peiris, H.V.; Verde, L. Is there evidence for additional neutrino species from cosmology? *J. Cosmol. Astropart. Phys.* **2013**, *4*, 036.
171. Olive, K.; Group, P.D. Review of particle physics. *Chin. Phys. C* **2014**, *38*, 090001.
172. Primack, J.R.; Gross, M.A.K. Hot Dark Matter in cosmology. In *Current Aspects of Neutrino Physics*; Caldwell, D.O., Ed.; Springer: Berlin, Germany, 2001; pp. 287–308.
173. Gerstein, G.; Zel'dovich, Y.B. Rest mass of the muonic neutrino and cosmology (English translation). *Lett. J. Exp. Theor. Phys.* **1966**, *4*, 120–122.
174. Marx, G.; Szalay, A.S. Cosmological limit on the neutretto rest mass. In Proceedings of the Neutrino 72, Balatonfured, Hungary, 11–17 June 1972; Volume 1, p. 123.
175. Cowsik, R.; McClelland, J. An upper limit on the neutrino rest mass. *Phys. Rev. Lett.* **1972**, *29*, 669–670.
176. Szalay, A.S.; Marx, G. Neutrino rest mass from cosmology. *Astron. Astrophys.* **1976**, *49*, 437–441.
177. Tegmark, M.; Eisenstein, D.J.; Strauss, M.A.; Weinberg, D.H.; Blanton, M.R.; Frieman, J.A.; Fukugita, M.; Gunn, J.E.; Hamilton, A.J.S.; Knapp, G.R.; et al. Cosmological constraints from the SDSS luminous red galaxies. *Phys. Rev. D* **2006**, *74*, 123507.
178. Shiraishi, M.; Ichikawa, K.; Ichiki, K.; Sugiyama, N.; Yamaguchi, M. Constraints on neutrino masses from WMAP5 and BBN in the lepton asymmetric universe. *J. Cosmol. Astropart. Phys.* **2009**, *7*, 5.
179. Tereno, I.; Schimd, C.; Uzan, J.P.; Kilbinger, M.; Vincent, F.H.; Fu, L. CFHTLS weak-lensing constraints on the neutrino masses. *Astron. Astrophys.* **2009**, *500*, 657–665.
180. Ichiki, K.; Takada, M.; Takahashi, T. Constraints on neutrino masses from weak lensing. *Phys. Rev. D* **2009**, *79*, 023520.
181. Xia, J.Q.; Granett, B.R.; Viel, M.; Bird, S.; Guzzo, L.; Haehnelt, M.G.; Coupon, J.; McCracken, H.J.; Mellier, Y. Constraints on massive neutrinos from the CFHTLS angular power spectrum. *J. Cosmol. Astropart. Phys.* **2012**, *6*, 010.
182. Olive, K.A. Inflation. *Phys. Rep.* **1990**, *190*, 307–403.
183. Guth, A.H. (Ed.) *The Inflationary Universe. The Quest for a New Theory of Cosmic Origins*; Addison-Wesley: Reading, MA, USA, 1997.
184. Peacock, J.A. *Cosmological Physics*; Cambridge University Press: Cambridge, UK, 1999.

185. Liddle, A.R.; Lyth, D.H. *Cosmological Inflation and Large-Scale Structure*; Cambridge University Press: Cambridge, UK, 2000.
186. Uzan, J.P. Inflation in the standard cosmological model. *C. R. Phys.* **2015**, *16*, 875–890.
187. Misner, C.W. Mixmaster universe. *Phys. Rev. Lett.* **1969**, *22*, 1071–1074.
188. Belinskij, V.A.; Khalatnikov, I.M.; Lifshits, E.M. Oscillatory approach to a singular point in the relativistic cosmology. *Adv. Phys.* **1970**, *19*, 525–573.
189. Starobinsky, A.A. Spectrum of relict gravitational radiation and the early state of the universe. *Sov. J. Exp. Theor. Phys. Lett.* **1979**, *30*, 682–685.
190. 't Hooft, G. Magnetic monopoles in unified gauge theories. *Nucl. Phys. B* **1974**, *79*, 276–284.
191. Preskill, J. Magnetic monopoles. *Annu. Rev. Nucl. Part. Sci.* **1984**, *34*, 461–530.
192. Zel'dovich, Y.B.; Khlopov, M.Y. On the concentration of relic monopoles in the universe. *Phys. Lett. B* **1979**, *79*, 239–241.
193. Preskill, J.P. Cosmological production of superheavy magnetic monopoles. *Phys. Rev. Lett.* **1979**, *43*, 1365–1368.
194. Guth, A.H. Inflationary universe: A possible solution to the horizon and flatness problems. *Phys. Rev. D* **1981**, *23*, 347–356.
195. Linde, A. *Inflation and Quantum Cosmology*; Academic Press: New York, NY, USA, 1989.
196. Lidsey, J.E.; Liddle, A.R.; Kolb, E.W.; Copeland, E.J.; Barreiro, T.; Abney, M. Reconstructing the inflaton potential—An overview. *Rev. Mod. Phys.* **1997**, *69*, 373–410.
197. Sato, K. Cosmological baryon-number domain structure and the first order phase transition of a vacuum. *Phys. Lett. B* **1981**, *99*, 66–70.
198. Einhorn, M.B.; Sato, K. Monopole production in the very early universe in a first-order phase transition. *Nucl. Phys. B* **1981**, *180*, 385–404.
199. Mukhanov, V.F.; Chibisov, G.V. Energy of vacuum and the large-scale structure of the universe. *Zhurnal Eksper. Teor. Fiz.* **1982**, *83*, 475–487.
200. Mukhanov, V.F.; Chibisov, G.V. Quantum fluctuations and a nonsingular universe. *Sov. J. Exp. Theor. Phys. Lett.* **1981**, *33*, 532–535.
201. Guth, A.H.; Pi, S.Y. Fluctuations in the new inflationary universe. *Phys. Rev. Lett.* **1982**, *49*, 1110–1113.
202. Hawking, S.W. The development of irregularities in a single bubble inflationary universe. *Phys. Lett. B* **1982**, *115*, 295–297.
203. Starobinsky, A.A. Dynamics of phase transition in the new inflationary universe scenario and generation of perturbations. *Phys. Lett. B* **1982**, *117*, 175–178.
204. Bardeen, J.M.; Steinhardt, P.J.; Turner, M.S. Spontaneous creation of almost scale-free density perturbations in an inflationary universe. *Phys. Rev. D* **1983**, *28*, 679–693.
205. Planck Collaboration. Planck 2015 Results. XX. Constraints on Inflation *Astron. Astrophys.* **2016**, *594*, A20.
206. Liddle, A.R.; Lyth, D.H. COBE, gravitational waves, inflation and extended inflation. *Phys. Lett. B* **1992**, *291*, 391–398.
207. Liddle, A.R.; Lyth, D.H. The cold Dark Matter density perturbation. *Phys. Rep.* **1993**, *231*, 1–105.
208. Linde, A.D. A new inflationary universe scenario: A possible solution of the horizon, flatness, homogeneity, isotropy and primordial monopole problems. *Phys. Lett. B* **1982**, *108*, 389–393.
209. Dodelson, S. *Modern Cosmology*; Academic Press: New York, NY, USA, 2003.
210. Kosowsky, A.; Turner, M.S. CBR anisotropy and the running of the scalar spectral index. *Phys. Rev. D* **1995**, *52*, 1739–1743.
211. Smoot, G.F.; Bennett, C.L.; Kogut, A.; Wright, E.L.; Aymon, J.; Boggess, N.W.; Cheng, E.S.; de Amici, G.; Gulkis, S.; Hauser, M.G.; et al.; Structure in the COBE differential microwave radiometer first-year maps. *Astrophys. J.* **1992**, *396*, L1–L5.
212. Fixsen, D.J.; Cheng, E.S.; Cottingham, D.A.; Eplee, R.E., Jr.; Isaacman, R.B.; Mather, J.C.; Meyer, S.S.; Noerdlinger, P.D.; Shafer, R.A.; Weiss, R.; et al. Cosmic microwave background dipole spectrum measured by the COBE FIRAS instrument. *Astrophys. J.* **1994**, *420*, 445–449.
213. Dwek, E.; Arendt, R.G.; Hauser, M.G.; Fixsen, D.; Kelsall, T.; Leisawitz, D.; Pei, Y.C.; Wright, E.L.; Mather, J.C.; Moseley, S.H.; et al. The COBE diffuse infrared background experiment search for the cosmic infrared background. IV. Cosmological implications. *Astrophys. J.* **1998**, *508*, 106–122.

214. Spergel, D.N.; Verde, L.; Peiris, H.V.; Komatsu, E.; Nolta, M.R.; Bennett, C.L.; Halpern, M.; Hinshaw, G.; Jarosik, N.; Kogut, A.; et al. First-year Wilkinson Microwave Anisotropy Probe (WMAP) observations: Determination of cosmological parameters. *Astrophys. J. Suppl.* **2003**, *148*, 175–194.
215. Peiris, H.V.; Komatsu, E.; Verde, L.; Spergel, D.N.; Bennett, C.L.; Halpern, M.; Hinshaw, G.; Jarosik, N.; Kogut, A.; Limon, M.; et al. First-year Wilkinson Microwave Anisotropy Probe (WMAP) observations: Implications for inflation. *Astrophys. J. Suppl.* **2003**, *148*, 213–231.
216. Spergel, D.N.; Bean, R.; Doré, O.; Nolta, M.R.; Bennett, C.L.; Dunkley, J.; Hinshaw, G.; Jarosik, N.; Komatsu, E.; Page, L.; et al. Three-year Wilkinson Microwave Anisotropy Probe (WMAP) observations: Implications for cosmology. *Astrophys. J. Suppl.* **2007**, *170*, 377–408.
217. Komatsu, E.; Dunkley, J.; Nolta, M.R.; Bennett, C.L.; Gold, B.; Hinshaw, G.; Jarosik, N.; Larson, D.; Limon, M.; Page, L.; et al. Five-year Wilkinson Microwave Anisotropy Probe observations: Cosmological interpretation. *Astrophys. J. Suppl.* **2009**, *180*, 330–376.
218. Komatsu, E.; Smith, K.M.; Dunkley, J.; Bennett, C.L.; Gold, B.; Hinshaw, G.; Jarosik, N.; Larson, D.; Nolta, M.R.; Page, L.; et al. Seven-year Wilkinson Microwave Anisotropy Probe (WMAP) observations: Cosmological interpretation. *Astrophys. J. Suppl.* **2011**, *192*, 18.
219. Hinshaw, G.; Larson, D.; Komatsu, E.; Spergel, D.N.; Bennett, C.L.; Dunkley, J.; Nolta, M.R.; Halpern, M.; Hill, R.S.; Odegard, N.; et al. Nine-year Wilkinson Microwave Anisotropy Probe (WMAP) observations: Cosmological parameter results. *Astrophys. J. Suppl.* **2013**, *208*, 19.
220. Planck Collaboration. Planck 2013 results. XVI. Cosmological parameters. *Astron. Astrophys.* **2014**, *571*, A16.
221. Planck Collaboration. Planck 2013 results. XXII. Constraints on inflation. *Astron. Astrophys.* **2014**, *571*, A22.
222. BICEP2 Collaboration. Detection of B-mode polarization at degree angular scales by BICEP2. *Phys. Rev. Lett.* **2014**, *112*, 241101.
223. Planck Collaboration. Planck intermediate results XXX. The angular power spectrum of polarized dust emission at intermediate and high Galactic latitudes. *Astron. Astrophys.* **2016**, *586*, A133.
224. Huang, Z.; Verde, L.; Vernizzi, F. Constraining inflation with future galaxy redshift surveys. *J. Cosmol. Astropart. Phys.* **2012**, *4*, 005.
225. Amendola, L.; Appleby, S.; Bacon, D.; Baker, T.; Baldi, M.; Bartolo, N.; Blanchard, A.; Bonvin, C.; Borgani, S.; Branchini, E.; et al. Cosmology and fundamental physics with the Euclid satellite. *Living Rev. Relativ.* **2013**, *16*, 6.
226. Namikawa, T.; Yamauchi, D.; Sherwin, B.; Nagata, R. Delensing cosmic microwave background B modes with the Square Kilometre Array Radio Continuum Survey. *Phys. Rev. D* **2016**, *93*, 043527.
227. Kapteyn, J.C. First attempt at a theory of the arrangement and motion of the sidereal system. *Astrophys. J.* **1922**, *55*, 302–328.
228. Oort, J.H. The force exerted by the stellar system in the direction perpendicular to the galactic plane and some related problems. *Bull. Astron. Inst. Neth.* **1932**, *6*, 249–287.
229. Zwicky, F. Die Rotverschiebung von extragalaktischen Nebeln. *Helv. Phys. Acta* **1933**, *6*, 110–127.
230. Zwicky, F. On the Masses of Nebulae and of Clusters of Nebulae. *Astrophys. J.* **1937**, *86*, 217–246.
231. Freese, K. Review of observational evidence for Dark Matter in the universe and in upcoming searches for dark stars. In Proceedings of the Dark Energy and Dark Matter: Observations, Experiments and Theories, Lyons, France, 7–11 July 2008; Volume 36, pp. 113–126.
232. Kahn, F.D.; Woltjer, L. Intergalactic matter and the galaxy. *Astrophys. J.* **1959**, *130*, 705–717.
233. Roberts, M.S.; Rots, A.H. Comparison of rotation curves of different galaxy types. *Astron. Astrophys.* **1973**, *26*, 483–485.
234. Einasto, J.; Kaasik, A.; Saar, E. Dynamic evidence on massive coronas of galaxies. *Nature* **1974**, *250*, 309–310.
235. Ostriker, J.P.; Peebles, P.J.E.; Yahil, A. The size and mass of galaxies, and the mass of the universe. *Astrophys. J.* **1974**, *193*, L1–L4.
236. Rubin, V.C.; Thonnard, N.; Ford, W.K., Jr. Extended rotation curves of high-luminosity spiral galaxies. IV—Systematic dynamical properties, SA through SC. *Astrophys. J.* **1978**, *225*, L107–L111.
237. Navarro, J.F.; Frenk, C.S.; White, S.D.M. The structure of cold Dark Matter halos. *Astrophys. J.* **1996**, *462*, 563.
238. Fukugita, M.; Hogan, C.J.; Peebles, P.J.E. The cosmic baryon budget. *Astrophys. J.* **1998**, *503*, 518–530.
239. Sachs, R.K.; Wolfe, A.M. Perturbations of a cosmological model and angular variations of the microwave background. *Astrophys. J.* **1967**, *147*, 73–90.

240. Eisenstein, D.J. Dark energy and cosmic sound. *New Astron. Rev.* **2005**, *49*, 360–365.
241. Bassett, B.; Hlozek, R. Baryon acoustic oscillations. In *Dark Energy: Observational and Theoretical Approaches*; Ruiz-Lapuente, P., Ed.; Cambridge University Press: Cambridge, UK, 2010; p. 246.
242. Schneider, P. Weak gravitational lensing. In *Gravitational Lensing: Strong, Weak and Micro, Saas-Fee Advanced Courses 33*; Schneider, P., Kochanek, C.S., Wambsganss, J., Eds.; Springer: Berlin/Heidelberg, Germany, 2006; p. 269.
243. Bacon, D.J.; Réfrégier, A.R.; Ellis, R.S. Detection of weak gravitational lensing by large-scale structure. *Mon. Not. R. Astron. Soc.* **2000**, *318*, 625–640.
244. Kaiser, N. A new shear estimator for weak-lensing observations. *Astrophys. J.* **2000**, *537*, 555–577.
245. Van Waerbeke, L.; Mellier, Y.; Erben, T.; Cuillandre, J.C.; Bernardeau, F.; Maoli, R.; Bertin, E.; Mc Cracken, H.J.; Le Fèvre, O.; Fort, B.; et al. Detection of correlated galaxy ellipticities from CFHT data: First evidence for gravitational lensing by large-scale structures. *Astron. Astrophys.* **2000**, *358*, 30–44.
246. Wittman, D.M.; Tyson, J.A.; Kirkman, D.; Dell'Antonio, I.; Bernstein, G. Detection of weak gravitational lensing distortions of distant galaxies by cosmic Dark Matter at large scales. *Nature* **2000**, *405*, 143–148.
247. Bertone, G.; Hooper, D.; Silk, J. Particle Dark Matter: Evidence, candidates and constraints. *Phys. Rep.* **2005**, *405*, 279–390.
248. Salati, P. The bestiary of Dark Matter species. In Proceedings of the Dark Energy and Dark Matter: Observations, Experiments and Theories, Lyons, France, 7–11 July 2008; Volume 36, pp. 175–186.
249. Garrett, K.; Dūda, G. Dark matter: A primer. *Adv. Astron.* **2011**, *2011*, 968283.
250. Kappl, R.; Winkler, M.W. New limits on Dark Matter from Super-Kamiokande. *Nucl. Phys. B* **2011**, *850*, 505–521.
251. Arina, C.; Hamann, J.; Wong, Y.Y.Y. A Bayesian view of the current status of Dark Matter direct searches. *J. Cosmol. Astropart. Phys.* **2011**, *9*, 022.
252. Porter, T.A.; Johnson, R.P.; Graham, P.W. Dark matter searches with astroparticle data. *Annu. Rev. Astron. Astrophys.* **2011**, *49*, 155–194.
253. Calore, F.; de Romeri, V.; Donato, F. Conservative upper limits on WIMP annihilation cross section from Fermi-LAT γ rays. *Phys. Rev. D* **2012**, *85*, 023004.
254. Klasen, M.; Pohl, M.; Sigl, G. Indirect and direct search for Dark Matter. *Prog. Part. Nucl. Phys.* **2015**, *85*, 1–32.
255. Mayet, F.; Green, A.M.; Battat, J.B.R.; Billard, J.; Bozorgnia, N.; Gelmini, G.B.; Gondolo, P.; Kavanagh, B.J.; Lee, S.K.; Loomba, D.; et al. A review of the discovery reach of directional Dark Matter detection. *Phys. Rep.* **2016**, *627*, 1–49.
256. Jungman, G.; Kamionkowski, M.; Griest, K. Supersymmetric Dark Matter. *Phys. Rep.* **1996**, *267*, 195–373.
257. IceCube Collaboration. Search for Dark Matter annihilations in the Sun with the 79-string IceCube detector. *Phys. Rev. Lett.* **2013**, *110*, 131302.
258. Aguirre, A.; Schaye, J.; Quataert, E. Problems for modified Newtonian dynamics in clusters and the Lyα forest? *Astrophys. J.* **2001**, *561*, 550–558.
259. Pointecouteau, E.; Silk, J. New constraints on modified Newtonian dynamics from galaxy clusters. *Mon. Not. R. Astron. Soc.* **2005**, *364*, 654–658.
260. Contaldi, C.R.; Wiseman, T.; Withers, B. TeVeS gets caught on caustics. *Phys. Rev. D* **2008**, *78*, 044034.
261. Lue, A.; Starkman, G.D. Squeezing MOND into a cosmological scenario. *Phys. Rev. Lett.* **2004**, *92*, 131102.
262. Dodelson, S. The real problem with MOND. *Int. J. Mod. Phys. D* **2011**, *20*, 2749–2753.
263. Ferreras, I.; Mavromatos, N.E.; Sakellariadou, M.; Yusaf, M.F. Confronting MOND and TeVeS with strong gravitational lensing over galactic scales: An extended survey. *Phys. Rev. D* **2012**, *86*, 083507.
264. Clowe, D.; Bradač, M.; Gonzalez, A.H.; Markevitch, M.; Randall, S.W.; Jones, C.; Zaritsky, D. A direct empirical proof of the existence of Dark Matter. *Astrophys. J.* **2006**, *648*, L109–L113.
265. Peebles, P.J.; Ratra, B. The cosmological constant and dark energy. *Rev. Mod. Phys.* **2003**, *75*, 559–606.
266. Slipher, V.M. Nebulae. *Proc. Am. Philos. Soc.* **1917**, *56*, 403–409.
267. Paal, G.; Horvath, I.; Lukacs, B. Inflation and compactification from galaxy redshifts? *Astrophys. Space Sci.* **1992**, *191*, 107–124.
268. Krauss, L.M. The end of the age problem, and the case for a cosmological constant revisited. *Astrophys. J.* **1998**, *501*, 461–466.

269. Räsänen, S. Structure formation as an alternative to dark energy and modified gravity. In Proceedings of the Dark Energy and Dark Matter: Observations, Experiments and Theories, Lyons, France, 7–11 July 2008; Volume 36, pp. 63–74.
270. Blasone, M.; Capolupo, A.; Capozziello, S.; Carloni, S.; Vitiello, G. Neutrino mixing contribution to the cosmological constant. *Phys. Lett. A* **2004**, *323*, 182–189.
271. Capolupo, A.; Capozziello, S.; Vitiello, G. Neutrino mixing as a source of dark energy. *Phys. Lett. A* **2007**, *363*, 53–56.
272. Riess, A.G.; Filippenko, A.V.; Challis, P.; Clocchiatti, A.; Diercks, A.; Garnavich, P.M.; Gilliland, R.L.; Hogan, C.J.; Jha, S.; Kirshner, R.P.; et al. Observational evidence from supernovae for an accelerating universe and a cosmological constant. *Astron. J.* **1998**, *116*, 1009–1038.
273. Perlmutter, S.; Aldering, G.; Goldhaber, G.; Knop, R.A.; Nugent, P.; Castro, P.G.; Deustua, S.; Fabbro, S.; Goobar, A.; Groom, D.E.; et al. The supernova cosmology project. Measurements of Omega and Lambda from 42 high-redshift supernovae. *Astrophys. J.* **1999**, *517*, 565–586.
274. Sahni, V.; Starobinsky, A. The case for a positive cosmological Λ-term. *Int. J. Mod. Phys. D* **2000**, *9*, 373–443.
275. Turner, M.S. Dark Matter and Dark Energy in the universe. In *The Third Stromlo Symposium: The Galactic Halo*; Gibson, B.K., Axelrod, R.S., Putman, M.E., Eds.; Astronomical Society of the Pacific Conference Series; The Astronomical Society of the Pacific: San Francisco, CA, USA 1999; Volume 165, p. 431.
276. Eisenstein, D.J.; Zehavi, I.; Hogg, D.W.; Scoccimarro, R.; Blanton, M.R.; Nichol, R.C.; Scranton, R.; Seo, H.J.; Tegmark, M.; Zheng, Z.; et al. Detection of the baryon acoustic peak in the large-scale correlation function of SDSS luminous red galaxies. *Astrophys. J.* **2005**, *633*, 560–574.
277. Frieman, J.A.; Turner, M.S.; Huterer, D. Dark Energy and the Accelerating Universe. *Annu. Rev. Astron. Astrophys.* **2008**, *46*, 385–432.
278. Astier, P.; Pain, R. Observational evidence of the accelerated expansion of the universe. *C. R. Phys.* **2012**, *13*, 521–538.
279. Enqvist, K. Lemaitre Tolman Bondi model and accelerating expansion. *Gen. Relativ. Gravit.* **2008**, *40*, 451–466.
280. Sotiriou, T.P.; Faraoni, V. $f(R)$ theories of gravity. *Rev. Mod. Phys.* **2010**, *82*, 451–497.
281. Felice, A.D.; Tsujikawa, S. $f(R)$ theories. *Living Rev. Relativ.* **2010**, *13*, 3.
282. Capozziello, S.; de Laurentis, M. Extended theories of gravity. *Phys. Rep.* **2011**, *509*, 167–321.
283. Khoury, J.; Weltman, A. Chameleon cosmology. *Phys. Rev. D* **2004**, *69*, 044026.
284. Durrer, R.; Maartens, R. Dark energy and dark gravity: Theory overview. *Gen. Relativ. Gravit.* **2008**, *40*, 301–328..
285. Linder, E.V. Exploring the expansion history of the universe. *Phys. Rev. Lett.* **2003**, *90*, 091301.
286. Chevallier, M.; Polarski, D. Accelerating Universes with scaling Dark Matter. *Int. J. Mod. Phys. D* **2001**, *10*, 213–223.
287. Linden, S.; Virey, J.M. Test of the Chevallier-Polarski-Linder parametrization for rapid dark energy equation of state transitions. *Phys. Rev. D* **2008**, *78*, 023526.
288. Wang, Y.; Freese, K. Probing dark energy using its density instead of its equation of state. *Phys. Lett. B* **2006**, *632*, 449–452.
289. Johri, V.B.; Rath, P.K. Parametrization of the dark energy equation of state. *Int. J. Mod. Phys. D* **2007**, *16*, 1581–1591.
290. Avelino, P.P.; Martins, C.J.A.P.; Nunes, N.J.; Olive, K.A. Reconstructing the dark energy equation of state with varying couplings. *Phys. Rev. D* **2006**, *74*, 083508.
291. Scherrer, R.J. Mapping the Chevallier-Polarski-Linder parametrization onto physical dark energy models. *Phys. Rev. D* **2015**, *92*, 043001.
292. Chongchitnan, S.; King, L. Imprints of dynamical dark energy on weak-lensing measurements. *Mon. Not. R. Astron. Soc.* **2010**, *407*, 1989–1997.
293. Astashenok, A.V.; Odintsov, S.D. Confronting dark energy models mimicking ΛCDM epoch with observational constraints: Future cosmological perturbations decay or future Rip? *Phys. Lett. B* **2013**, *718*, 1194–1202.
294. Debono, I. Bayesian model selection for dark energy using weak lensing forecasts. *Mon. Not. R. Astron. Soc.* **2014**, *437*, 887–897.
295. Bardeen, J.M. Gauge-invariant cosmological perturbations. *Phys. Rev. D* **1980**, *22*, 1882–1905.

296. Lifshitz, E.M. On the gravitational stability of the expanding universe. *J. Phys. (USSR)* **1946**, *46*, 587–602.
297. Bartelmann, M.; Schneider, P. Weak gravitational lensing. *Phys. Rep.* **2001**, *340*, 291–472.
298. Jeans, J.H. The stability of a spherical nebula. *Philos. Trans. R. Soc. Lond. A Math. Phys. Eng. Sci.* **1902**, *199*, 1–53.
299. Silk, J. Cosmic black-body radiation and galaxy formation. *Astrophys. J.* **1968**, *151*, 459–471.
300. Bond, J.R.; Efstathiou, G.; Silk, J. Massive neutrinos and the large-scale structure of the universe. *Phys. Rev. Lett.* **1980**, *45*, 1980–1984.
301. Silk, J.; Mamon, G.A. The current status of galaxy formation. *Res. Astron. Astrophys.* **2012**, *12*, 917–946.
302. Linder, E.V.; Jenkins, A. Cosmic structure growth and dark energy. *Mon. Not. R. Astron. Soc.* **2003**, *346*, 573–583.
303. Bernardeau, F.; Colombi, S.; Gaztañaga, E.; Scoccimarro, R. Large-scale structure of the Universe and cosmological perturbation theory. *Phys. Rep.* **2002**, *367*, 1–248.
304. Percival, W.J. Cosmological structure formation in a homogeneous dark energy background. *Astron. Astrophys.* **2005**, *443*, 819–830.
305. Hamilton, A.J.S.; Kumar, P.; Lu, E.; Matthews, A. Reconstructing the primordial spectrum of fluctuations of the universe from the observed nonlinear clustering of galaxies. *Astrophys. J.* **1991**, *374*, L1–L4.
306. Jain, B.; Mo, H.J.; White, S.D.M. The evolution of correlation functions and power spectra in gravitational clustering. *Mon. Not. R. Astron. Soc.* **1995**, *276*, L25–L29.
307. Peacock, J.A.; Dodds, S.J. Non-linear evolution of cosmological power spectra. *Mon. Not. R. Astron. Soc.* **1996**, *280*, L19–L26.
308. Smith, R.E.; Peacock, J.A.; Jenkins, A.; White, S.D.M.; Frenk, C.S.; Pearce, F.R.; Thomas, P.A.; Efstathiou, G.; Couchman, H.M.P. Stable clustering, the halo model and non-linear cosmological power spectra. *Mon. Not. R. Astron. Soc.* **2003**, *341*, 1311–1332.
309. Seljak, U. Analytic model for galaxy and Dark Matter clustering. *Mon. Not. R. Astron. Soc.* **2000**, *318*, 203–213.
310. Peacock, J.A.; Smith, R.E. Halo occupation numbers and galaxy bias. *Mon. Not. R. Astron. Soc.* **2000**, *318*, 1144–1156.
311. Huterer, D.; Takada, M. Calibrating the nonlinear matter power spectrum: Requirements for future weak lensing surveys. *Astropart. Phys.* **2005**, *23*, 369–376.
312. Van Daalen, M.P.; Schaye, J.; Booth, C.M.; Dalla Vecchia, C. The effects of galaxy formation on the matter power spectrum: A challenge for precision cosmology. *Mon. Not. R. Astron. Soc.* **2011**, *415*, 3649–3665.
313. Semboloni, E.; Hoekstra, H.; Schaye, J.; van Daalen, M.P.; McCarthy, I.G. Quantifying the effect of baryon physics on weak lensing tomography. *Mon. Not. R. Astron. Soc.* **2011**, *417*, 2020–2035.
314. Bird, S.; Viel, M.; Haehnelt, M.G. Massive neutrinos and the non-linear matter power spectrum. *Mon. Not. R. Astron. Soc.* **2012**, *420*, 2551–2561.
315. Hearin, A.P.; Zentner, A.R.; Ma, Z. General requirements on matter power spectrum predictions for cosmology with weak lensing tomography. *J. Cosmol. Astropart. Phys.* **2012**, *4*, 034.
316. Takahashi, R.; Sato, M.; Nishimichi, T.; Taruya, A.; Oguri, M. Revising the halofit model for the nonlinear matter power spectrum. *Astrophys. J.* **2012**, *761*, 152.
317. Villaescusa-Navarro, F.; Viel, M.; Datta, K.K.; Choudhury, T.R. Modeling the neutral hydrogen distribution in the post-reionization Universe: Intensity mapping. *J. Cosmol. Astropart. Phys.* **2014**, *9*, 050.
318. Bentivegna, E.; Bruni, M. Effects of nonlinear inhomogeneity on the cosmic expansion with numerical relativity. *Phys. Rev. Lett.* **2016**, *116*, 251302.
319. Mertens, J.B.; Giblin, J.T.; Starkman, G.D. Integration of inhomogeneous cosmological spacetimes in the BSSN formalism. *Phys. Rev. D* **2016**, *93*, 124059.
320. Giblin, J.T.; Mertens, J.B.; Starkman, G.D. Departures from the Friedmann-Lemaitre-Robertston-Walker cosmological model in an inhomogeneous universe: A numerical examination. *Phys. Rev. Lett.* **2016**, *116*, 251301.
321. Harrison, E.R. Fluctuations at the threshold of classical cosmology. *Phys. Rev. D* **1970**, *1*, 2726–2730.
322. Zel'dovich, Y.B. A hypothesis, unifying the structure and the entropy of the Universe. *Mon. Not. R. Astron. Soc.* **1972**, *160*, 1–3.
323. Peebles, P.J.E.; Yu, J.T. Primeval adiabatic perturbation in an expanding universe. *Astrophys. J.* **1970**, *162*, 815–836.

324. Mukherjee, P.; Wang, Y. Model-independent reconstruction of the primordial power spectrum from Wilkinson Microwave Anistropy Probe data. *Astrophys. J.* **2003**, *599*, 1–6.
325. Trotta, R. Forecasting the Bayes factor of a future observation. *Mon. Not. R. Astron. Soc.* **2007**, *378*, 819–824.
326. Trotta, R. Applications of Bayesian model selection to cosmological parameters. *Mon. Not. R. Astron. Soc.* **2007**, *378*, 72–82.
327. Bridges, M.; Feroz, F.; Hobson, M.P.; Lasenby, A.N. Bayesian optimal reconstruction of the primordial power spectrum. *Mon. Not. R. Astron. Soc.* **2009**, *400*, 1075–1084.
328. Vázquez, J.A.; Bridges, M.; Hobson, M.P.; Lasenby, A.N. Model selection applied to reconstruction of the Primordial Power Spectrum. *J. Cosmol. Astropart. Phys.* **2012**, *6*, 006.
329. Hoyle, C.D.; Kapner, D.J.; Heckel, B.R.; Adelberger, E.G.; Gundlach, J.H.; Schmidt, U.; Swanson, H.E. Submillimeter tests of the gravitational inverse-square law. *Phys. Rev. D* **2004**, *70*, 042004.
330. Kapner, D.J.; Cook, T.S.; Adelberger, E.G.; Gundlach, J.H.; Heckel, B.R.; Hoyle, C.D.; Swanson, H.E. Tests of the gravitational inverse-square law below the dark-energy length scale. *Phys. Rev. Lett.* **2007**, *98*, 021101.
331. Baker, T.; Psaltis, D.; Skordis, C. Linking tests of gravity on all scales: From the strong-field regime to cosmology. *Astrophys. J.* **2015**, *802*, 63.
332. Eddington, A.S. *The Mathematical Theory of Relativity*; Cambridge University Press: Cambridge, UK, 1923.
333. Shapiro, I.I. Solar system tests of general relativity: Recent results and present plans. In *General Relativity and Gravitation, 1989*; Ashby, N., Bartlett, D.F., Wyss, W., Eds.; Cambridge University Press: Cambridge, UK, 2005; p. 313.
334. Williams, J.G.; Turyshev, S.G.; Boggs, D.H. Progress in lunar laser ranging tests of relativistic gravity. *Phys. Rev. Lett.* **2004**, *93*, 261101.
335. Williams, J.G.; Turyshev, S.G.; Boggs, D.H. Lunar laser ranging tests of the equivalence principle with the Earth and Moon. *Int. J. Mod. Phys. D* **2009**, *18*, 1129–1175.
336. Shapiro, S.S.; Davis, J.L.; Lebach, D.E.; Gregory, J.S. Measurement of the solar gravitational deflection of radio waves using geodetic very-long-baseline interferometry data, 1979–1999. *Phys. Rev. Lett.* **2004**, *92*, 121101.
337. Pitjeva, E.V.; Pitjev, N.P. Relativistic effects and Dark Matter in the Solar system from observations of planets and spacecraft. *Mon. Not. R. Astron. Soc.* **2013**, *432*, 3431–3437.
338. Fienga, A.; Laskar, J.; Kuchynka, P.; Manche, H.; Desvignes, G.; Gastineau, M.; Cognard, I.; Theureau, G. The INPOP10a planetary ephemeris and its applications in fundamental physics. *Celest. Mech. Dyn. Astron.* **2011**, *111*, 363–385.
339. Iorio, L.; Lichtenegger, H.I.M.; Ruggiero, M.L.; Corda, C. Phenomenology of the Lense-Thirring effect in the solar system. *Astrophys. Space Sci.* **2011**, *331*, 351–395.
340. Hughes, V.W.; Robinson, H.G.; Beltran-Lopez, V. Upper Limit for the anisotropy of inertial mass from nuclear resonance experiments. *Phys. Rev. Lett.* **1960**, *4*, 342–344.
341. Drever, R.W.P. A search for anisotropy of inertial mass using a free precession technique. *Philos. Mag.* **1961**, *6*, 683–687.
342. Allmendinger, F.; Heil, W.; Karpuk, S.; Kilian, W.; Scharth, A.; Schmidt, U.; Schnabel, A.; Sobolev, Y.; Tullney, K. New limit on Lorentz-invariance- and CPT-violating neutron spin interactions using a free-spin-precession He3-Xe129 comagnetometer. *Phys. Rev. Lett.* **2014**, *112*, 110801.
343. Williams, J.G.; Turyshev, S.G.; Boggs, D.H. Lunar laser ranging tests of the equivalence principle. *Class. Quantum Gravity* **2012**, *29*, 184004.
344. Delva, P.; Hees, A.; Bertone, S.; Richard, E.; Wolf, P. Test of the gravitational redshift with stable clocks in eccentric orbits: Application to Galileo satellites 5 and 6. *Class. Quantum Gravity* **2015**, *32*, 232003.
345. Iorio, L. Gravitational anomalies in the solar system? *Int. J. Mod. Phys. D* **2015**, *24*, 1530015.
346. Uzan, J.P. Varying constants, gravitation and cosmology. *Living Rev. Relativ.* **2011**, *14*, 2.
347. Flambaum, V.V. Variation of the fundamental constants: Theory and observations. *Int. J. Mod. Phys. A* **2007**, *22*, 4937–4950.
348. Lea, S.N. Limits to time variation of fundamental constants from comparisons of atomic frequency standards. *Rep. Prog. Phys.* **2007**, *70*, 1473–1523.
349. Rich, J. Which fundamental constants for cosmic microwave background and baryon-acoustic oscillation? *Astron. Astrophys.* **2015**, *584*, A69.

350. Barrow, J.D.; Sandvik, H.B.; Magueijo, J.A. Behavior of varying-alpha cosmologies. *Phys. Rev. D* **2002**, *65*, 063504.
351. Barrow, J.D.; Graham, A.A.H. General dynamics of varying-alpha universes. *Phys. Rev. D* **2013**, *88*, 103513.
352. Fujii, Y.; Iwamoto, A.; Fukahori, T.; Ohnuki, T.; Nakagawa, M.; Hidaka, H.; Oura, Y.; Möller, P. The nuclear interaction at Oklo 2 billion years ago. *Nucl. Phys. B* **2000**, *573*, 377–401.
353. Uzan, J.P. The fundamental constants and their variation: Observational and theoretical status. *Rev. Mod. Phys.* **2003**, *75*, 403–455.
354. Lamoreaux, S.K.; Torgerson, J.R. Neutron moderation in the Oklo natural reactor and the time variation of α. *Phys. Rev. D* **2004**, *69*, 121701.
355. Khatri, R.; Wandelt, B.D. 21-cm radiation: A new probe of variation in the fine-structure constant. *Phys. Rev. Lett.* **2007**, *98*, 111301.
356. Nakashima, M.; Ichikawa, K.; Nagata, R.; Yokoyama, J. Constraining the time variation of the coupling constants from cosmic microwave background: Effect of ΛQCD. *J. Cosmol. Astropart. Phys.* **2010**, *2010*, 030.
357. Martins, C.J.A.P.; Menegoni, E.; Galli, S.; Mangano, G.; Melchiorri, A. Varying couplings in the early universe: Correlated variations of α and G. *Phys. Rev. D* **2010**, *82*, 023532.
358. Anderson, J.D.; Schubert, G.; Trimble, V.; Feldman, M.R. Measurements of Newton's gravitational constant and the length of day. *Europhys. Lett.* **2015**, *110*, 10002.
359. Pitkin, M. Comment on "Measurements of Newton's gravitational constant and the length of day". *Europhys. Lett.* **2015**, *111*, 30002.
360. Anderson, J.D.; Schubert, G.; Trimble, V.; Feldman, M.R. Reply to the comment by M. Pitkin. *Europhys. Lett.* **2015**, *111*, 30003.
361. Iorio, L. Does Newton's gravitational constant vary sinusoidally with time? Orbital motions say no. *Class. Quantum Gravity* **2016**, *33*, 045004.
362. Feldman, M.R.; Anderson, J.D.; Schubert, G.; Trimble, V.; Kopeikin, S.M.; Lämmerzahl, C. Deep space experiment to measure G. *Class. Quantum Gravity* **2016**, *33*, 125013.
363. Lahav, O. Observational tests of FRW world models. *Class. Quantum Gravity* **2002**, *19*, 3517–3526.
364. Hansen, F.K.; Banday, A.J.; Górski, K.M. Testing the cosmological principle of isotropy: Local power-spectrum estimates of the WMAP data. *Mon. Not. R. Astron. Soc.* **2004**, *354*, 641–665.
365. Schwarz, D.J.; Bacon, D.; Chen, S.; Clarkson, C.; Huterer, D.; Kunz, M.; Maartens, R.; Raccanelli, A.; Rubart, M.; Starck, J.L. Testing foundations of modern cosmology with SKA all-sky surveys. In Proceedings of the Advancing Astrophysics with the Square Kilometre Array (AASKA14), Giardini Naxos, Italy, 9–13 June 2014; p. 32.
366. Segre, M. Galileo, Viviani and the tower of Pisa. *Stud. Hist. Philos. Sci. Part A* **1989**, *20*, 435–451.
367. Eötvos, R. Über die Anziehung der Erde auf Verschiedene Substanzen. *Math. Naturwissenschaft. Ber. Ung.* **1890**, *8*, 65–68.
368. Gundlach, J.H.; Smith, G.L.; Adelberger, E.G.; Heckel, B.R.; Swanson, H.E. Short-range test of the equivalence principle. *Phys. Rev. Lett.* **1997**, *78*, 2523–2526.
369. Schlamminger, S.; Choi, K.Y.; Wagner, T.A.; Gundlach, J.H.; Adelberger, E.G. Test of the equivalence principle using a rotating torsion balance. *Phys. Rev. Lett.* **2008**, *100*, 041101.
370. Adelberger, E.; Gundlach, J.; Heckel, B.; Hoedl, S.; Schlamminger, S. Torsion balance experiments: A low-energy frontier of particle physics. *Prog. Part. Nucl. Phys.* **2009**, *62*, 102–134.
371. Wagner, T.A.; Schlamminger, S.; Gundlach, J.H.; Adelberger, E.G. Torsion-balance tests of the weak equivalence principle. *Class. Quantum Gravity* **2012**, *29*, 184002.
372. Dickey, J.O.; Bender, P.L.; Faller, J.E.; Newhall, X.X.; Ricklefs, R.L.; Ries, J.G.; Shelus, P.J.; Veillet, C.; Whipple, A.L.; Wiant, J.R.; et al. Lunar Laser ranging: A continuing legacy of the Apollo program. *Science* **1994**, *265*, 482–490.
373. Murphy, T.W. Lunar laser ranging: The millimeter challenge. *Rep. Prog. Phys.* **2013**, *76*, 076901.
374. Pearlman, M.R.; Degnan, J.J.; Bosworth, J.M. The international laser ranging service. *Adv. Space Res.* **2002**, *30*, 135–143.
375. Appleby, G.; Rodríguez, J.; Altamimi, Z. Assessment of the accuracy of global geodetic satellite laser ranging observations and estimated impact on ITRF scale: Estimation of systematic errors in LAGEOS observations 1993–2014. *J. Geod.* **2016**, doi:10.1007/s00190-016-0929-2.

376. Degnan, J.J. Laser transponders for high-accuracy interplanetary laser ranging and time transfer. In *Lasers, Clocks and Drag-Free Control: Exploration of Relativistic Gravity in Space*; Dittus, H., Lammerzahl, C., Turyshev, S.G., Eds.; Astrophysics and Space Science Library; Springer: Berlin/Heidelberg, Germany, 2008; Volume 349, p. 231.
377. Iorio, L. Effects of standard and modified gravity on interplanetary ranges. *Int. J. Mod. Phys. D* **2011**, *20*, 181–232.
378. Dirkx, D.; Noomen, R.; Visser, P.N.A.M.; Bauer, S.; Vermeersen, L.L.A. Comparative analysis of one- and two-way planetary laser ranging concepts. *Planet. Space Sci.* **2015**, *117*, 159–176.
379. Smith, D.; Zuber, M.; Sun, X.; Neumann, G.; Cavanaugh, J.; McGarry, J.; Zagwodzki, T. Two-way laser link over interplanetary distance. *Science* **2006**, *311*, 53.
380. Chen, Y.; Birnbaum, K.M.; Hemmati, H. Active laser ranging over planetary distances with millimeter accuracy. *Appl. Phys. Lett.* **2013**, *102*, 241107.
381. Margot, J.L.; Giorgini, J.D. Probing general relativity with radar astrometry in the inner solar system. *Proc. Int. Astron. Union* **2010**, *261*, 183–188.
382. Fienga, A.; Laskar, J.; Manche, H.; Gastineau, M. Tests of GR with INPOP15a Planetary Ephemerides: Estimations of Possible Supplementary Advances of Perihelia for Mercury and Saturn. 2016, arXiv:1601.00947.
383. Iorio, L. The Solar Lense-Thirring Effect: Perspectives for a Future Measurement. 2016, arXiv:1601.01382.
384. Anderson, J.D.; Laing, P.A.; Lau, E.L.; Liu, A.S.; Nieto, M.M.; Turyshev, S.G. Study of the anomalous acceleration of Pioneer 10 and 11. *Phys. Rev. D* **2002**, *65*, 082004.
385. Nieto, M.M.; Anderson, J.D. Using early data to illuminate the Pioneer anomaly. *Class. Quantum Gravity* **2005**, *22*, 5343–5354.
386. Turyshev, S.G.; Toth, V.T. The pioneer anomaly. *Living Rev. Relativ.* **2010**, *13*, 4.
387. Nieto, M.M. New Horizons and the onset of the Pioneer anomaly. *Phys. Lett. B* **2008**, *659*, 483–485.
388. Iorio, L. Perspectives on effectively constraining the location of a massive trans-Plutonian object with the New Horizons spacecraft: A sensitivity analysis. *Celest. Mech. Dyn. Astron.* **2013**, *116*, 357–366.
389. Iorio, L.; Ruggiero, M.L.; Radicella, N.; Saridakis, E.N. Constraining the Schwarzschild-de Sitter solution in models of modified gravity. *Phys. Dark Universe* **2016**, *13*, 111–120.
390. Damour, T.; Taylor, J.H. Strong-field tests of relativistic gravity and binary pulsars. *Phys. Rev. D* **1992**, *45*, 1840–1868.
391. Lyne, A.G.; Burgay, M.; Kramer, M.; Possenti, A.; Manchester, R.N.; Camilo, F.; McLaughlin, M.A.; Lorimer, D.R.; D'Amico, N.; Joshi, B.C.; et al. A double-pulsar system: A rare laboratory for relativistic gravity and plasma physics. *Science* **2004**, *303*, 1153–1157.
392. Kramer, M.; Wex, N. The double pulsar system: A unique laboratory for gravity. *Class. Quantum Gravity* **2009**, *26*, 073001.
393. McGaugh, S.S.; Schombert, J.M.; Bothun, G.D.; de Blok, W.J.G. The baryonic tully-fisher relation. *Astrophys. J.* **2000**, *533*, L99–L102.
394. Famaey, B.; McGaugh, S.S. Modified Newtonian Dynamics (MOND): Observational phenomenology and relativistic extensions. *Living Rev. Relativ.* **2012**, *15*, 10.
395. Adams, F.C.; Laughlin, G. Relativistic effects in extrasolar planetary systems. *Int. J. Mod. Phys. D* **2006**, *15*, 2133–2140.
396. Adams, F.C.; Laughlin, G. Effects of secular interactions in extrasolar planetary systems. *Astrophys. J.* **2006**, *649*, 992–1003.
397. Adams, F.C.; Laughlin, G. Long-term evolution of close planets including the effects of secular interactions. *Astrophys. J.* **2006**, *649*, 1004–1009.
398. Iorio, L. Are we far from testing general relativity with the transitting extrasolar planet HD 209458b 'Osiris'? *New Astron.* **2006**, *11*, 490–494.
399. Jordán, A.; Bakos, G.Á. Observability of the general relativistic precession of periastra in exoplanets. *Astrophys. J.* **2008**, *685*, 543–552.
400. Pál, A.; Kocsis, B. Periastron precession measurements in transiting extrasolar planetary systems at the level of general relativity. *Mon. Not. R. Astron. Soc.* **2008**, *389*, 191–198.
401. Iorio, L. Classical and relativistic node precessional effects in WASP-33b and perspectives for detecting them. *Astrophys. Space Sci.* **2011**, *331*, 485–496.

402. Iorio, L. Classical and relativistic long-term time variations of some observables for transiting exoplanets. *Mon. Not. R. Astron. Soc.* **2011**, *411*, 167–183.
403. Iorio, L. Accurate characterization of the stellar and orbital parameters of the exoplanetary system WASP-33 b from orbital dynamics. *Mon. Not. R. Astron. Soc.* **2016**, *455*, 207–213.
404. Iorio, L. Post-Keplerian corrections to the orbital periods of a two-body system and their measurability. *Mon. Not. R. Astron. Soc.* **2016**, *460*, 2445–2452.
405. Henry, R.C. Kretschmann scalar for a kerr-newman black hole. *Astrophys. J.* **2000**, *535*, 350–353.
406. Will, C.M. Focus Issue: Gravity Probe B. *Class. Quantum Gravity* **2015**, *32*, 220301.
407. Everitt, C.W.F.; Muhlfelder, B.; DeBra, D.B.; Parkinson, B.W.; Turneaure, J.P.; Silbergleit, A.S.; Acworth, E.B.; Adams, M.; Adler, R.; Bencze, W.J.; et al. The Gravity Probe B test of general relativity. *Class. Quantum Gravity* **2015**, *32*, 224001.
408. Hulse, R.A.; Taylor, J.H. Discovery of a pulsar in a binary system. *Astrophys. J.* **1975**, *195*, L51–L53.
409. Psaltis, D. Probes and tests of strong-field gravity with observations in the electromagnetic spectrum. *Living Rev. Relativ.* **2008**, *11*, 9.
410. Psaltis, D. Two approaches to testing general relativity in the strong-field regime. *J. Phys. Conf. Ser.* **2009**, *189*, 012033.
411. Johannsen, T. Testing general relativity in the strong-field regime with observations of black holes in the electromagnetic spectrum. *Publ. Astron. Soc. Pac.* **2012**, *124*, 1133–1134.
412. Kramer, M. Precision tests of theories of gravity using pulsars. In *Thirteenth Marcel Grossmann Meeting: On Recent Developments in Theoretical and Experimental General Relativity, Astrophysics and Relativistic Field Theories*; Rosquist, K., Ed.; World Scientific: Singapore, 2015; pp. 315–332.
413. Weisberg, J.M.; Taylor, J.H. The Relativistic Binary Pulsar B1913+16: Thirty Years of Observations and Analysis. *Astron. Soc. Pac. Conf. Ser.* **2005**, *328*, 25–32.
414. Bailes, M.; Ord, S.M.; Knight, H.S.; Hotan, A.W. Self-consistency of relativistic observables with general relativity in the white dwarf-neutron star binary PSR J1141-6545. *Astrophys. J.* **2003**, *595*, L49–L52.
415. Stairs, I.H.; Thorsett, S.E.; Taylor, J.H.; Wolszczan, A. Studies of the relativistic binary pulsar PSR B1534+12. I. Timing analysis. *Astrophys. J.* **2002**, *581*, 501–508.
416. Kramer, M.; Lorimer, D.R.; Lyne, A.G.; McLaughlin, M.; Burgay, M.; D'Amico, N.; Possenti, A.; Camilo, F.; Freire, P.C.C.; Joshi, B.C.; et al. Testing GR with the double pulsar: Recent results. In Proceedings of the 22nd Texas Symposium on Relativistic Astrophysics, Stanford, CA, USA , 13–17 December 2004; pp. 142–148.
417. Kramer, M.; Stairs, I.H.; Manchester, R.N.; McLaughlin, M.A.; Lyne, A.G.; Ferdman, R.D.; Burgay, M.; Lorimer, D.R.; Possenti, A.; D'Amico, N.; et al. Tests of general relativity from timing the double pulsar. *Science* **2006**, *314*, 97–102.
418. Desai, S.; Kahya, E.O. Galactic one-way Shapiro delay to PSR B1937+21. *Mod. Phys. Lett. A* **2016**, *31*, 1650083.
419. Huang, Y.; Weisberg, J.M. Timing the relativistic binary pulsar PSR B1913+16. In Proceedings of the 228th American Astronomical Society Meeting, San Diego, CA, USA, 12–16 June 2016.
420. Iorio, L. Prospects for measuring the moment of inertia of pulsar J0737-3039A. *New Astron.* **2009**, *14*, 40–43.
421. Kehl, M.S.; Wex, N.; Kramer, M.; Liu, K. Future Measurements of the Lense-Thirring Effect in the Double Pulsar. 2016, arXiv:1605.00408.
422. Iorio, L. Constraints from orbital motions around the Earth of the environmental fifth-force hypothesis for the OPERA superluminal neutrino phenomenology. *J. High Energy Phys.* **2012**, *5*, 73.
423. Deng, X.M.; Xie, Y. Yukawa effects on the clock onboard a drag-free satellite. *Mon. Not. R. Astron. Soc.* **2013**, *431*, 3236–3239.
424. Li, Z.W.; Yuan, S.F.; Lu, C.; Xie, Y. New upper limits on deviation from the inverse-square law of gravity in the solar system: A Yukawa parameterization. *Res. Astron. Astrophys.* **2014**, *14*, 139–143.
425. Khoury, J.; Weltman, A. Chameleon fields: Awaiting surprises for tests of gravity in space. *Phys. Rev. Lett.* **2004**, *93*, 171104.
426. Tully, R.B.; Fisher, J.R. A new method of determining distances to galaxies. *Astron. Astrophys.* **1977**, *54*, 661–673.
427. Dunsby, P.; Goheer, N.; Osano, B.; Uzan, J.P. How close can an inhomogeneous universe mimic the concordance model? *J. Cosmol. Astropart. Phys.* **2010**, *6*, 017.

428. Uzan, J.P.; Bernardeau, F. Lensing at cosmological scales: A test of higher dimensional gravity. *Phys. Rev. D* **2001**, *64*, 083004.
429. Peebles, P.J.E. *The Large-Scale Structure of the Universe*; Princeton Series in Physics: Princeton, NJ, USA, 1980.
430. Rich, J. *Fundamentals of Cosmology*; Springer: Berlin, Germany, 2001.
431. Uzan, J.P. The acceleration of the universe and the physics behind it. *Gen. Relativ. Gravit.* **2007**, *39*, 307–342.
432. Uzan, J.P. Tests of general relativity on astrophysical scales. *Gen. Relativ. Gravit.* **2010**, *42*, 2219–2246.
433. Lue, A.; Scoccimarro, R.; Starkman, G.D. Probing Newton's constant on vast scales: Dvali-Gabadadze-Porrati gravity, cosmic acceleration, and large scale structure. *Phys. Rev. D* **2004**, *69*, 124015.
434. Song, Y.S. Large scale structure formation of the normal branch in the DGP brane world model. *Phys. Rev. D* **2008**, *77*, 124031.
435. Benabed, K.; Bernardeau, F. Testing quintessence models with large-scale structure growth. *Phys. Rev. D* **2001**, *64*, 083501.
436. Koivisto, T.; Mota, D.F. Gauss-Bonnet quintessence: Background evolution, large scale structure, and cosmological constraints. *Phys. Rev. D* **2007**, *75*, 023518.
437. Tsujikawa, S. Quintessence: A review. *Class. Quantum Gravity* **2013**, *30*, 214003.
438. Baldi, M.; Pettorino, V.; Amendola, L.; Wetterich, C. Oscillating non-linear large-scale structures in growing neutrino quintessence. *Mon. Not. R. Astron. Soc.* **2011**, *418*, 214–229.
439. Koivisto, T.S.; Saridakis, E.N.; Tamanini, N. Scalar-fluid theories: Cosmological perturbations and large-scale structure. *J. Cosmol. Astropart. Phys.* **2015**, *2015*, 047.
440. Schimd, C.; Uzan, J.P.; Riazuelo, A. Weak lensing in scalar-tensor theories of gravity. *Phys. Rev. D* **2005**, *71*, 083512.
441. Rodríguez-Meza, M.A. Power spectrum of large-scale structure cosmological models in the framework of scalar-tensor theories. *J. Phys. Conf. Ser.* **2010**, *229*, 012063.
442. Goenner, H. Some remarks on the genesis of scalar-tensor theories. *Gen. Relativ. Gravit.* **2012**, *44*, 2077–2097.
443. Takushima, Y.; Terukina, A.; Yamamoto, K. Bispectrum of cosmological density perturbations in the most general second-order scalar-tensor theory. *Phys. Rev. D* **2014**, *89*, 104007.
444. Dossett, J.N.; Ishak, M.; Parkinson, D.; Davis, T.M. Constraints and tensions in testing general relativity from Planck and CFHTLenS data including intrinsic alignment systematics. *Phys. Rev. D* **2015**, *92*, 023003.
445. Di Valentino, E.; Melchiorri, A.; Silk, J. Cosmological hints of modified gravity? *Phys. Rev. D* **2016**, *93*, 023513.
446. Hu, B.; Liguori, M.; Bartolo, N.; Matarrese, S. Parametrized modified gravity constraints after Planck. *Phys. Rev. D* **2013**, *88*, 123514.
447. Pettorino, V. Testing modified gravity with Planck: The case of coupled dark energy. *Phys. Rev. D* **2013**, *88*, 063519.
448. Planck Collaboration. Planck 2015 Results. XIV. Dark Energy and Modified Gravity. *Astron. Astrophys.* **2016**, *594*, A14.
449. Räsänen, S. Dark energy from back-reaction. *J. Cosmol. Astropart. Phys.* **2004**, *2*, 003.
450. Wiltshire, D.L. Gravitational energy as dark energy: Towards concordance cosmology without Lambda. In Proceedings of the Dark Energy and Dark Matter: Observations, Experiments and Theories, Lyons, France, 7–11 July 2008; Volume 36, pp. 91–98.
451. Räsänen, S. Backreaction: Directions of progress. *Class. Quantum Gravity* **2011**, *28*, 164008.
452. Buchert, T.; Carfora, M.; Ellis, G.F.R.; Kolb, E.W.; MacCallum, M.A.H.; Ostrowski, J.J.; Räsänen, S.; Roukema, B.F.; Andersson, L.; Coley, A.A.; et al. Is there proof that backreaction of inhomogeneities is irrelevant in cosmology? *Class. Quantum Gravity* **2015**, *32*, 215021.
453. Räsänen, S.; Bolejko, K.; Finoguenov, A. New test of the Friedmann-Lemaître-Robertson-Walker metric using the distance sum rule. *Phys. Rev. Lett.* **2015**, *115*, 101301.
454. Räsänen, S.; Väliviita, J.; Kosonen, V. Testing distance duality with CMB anisotropies. *J. Cosmol. Astropart. Phys.* **2016**, *4*, 050.
455. Boehm, C.; Räsänen, S. Violation of the FRW consistency condition as a signature of backreaction. *J. Cosmol. Astropart. Phys.* **2013**, *9*, 003.
456. Buchert, T. On average properties of inhomogeneous fluids in general relativity: Dust cosmologies. *Gen. Relativ. Gravit.* **2000**, *32*, 105–126.

457. Buchert, T. Dark Energy from structure: A status report. *Gen. Relativ. Gravit.* **2008**, *40*, 467–527.
458. Banks, T. Supersymmetry breaking and the cosmological constant. *Int. J. Mod. Phy. A* **2014**, *29*, 1430010.
459. Arkani-Hamed, N.; Dimopoulos, S. Supersymmetric unification without low energy supersymmetry and signatures for fine-tuning at the LHC. *J. High Energy Phys.* **2005**, *6*, 073.
460. Susskind, L. The anthropic landscape of string theory. In *Universe or Multiverse?* Carr, B., Ed.; Cambridge University Press: Cambridge, UK, 2007.
461. Bauer, F.; Solà, J.; Štefancić, H. Dynamically avoiding fine-tuning the cosmological constant: The "Relaxed Universe". *J. Cosmol. Astropart. Phys.* **2010**, *12*, 029.
462. Bombelli, L.; Lee, J.; Meyer, D.; Sorkin, R.D. Space-time as a causal set. *Phys. Rev. Lett.* **1987**, *59*, 521–524.
463. Ahmed, M.; Dodelson, S.; Greene, P.B.; Sorkin, R. Everpresent Λ. *Phys. Rev. D* **2004**, *69*, 103523.
464. Weinberg, S. Anthropic bound on the cosmological constant. *Phys. Rev. Lett.* **1987**, *59*, 2607–2610.
465. Garriga, J.; Linde, A.; Vilenkin, A. Dark energy equation of state and anthropic selection. *Phys. Rev. D* **2004**, *69*, 063521.
466. Armendariz-Picon, C.; Mukhanov, V.; Steinhardt, P.J. Essentials of k-essence. *Phys. Rev. D* **2001**, *63*, 103510.
467. Kamenshchik, A.; Moschella, U.; Pasquier, V. An alternative to quintessence. *Phys. Lett. B* **2001**, *511*, 265–268.
468. de Putter, R.; Linder, E.V. Kinetic k-essence and quintessence. *Astropart. Phys.* **2007**, *28*, 263–272.
469. Durrer, R.; Maartens, R. Dark Energy and Modified Gravity. In *Dark Energy: Observational and Theoretical Approaches*; Ruiz-Lapuente, P., Ed.; Cambridge University Press: Cambridge, UK, 2008; pp. 48–91.
470. Deffayet, C.; Dvali, G.; Gabadadze, G. Accelerated universe from gravity leaking to extra dimensions. *Phys. Rev. D* **2002**, *65*, 044023.
471. Amendola, L.; Polarski, D.; Tsujikawa, S. Are f(R) Dark Energy Models Cosmologically Viable? *Phys. Rev. Lett.* **2007**, *98*, 131302.
472. Turyshev, S.G.; Toth, V.T.; Kinsella, G.; Lee, S.C.; Lok, S.M.; Ellis, J. Support for the thermal origin of the pioneer anomaly. *Phys. Rev. Lett.* **2012**, *108*, 241101.
473. Adler, C.G.; Coulter, B.L. Galileo and the Tower of Pisa experiment. *Am. J. Phys.* **1978**, *46*, 199–201.
474. Bergé, J.; Touboul, P.; Rodrigues, M.; The MICROSCOPE Team. Status of MICROSCOPE, a mission to test the Equivalence Principle in space. *J. Phys. Conf. Ser.* **2015**, *610*, 012009.
475. Overduin, J.; Everitt, F.; Worden, P.; Mester, J. STEP and fundamental physics. *Class. Quantum Gravity* **2012**, *29*, 184012.
476. Nobili, A.M.; Comandi, G.L.; Doravari, S.; Bramanti, D.; Kumar, R.; Maccarrone, F.; Polacco, E.; Turyshev, S.G.; Shao, M.; Lipa, J.; et al. "Galileo Galilei" (GG) a small satellite to test the equivalence principle of Galileo, Newton and Einstein. *Exp. Astron.* **2009**, *23*, 689–710.
477. Muller, H.; Peters, A.; Chu, S. A precision measurement of the gravitational redshift by the interference of matter waves. *Nature* **2010**, *463*, 926–929.
478. Bardeen, J.M.; Petterson, J.A. The lense-thirring effect and accretion disks around kerr black holes. *Astrophys. J.* **1975**, *195*, L65.
479. Pound, R.V.; Rebka, G.A. Apparent weight of photons. *Phys. Rev. Lett.* **1960**, *4*, 337–341.
480. Turyshev, S.G.; Yu, N.; Toth, V.T. General relativistic observables for the ACES experiment. *Phys. Rev. D* **2016**, *93*, 045027.
481. Vessot, R.F.C.; Levine, M.W.; Mattison, E.M.; Blomberg, E.L.; Hoffman, T.E.; Nystrom, G.U.; Farrel, B.F.; Decher, R.; Eby, P.B.; Baugher, C.R. Test of relativistic gravitation with a space-borne hydrogen maser. *Phys. Rev. Lett.* **1980**, *45*, 2081–2084.
482. Schiller, S.; Tino, G.M.; Gill, P.; Salomon, C.; Sterr, U.; Peik, E.; Nevsky, A.; Görlitz, A.; Svehla, D.; Ferrari, G.; et al. Einstein Gravity Explorer—A medium-class fundamental physics mission. *Exp. Astron.* **2009**, *23*, 573–610.
483. Antoniadis, J.; Freire, P.C.C.; Wex, N.; Tauris, T.M.; Lynch, R.S.; van Kerkwijk, M.H.; Kramer, M.; Bassa, C.; Dhillon, V.S.; Driebe, T.; et al. A massive pulsar in a compact relativistic binary. *Science* **2013**, *340*, 448.
484. Iorio, L. Constraining the preferred-frame α_1, α_2 parameters from solar system planetary precessions. *Int. J. Mod. Phys. D* **2014**, *23*, 1450006.
485. Fienga, A.; Laskar, J.; Exertier, P.; Manche, H.; Gastineau, M. Numerical estimation of the sensitivity of INPOP planetary ephemerides to general relativity parameters. *Celest. Mech. Dyn. Astron.* **2015**, *123*, 325–349.

486. Everitt, C.W.F.; Debra, D.B.; Parkinson, B.W.; Turneaure, J.P.; Conklin, J.W.; Heifetz, M.I.; Keiser, G.M.; Silbergleit, A.S.; Holmes, T.; Kolodziejczak, J.; et al. Gravity Probe B: Final results of a space experiment to test general relativity. *Phys. Rev. Lett.* **2011**, *106*, 221101.
487. Kramer, M. Pulsars as probes of gravity and fundamental physics. *Int. J. Mod. Phys. D* **2016**, *25*, 1630029.
488. Bertotti, B.; Ciufolini, I.; Bender, P.L. New test of general relativity—Measurement of de Sitter geodetic precession rate for lunar perigee. *Phys. Rev. Lett.* **1987**, *58*, 1062–1065.
489. Shapiro, I.I.; Reasenberg, R.D.; Chandler, J.F.; Babcock, R.W. Measurement of the de Sitter precession of the moon—A relativistic three-body effect. *Phys. Rev. Lett.* **1988**, *61*, 2643–2646.
490. Williams, J.G.; Newhall, X.X.; Dickey, J.O. Relativity parameters determined from lunar laser ranging. *Phys. Rev. D* **1996**, *53*, 6730–6739.
491. Merkowitz, S.M. Tests of gravity using lunar laser ranging. *Living Rev. Relativ.* **2010**, *13*, 7.
492. Martini, M.; Dell'Agnello, S.; Currie, D.; Delle Monache, G.O.; Vittori, R.; Berardi, S.; Boni, A.; Cantone, C.; Ciocci, E.; Lops, C.; et al. Moonlight: A new lunar laser ranging retroreflector and the lunar geodetic precession. *Acta Polytech.* **2013**, *53*, 745–750.
493. Kramer, M. Determination of the geometry of the PSR B1913 + 16 System by geodetic precession. *Astrophys. J.* **1998**, *509*, 856–860.
494. Breton, R.P.; Kaspi, V.M.; Kramer, M.; McLaughlin, M.A.; Lyutikov, M.; Ransom, S.M.; Stairs, I.H.; Ferdman, R.D.; Camilo, F.; Possenti, A. Relativistic spin precession in the double pulsar. *Science* **2008**, *321*, 104–107.
495. Lucchesi, D.M.; Anselmo, L.; Bassan, M.; Pardini, C.; Peron, R.; Pucacco, G.; Visco, M. Testing the gravitational interaction in the field of the Earth via satellite laser ranging and the Laser Ranged Satellites Experiment (LARASE). *Class. Quantum Gravity* **2015**, *32*, 155012.
496. Ingram, A.; van der Klis, M.; Middleton, M.; Done, C.; Altamirano, D.; Heil, L.; Uttley, P.; Axelsson, M. A quasi-periodic modulation of the iron line centroid energy in the black hole binary H 1743-322. *Mon. Not. R. Astron. Soc.* **2016**, *461*, 1967–1980.
497. Bosi, F.; Cella, G.; di Virgilio, A.; Ortolan, A.; Porzio, A.; Solimeno, S.; Cerdonio, M.; Zendri, J.P.; Allegrini, M.; Belfi, J.; et al. Measuring gravitomagnetic effects by a multi-ring-laser gyroscope. *Phys. Rev. D* **2011**, *84*, 122002.
498. Stone, N.; Loeb, A. Observing lense-thirring precession in tidal disruption flares. *Phys. Rev. Lett.* **2012**, *108*, 061302.
499. Franchini, A.; Lodato, G.; Facchini, S. Lense-Thirring precession around supermassive black holes during tidal disruption events. *Mon. Not. R. Astron. Soc.* **2016**, *455*, 1946–1956.
500. Martynov, D.V.; Hall, E.D.; Abbott, B.P.; Abbott, R.; Abbott, T.D.; Adams, C.; Adhikari, R.X.; Anderson, R.A.; Anderson, S.B.; Arai, K.; et al. Sensitivity of the advanced LIGO detectors at the beginning of gravitational wave astronomy. *Phys. Rev. D* **2016**, *93*, 112004.
501. LIGO Scientific Collaboration and Virgo Collaboration. GW151226: Observation of gravitational waves from a 22-solar-mass binary black hole coalescence. *Phys. Rev. Lett.* **2016**, *116*, 241103.
502. Johannsen, T. Sgr A* and general relativity. *Class. Quantum Gravity* **2016**, *33*, 113001.
503. Soffel, M.; Wirrer, R.; Schastok, J.; Ruder, H.; Schneider, M. Relativistic effects in the motion of artificial satellites: The oblateness of the central body I. *Celest. Mech.* **1987**, *42*, 81–89.
504. Heimberger, J.; Soffel, M.; Ruder, H. Relativistic effects in the motion of artificial satellites: The oblateness of the central body II. *Celest. Mech. Dyn. Astron.* **1989**, *47*, 205–217.
505. Soffel, M. *Relativity in Astrometry, Celestial Mechanics and Geodesy*; Springer-Verlag: Berlin, Germany, 1989.
506. Iorio, L. Post-Newtonian direct and mixed orbital effects due to the oblateness of the central body. *Int. J. Mod. Phys. D* **2015**, *24*, 1550067.
507. Iorio, L. A possible new test of general relativity with Juno. *Class. Quantum Gravity* **2013**, *30*, 195011.
508. Starkman, G.D.; Trotta, R.; Vaudrevange, P.M. The virtues of frugality—Why cosmological observers should release their data slowly. *Mon. Not. R. Astron. Soc.* **2010**, *401*, L15–L18.
509. Domcke, V.; Pieroni, M.; Binétruy, P. Primordial gravitational waves for universality classes of pseudoscalar inflation. *J. Cosmol. Astropart. Phys.* **2016**, *6*, 031.
510. Ito, A.; Soda, J. MHz gravitational waves from short-term anisotropic inflation. *J. Cosmol. Astropart. Phys.* **2016**, *4*, 035.
511. Fidler, C.; Pettinari, G.W.; Beneke, M.; Crittenden, R.; Koyama, K.; Wands, D. The intrinsic B-mode

polarisation of the Cosmic Microwave Background. *J. Cosmol. Astropart. Phys.* **2014**, *7*, 011.

512. Namikawa, T.; Nagata, R. Non-Gaussian structure of B-mode polarization after delensing. *J. Cosmol. Astropart. Phys.* **2015**, *10*, 004.
513. Abazajian, K.N.; Arnold, K.; Austermann, J.; Benson, B.A.; Bischoff, C.; Bock, J.; Bond, J.R.; Borrill, J.; Calabrese, E.; Carlstrom, J.E.; et al. Neutrino physics from the cosmic microwave background and large scale structure. *Astropart. Phys.* **2015**, *63*, 66–80.
514. Trotta, R. Bayes in the sky: Bayesian inference and model selection in cosmology. *Contemp. Phys.* **2008**, *49*, 71–104.
515. Hobson, M.P.; Jaffe, A.H.; Liddle, A.R.; Mukeherjee, P.; Parkinson, D. *Bayesian Methods in Cosmology*; Cambridge University Press: Cambridge, UK, 2010.
516. Liddle, A.R. How many cosmological parameters? *Mon. Not. R. Astron. Soc.* **2004**, *351*, L49–L53.
517. Metcalf, R.B.; Silk, J. On breaking cosmic degeneracy. *Astrophys. J.* **1998**, *492*, L1–L4.
518. Debono, I.; Rassat, A.; Réfrégier, A.; Amara, A.; Kitching, T.D. Weak lensing forecasts for dark energy, neutrinos and initial conditions. *Mon. Not. R. Astron. Soc.* **2010**, *404*, 110–119.
519. Howlett, C.; Lewis, A.; Hall, A.; Challinor, A. CMB power spectrum parameter degeneracies in the era of precision cosmology. *J. Cosmol. Astropart. Phys.* **2012**, *4*, 027.
520. Joyce, A.; Jain, B.; Khoury, J.; Trodden, M. Beyond the cosmological standard model. *Phys. Rep.* **2015**, *568*, 1–98.
521. Jaffe, A. H_0 and odds on cosmology. *Astrophys. J.* **1996**, *471*, 24.
522. Mukherjee, P.; Parkinson, D.; Liddle, A.R. A nested sampling algorithm for cosmological model selection. *Astrophys. J.* **2006**, *638*, L51–L54.
523. Kilbinger, M.; Wraith, D.; Robert, C.P.; Benabed, K.; Cappé, O.; Cardoso, J.F.; Fort, G.; Prunet, S.; Bouchet, F.R. Bayesian model comparison in cosmology with Population Monte Carlo. *Mon. Not. R. Astron. Soc.* **2010**, *405*, 2381–2390.
524. Wraith, D.; Kilbinger, M.; Benabed, K.; Cappé, O.; Cardoso, J.; Fort, G.; Prunet, S.; Robert, C.P. Estimation of cosmological parameters using adaptive importance sampling. *Phys. Rev. D* **2009**, *80*, 023507.
525. Del Pozzo, W.; Veitch, J.; Vecchio, A. Testing general relativity using Bayesian model selection: Applications to observations of gravitational waves from compact binary systems. *Phys. Rev. D* **2011**, *83*, 082002.
526. Ellis, G.F.R. *Dark Matter and Dark Energy Proposals: Maintaining Cosmology as a True Science*; EAS Publications Series; European Astronomical Society: Versoix, Switzerland, 2009; Volume 36, pp. 325–336.
527. Ellis, G.F.R. Issues in the Philosophy of Cosmology. In *Philosophy of Physics*; Butterfield, J., Earman, J., Eds.; Handbook of the Philosophy of Science; Elsevier: Amsterdam, The Netherlands, 2006; pp. 1183–1285.
528. Jackman, S. *Bayesian Analysis for the Social Sciences*; Wiley: Hoboken, NJ, USA, 2009.
529. Niemack, M.D.; Ade, P.A.R.; Aguirre, J.; Barrientos, F.; Beall, J.A.; Bond, J.R.; Britton, J.; Cho, H.M.; Das, S.; Devlin, M.J.; et al. ACTPol: A polarization-sensitive receiver for the Atacama Cosmology Telescope. *Proc. SPIE* **2010**, *7741*, 77411S.
530. Beynon, E.; Bacon, D.J.; Koyama, K. Weak lensing predictions for modified gravities at non-linear scales. *Mon. Not. R. Astron. Soc.* **2010**, *403*, 353–362.
531. Harnois-Déraps, J.; Munshi, D.; Valageas, P.; van Waerbeke, L.; Brax, P.; Coles, P.; Rizzo, L. Testing modified gravity with cosmic shear. *Mon. Not. R. Astron. Soc.* **2015**, *454*, 2722–2735.
532. Carlstrom, J.E.; CMB-S4 Collaboration. The next generation ground-based CMB experiment, CMB-S4. In Proceedings of the 228th AAS Meeting, San Diego, CA, USA, 12–16 June 2016.
533. Santos, M.; Bull, P.; Alonso, D.; Camera, S.; Ferreira, P.; Bernardi, G.; Maartens, R.; Viel, M.; Villaescusa-Navarro, F.; Abdalla, F.B.; et al. Cosmology from a SKA HI intensity mapping survey. In Proceedings of the Advancing Astrophysics with the Square Kilometre Array (AASKA14), Giardini Naxos, Italy, 9–13 June 2014.
534. Hall, A.; Bonvin, C.; Challinor, A. Testing general relativity with 21-cm intensity mapping. *Phys. Rev. D* **2013**, *87*, 064026.
535. Lahav, O.; Suto, Y. Measuring our Universe from Galaxy Redshift Surveys. *Living Rev. Relativ.* **2004**, *7*, 8.
536. Dupé, F.X.; Rassat, A.; Starck, J.L.; Fadili, M.J. Measuring the integrated Sachs-Wolfe effect. *Astron. Astrophys.* **2011**, *534*, A51.
537. Stabenau, H.F.; Jain, B. N-body simulations of alternative gravity models. *Phys. Rev. D* **2006**, *74*, 084007.

538. Adamek, J.; Daverio, D.; Durrer, R.; Kunz, M. General relativistic N-body simulations in the weak field limit. *Phys. Rev. D* **2013**, *88*, 103527.
539. He, J.H.; Li, B.; Hawken, A.J. Effective Dark Matter power spectra in $f(R)$ gravity. *Phys. Rev. D* **2015**, *92*, 103508.
540. Winther, H.A.; Schmidt, F.; Barreira, A.; Arnold, C.; Bose, S.; Llinares, C.; Baldi, M.; Falck, B.; Hellwing, W.A.; Koyama, K.; et al. Modified gravity N-body code comparison project. *Mon. Not. R. Astron. Soc.* **2015**, *454*, 4208–4234.
541. Eingorn, M. First-order cosmological perturbations engendered by point-like masses. *Astrophys. J.* **2016**, *825*, 84.
542. Hahn, O.; Paranjape, A. General Relativistic "Screening" in Cosmological Simulations. 2016, arXiv:1602.07699.
543. Kamdar, H.M.; Turk, M.J.; Brunner, R.J. Machine learning and cosmological simulations—I. Semi-analytical models. *Mon. Not. R. Astron. Soc.* **2016**, *455*, 642–658.
544. Kamdar, H.M.; Turk, M.J.; Brunner, R.J. Machine learning and cosmological simulations—II. Hydrodynamical simulations. *Mon. Not. R. Astron. Soc.* **2016**, *457*, 1162–1179.
545. Smoot, G.F. See Saw Inflation/Dark Matter/Dark Energy/Baryogenesis. 2014, arXiv:1405.2776.
546. Rickles, D. A "Second Superstring Revolution" and the Future of String Theory. In *A Brief History of String Theory: From Dual Models to M-Theory*; Springer: Berlin, Germany, 2014; pp. 207–242.
547. Maartens, R.; Koyama, K. Brane-world gravity. *Living Rev. Relativ.* **2010**, *13*, 5.
548. Woodard, R.P. How far are we from the quantum theory of gravity? *Rep. Prog. Phys.* **2009**, *72*, 126002.
549. Norris, R.P. Data challenges for next-generation radio telescopes. In Proceedings of the Sixth IEEE International Conference on eScience, Brisbane, Australia, 7–10 December 2010; pp. 21–24.
550. Planck Collaboration. Planck 2015 Results. II. Low Frequency Instrument Data Processing. *Astron. Astrophys.* **2016**, *594*, A2.
551. Planck Collaboration. Planck 2015 Results. VIII. High Frequency Instrument Data Processing: Calibration and Maps. *Astron. Astrophys.* **2016**, *594*, A8.
552. Jurić, M.; Kantor, J.; Lim, K.; Lupton, R.H.; Dubois-Felsmann, G.; Jenness, T.; Axelrod, T.S.; Aleksić, J.; Allsman, R.A.; AlSayyad, Y.; et al. The LSST Data Management System. 2015, arXiv:1512.07914.
553. Desai, S.; Mohr, J.J.; Henderson, R.; Kümmel, M.; Paech, K.; Wetzstein, M. CosmoDM and its application to Pan-STARRS data. *J. Instrum.* **2015**, *10*, C06014.
554. Dodson, R.; Vinsen, K.; Wu, C.; Popping, A.; Meyer, M.; Wicenec, A.; Quinn, P.; van Gorkom, J.; Momjian, E. Imaging SKA-scale data in three different computing environments. *Astron. Comput.* **2016**, *14*, 8–22.
555. Arviset, C.; González, J.; Gutiérrez, R.; Hernández, J.; Salgado, J.; Segovia, J.C. The Gaia archive: VO in action in the big data era. In Proceedings of the International Symposium Dedicated to the 50th Anniversary of the Markarian Survey and the 10th Anniversary of the Armenian Virtual Observatory, Byurakan, Armenia, 5–8 October 2015; Volume 505, p. 218.

8

What is the Validity Domain of Einstein's Equations? Distributional Solutions over Singularities and Topological Links in Geometrodynamics

Elias Zafiris

Parmenides Foundation, Center for the Conceptual Foundations of Science, Kirchplatz 1, Pullach, 82049 Munich, Germany; elias.zafiris@parmenides-foundation.org

Academic Editors: Stephon Alexander, Jean-Michel Alimi, Elias C. Vagenas and Lorenzo Iorio

Abstract: The existence of singularities alerts that one of the highest priorities of a centennial perspective on general relativity should be a careful re-thinking of the validity domain of Einstein's field equations. We address the problem of constructing distinguishable extensions of the smooth spacetime manifold model, which can incorporate singularities, while retaining the form of the field equations. The sheaf-theoretic formulation of this problem is tantamount to extending the algebra sheaf of smooth functions to a distribution-like algebra sheaf in which the former may be embedded, satisfying the pertinent cohomological conditions required for the coordinatization of all of the tensorial physical quantities, such that the form of the field equations is preserved. We present in detail the construction of these distribution-like algebra sheaves in terms of residue classes of sequences of smooth functions modulo the information of singular loci encoded in suitable ideals. Finally, we consider the application of these distribution-like solution sheaves in geometrodynamics by modeling topologically-circular boundaries of singular loci in three-dimensional space in terms of topological links. It turns out that the Borromean link represents higher order wormhole solutions.

Keywords: general relativity; sheaf cohomology; abstract differential geometry; singularities; geometrodynamics; distributions; generalized functions; nowhere dense algebras; algebra sheaves; topological links; wormholes; Borromean rings

---

1. Introduction

One hundred years after Einstein's initial conception and formulation of the general theory of relativity, it still remains a vibrant subject of intense research and formidable depth. In this way, during all of these years, our understanding of gravitation in differential geometric terms is being continuously refined. We believe that one of the highest priorities of a centennial perspective on general relativity should be a careful re-examination of the validity domain of Einstein's field equations. These equations constitute the irreducible kernel of general relativity and the possibility of retaining the form of Einstein's equations, while concurrently extending their domain of validity is promising for shedding new light on old problems and guiding toward their effective resolution. These problems are primarily related to the following perennial issues: (a) the smooth manifold background of the theory; (b) the existence of singular loci in spacetime where the metric breaks down or the curvature blows up; and (c) the non-geometric nature of the second part of Einstein's equations involving the energy-momentum tensor. It turns out that these problems are intrinsically related to each other and require a critical re-thinking of the initial assumptions referring to the domain of validity of Einstein's equations.

In this communication, first of all, we would like to consider the problem of constructing distinguishable extensions of the smooth spacetime manifold solution space of Einstein's equations incorporating singularities by taking into account recent developments in differential geometry. These developments pertain to the possible generalization of the technical framework of differential manifolds, on which the formalism and interpretation of general relativity is based on, to non-smooth or singular topological spaces by applying concepts and methods of sheaf theory and sheaf cohomology. In a nutshell, it turns out that all of the usual local constructions of differential geometry, re-interpreted sheaf-theoretically, do not require the notion of a global smooth manifold, but are based on much weaker conditions of an essentially cohomological nature. The physical interpretation of these findings, referring to appropriate extensions of Einstein's equations over singular domains, is tantamount to the viable possibility of extending the covariant formulation of Einstein's equations using continuous distribution-like or even non-smooth sheaves of coefficients for all of the involved tensorial physical quantities.

Second, we would like to show explicitly how certain generalized distribution-like solutions of partial differential equations, which fit appropriately in the above-mentioned sheaf-theoretic framework of differential geometry, bear significance in relation to obtaining singularity-free solutions of Einstein's equations in extended domains. We scrutinize the generation of these distribution-like algebra sheaves of coefficients from a physical perspective and explain the means of their construction in terms of residue classes of sequences of smooth functions modulo the information of singular loci encoded in suitable ideals.

Finally, we consider the application of these distribution-like solution sheaves in geometrodynamics. The geometrodynamical formalism is very instructive in relation to the proposed extensions because it leads to the conclusion that active positive gravitational mass may emerge from purely topological considerations taking into account the constraints imposed by Einstein's field equations in the vacuum. In this manner, we may re-assessfruitfully Wheeler's insights referring to "mass without mass" and "charge without charge", as well as re-evaluate the notion of wormhole solutions from a cohomological point of view. In this context, we propose to model topologically-circular boundaries of singular loci in three-dimensional space in terms of topological links. It turns out that there exists a universal topological link bearing the connectivity property of the Borromean rings. The cohomological expression of the Borromean link points to its physical interpretation as a higher order wormhole solution of the field equations.

2. General Relativity from the Perspective of Sheaf Theory

In the standard formulation of general relativity, the spacetime event structure is represented by means of a connected, four-dimensional real smooth manifold X. The chronogeometric relations on the event manifold X are expressed in terms of a pseudo-Riemannian metric of the Lorentzian signature, called the spacetime metric. The chronogeometric relations are not fixed kinematically a priori, like in all predecessor classical field theories, but they should be obtained dynamically in terms of the metric as a solution of Einstein's field equations depending on the energy-momentum matter field distributions. In this manner, all of the pertinent chronogeometric relations defined on a four-dimensional smooth manifold, endowing it with the structure of a spacetime manifold, become variable. The dynamical constitution of these relations by means of the field equations requires the imposition of a compatibility requirement relating the metric tensor, which represents the spacetime geometry, with the affine connection, which represents the differential evolution of the gravitational field. A spacetime manifold is considered to be without singularities if the coefficients of the metric tensor field are smooth and the manifold X is geodesically complete with respect to the metric. In this case, all timelike geodesic curves can be extended to arbitrary length in the smooth spacetime manifold X. From a physical viewpoint, according to the above requirements, the notion of localization at a spacetime point-event is sensible only if the coefficients of the metric tensor field are smooth in an open neighborhood of this point.

Algebraically speaking, a real smooth manifold X can be reconstructed entirely from the $\mathbb{R}$-algebra $\mathbb{C}^{\infty}(X)$ of smooth real-valued functions on it, and in particular, the points of X are derived from the algebra $\mathbb{C}^\infty(X)$ as the $\mathbb{R}$-algebra homomorphisms $\mathbb{C}^\infty(X) \to \mathbb{R}$. This important observation in relation to general relativity has been first proposed and explicated by Geroch in the form of Einstein algebras [1]. From a modern mathematical perspective, it is a consequence of the Gelfand representation theorem applied to the case of smooth manifolds [2,3]. In this way, manifold points constitute the $\mathbb{R}$-spectrum of the algebra of smooth functions $\mathbb{C}^\infty(X)$, being isomorphic with the maximal ideals of this algebra. Notice that the $\mathbb{R}$-algebra $\mathbb{C}^\infty(X)$ is a commutative topological algebra that contains the field of real numbers $\mathbb{R}$ as a distinguished subalgebra, encapsulating the predominant physical assumption that our means of characterizing events is conducted by evaluations in the field of real numbers $\mathbb{R}$.

The algebraic viewpoint is instructive because it makes clear that in the standard differential geometric setting of general relativity, all of the tensorial physical quantities are coordinatized by means of the commutative $\mathbb{R}$-algebra of globally-defined smooth real-valued functions $\mathbb{C}^\infty(X)$. Hence, the background of the theory remains fixed as the $\mathbb{R}$-spectrum of the commutative topological algebra $\mathbb{C}^\infty(X)$, supplying smooth coefficients for the coordinatization of physical quantities. The points of the manifold X, although not dynamically localizable degrees of freedom in general relativity, serve as the semantic information carriers of spacetime events. More precisely, the points are marked on a smooth manifold in terms of global evaluations of the smooth algebra $\mathbb{C}^\infty(X)$ in the field of real numbers. The subtlety of general relativity is exactly that manifold points are not dynamically localizable entities in the theory. More precisely, manifold points assume an indirect reference as indicators of spacetime events, only after the dynamical specification of chronogeometrical relations among them, as particular solutions of the generally covariant field equations. Clearly, the existence of singular loci in spacetime where the metric breaks down in terms of smooth function coefficients forbids the association of smooth manifold points with spacetime events. What remains is an emergent notion of an event horizon of a singular locus where spacetime information may be encoded appropriately.

The dynamical variability of the coefficients coordinatizing all tensorial physical quantities requires the action of a covariant differential operator to be applied upon them. This takes place via the notion of an affine connection, which is expressed as a covariant derivative acting on these smooth coefficients. The result of differentiation is encoded in $\mathbb{C}^\infty(X)$-modules over the algebra $\mathbb{C}^\infty(X)$, called modules of differential forms Ω and their duals $\Xi = Hom(\Omega, \mathbb{C}^\infty(X))$, as well as their higher powers constructed by means of exterior algebra.

In the same algebraic context, the role of a metric geometry on a smooth manifold, as related to the above modules of differential forms and their duals in general relativity, pertains to the representability of spacetime events by points of the manifold, which in turn necessitates their coordinatization in terms of real numbers. This is tantamount to the requirement that all types of differentially-variable quantities should possess uniquely-defined dual types, such that their point-event representability can be made possible by means of real numbers. This is precisely the role of a geometry induced by a metric. Concretely, the spacetime metric assigns a unique dual to each differentially-variable quantity, by effecting an isomorphism between the modules Ω and $\Xi := Hom(\Omega, \mathbb{C}^\infty(X))$, that is $g : \Omega \simeq \Xi$, such that $df \mapsto v_f := g(df)$.

The important thing to notice is that all of these constructions can be performed strictly locally, that is by using only sections defined in the neighborhood of points. This is an implication that differential geometric constructions should be expressed not in terms of global algebra coefficients, but in terms of sheaves of coefficients defined locally. Then, the task is to study the maximal extendibility of these constructions from the local to the global level, which is technically expressed via the theory of sheaf cohomology.

In the context of general relativity, the modeling of the dynamical variability, caused by the gravitational field by means of the Levi–Civita connection, from a local sheaf-theoretic perspective, is becoming even more relevant in view of the spacetime metric compatibility of this connection and the associated solution space of the theory. Einstein's equations are formulated in terms of non-linear partial differential equations involving smooth functions, playing the role of local coefficients coordinatizing the metric tensor, the Ricci tensor and the scalar curvature. The solution of these equations in terms of the spacetime metric determines the local metrical properties of the spacetime manifold around any point, depending on the energy-momentum tensor. Notwithstanding this, all of the global cosmological predictions of the theory are obtained not from these local solutions of the field equations per se, but from the possibility of the continuation of some local solution to an extended region. The method of the continuation or extension of some solution from the local to the global level is mathematically of a sheaf-theoretic nature.

In view of the problem of singularities in general relativity, this is a clear warning that distribution-like sheaves of coefficients may be more appropriate for the continuation of some local solution over extended regions when the smooth ones become ill-defined over singular loci. It is a natural requirement that these sheaves of coefficients contain the standard smooth ones as a subalgebra, or equivalently, there is an algebra sheaf embedding of the smooth coefficients into the generalized ones. It is expected that distribution-like sheaves of coefficients can prevent the breaking down of the metric at singularities and, therefore, provide the means to extend the domain of validity of the field equations, under the proviso that the same tensorial equations can be re-expressed covariantly in terms of these generalized sheaves of coefficients.

3. Cohomological Conditions for Extending the Smooth Sheaf of Coefficients in General Relativity

Cohomology theory constitutes a sophisticated algebraic-topological method of assigning global invariants to a topological space in a homotopy-invariant way. The cohomology groups measure the global obstructions for extending sections from the local to the global level, for instance extending local solutions of a differential equation to a global solution. The differential geometric mechanism of smooth manifolds is essentially based on the set-up of the de Rham complex in terms of locally-defined smooth coefficients. In particular, de Rham cohomology measures the extent that closed differential forms fail to be exact and, thus, the obstruction to integrability. In this context, the central role is played by the lemma of Poincaré, according to which every closed differential form is locally exact in terms of smooth coefficients. The de Rham theorem asserts that the homomorphism from the de Rham cohomology ring to the differentiable singular cohomology ring, given by the integration of closed forms over differentiable singular cycles, is a ring isomorphism. The sheaf-theoretic understanding of this deep result came after the realization that both the de Rham cohomology and the differentiable singular cohomology are actually special isomorphic cases of sheaf cohomology. In particular, it has been also crystallized that the de Rham cohomology of a differential manifold depends only on the property of paracompactness of the underlying topological space. In turn, the paracompactness property, which is required in the definition of a differential manifold, can be also characterized cohomologically via the acyclic behavior of soft sheaves, like the sheaf of smooth functions. In other words, soft sheaves, namely sheaves whose sections over any closed subset can be extended to a global section, are acyclic over a paracompact topological space.

The re-interpretation and generalization of the standard de Rham cohomology theory on manifolds in sheaf-cohomological terms is physically significant, because it provides an intrinsic way to set up and solve differential equations expressing the dynamical variability of physical quantities. The concepts and technical tools of sheaf cohomology have been developed through the ground-breaking work of Grothendieck in geometry [4,5]. What should be initially kept in mind for physical applications is that the natural argument of a cohomology theory is a pair consisting of a topological space together with a sheaf of commutative algebras defined over it, rather than just a space.

It is instructive to include the basic definition characterizing the notion of a sheaf of sets on a topological space X, which also gives rise in a direct way to the notion of a sheaf of commutative algebras over X that we will employ in the sequel:

A presheaf $\mathbb{F}$ of sets on a topological space X, consists of the following information:

(I) For every open set U of X, a set denoted by $\mathbb{F}(U)$, and

(II) For every inclusion $V \hookrightarrow U$ of open sets of X, a restriction morphism of sets in the opposite direction:

$$r(U|V) : \mathbb{F}(U) \to \mathbb{F}(V) \tag{1}$$

such that:

(a) $r(U|U)$ = identity at $\mathbb{F}(U)$ for all open sets U of X.
(b) $r(V|W) \circ r(U|V) = r(U|W)$ for all open sets $W \hookrightarrow V \hookrightarrow U$. Usually, the following simplifying notation is used: $r(U|V)(s) := s|_V$.

A presheaf $\mathbb{F}$ of sets on a topological space X is defined to be a sheaf if it satisfies the following two conditions, for every family V_a, $a \in I$, of local open covers of V, where V open set in X, such that $V = \cup_a V_a$:

(1) Local identity axiom of sheaf:

Given $s, t \in \mathbb{F}(V)$ with $s|_{V_a} = t|_{V_a}$ for all $a \in I$, then $s = t$.

(2) Gluing axiom of sheaf:

Given $s_a \in \mathbb{F}(V_a)$, $s_b \in \mathbb{F}(V_b)$, $a, b \in I$, such that:

$$s_a|_{V_a \cap V_b} = s_b|_{V_a \cap V_b}, \tag{2}$$

for all $a, b \in I$, then there exists a unique $s \in \mathbb{F}(V)$, such that: $s|_{V_a} = s_a \in F(V_a)$ and $s|_{V_b} = s_b \in F(V_b)$.

As a basic example, if $\mathbb{F}$ denotes the presheaf that assigns to each open set $U \subset X$ the commutative algebra of all real-valued continuous functions on U, then $\mathbb{F}$ is actually a sheaf. This is intuitively clear since the specification of a topology on X is solely used for the definition of the continuous functions on X. Thus, the continuity of each function can be determined locally. This means that continuity respects the operation of restriction to open sets and, moreover, that continuous functions can be amalgamated together in a unique manner, as is required for the satisfaction of the sheaf condition.

The realization that the natural argument of a cohomology theory is not only a space, but it is actually a pair consisting of a topological space together with a sheaf of commutative algebras localized over it, has given rise to the notion of a commutative locally $\mathbb{R}$-algebraized space, defined by means of a pair $(X, \mathbb{A})$ consisting of a topological space X and a sheaf of commutative $\mathbb{R}$-algebras $\mathbb{A}$ on X, such that the restriction $\mathbb{A}_x$ is a local commutative $\mathbb{R}$-algebra for any point $x \in X$. Regarding the possibility of extending consistently all of the standard local differential geometric constructions in the context of smooth manifolds to singular spaces, in terms of locally $\mathbb{R}$-algebraized spaces, where a suitable sheaf of commutative $\mathbb{R}$-algebras $\mathbb{A}$ on X substitutes the smooth sheaf of $\mathbb{R}$-algebras $\mathbb{C}^\infty(X)$), a full-grown theory has been recently developed, called Abstract Differential Geometry (ADG). This theory has shown that the standard differential-analytic tools of locally-Euclidean spaces and smooth manifolds leading to the formulation and solution of differential equations can be actually re-produced and generalized to non-smooth or singular topological spaces by means of sheaf cohomology. Equivalently, the suitability of a sheaf of commutative $\mathbb{R}$-algebras $\mathbb{A}$ on an abstract topological space X for expressing the differential geometric mechanism in terms of these coefficients instead of the smooth ones is entirely determined only by the satisfaction of precise cohomological conditions pertaining to the characterization of the algebra sheaf $\mathbb{A}$. We note, in passing, that for the economy of symbols, we denote algebra sheaves by the same symbols as we used for the algebras before, since the difference is clear from the context.

The mathematical theory of ADG has been built rigorously by Mallios [6,7] (see also [8]), based on critical prior work of Selesnick [9]. The significance of ADG for physics has been also shown by an explicit reconstruction and generalization of the framework of the Maxwell and Yang–Mills gauge field theories in sheaf cohomological terms [10,11]; see also [12–14]. An exposition of the basic didactics of ADG in relation to its physical applications has been presented by Raptis in [15]. The basic method introduced for the generalization of the standard analytic tools of Classical Differential Geometry (CDG) consists of the following: Initially, a concept of CDG is suitable for the extension to a broader differential context (beyond the context of smooth manifolds) if it is liable to a process of sheaf-theoretic localization [16]. In CDG, all of the differential geometric constructions require that the base space is a smooth manifold. The underlying reason is that the means of differentiation are lifted locally from the structure of a Euclidean space. In this way, the de Rham complex is fixed with respect to smooth coefficients, and all tensorial quantities are coordinatized in smooth terms. In ADG, the base space provides merely a topological basis of sheaf-theoretic localization, such that all of the pertinent differential geometric constructions can take place locally, whereas the latter are not subordinate to this topological basis, meaning that they are not dependent on any particular localization basis. Thus, the object of primary significance in ADG is not the base space itself, but the algebra sheaf of coefficients localized over it. The differentiation structure is built in the algebra sheaf of coefficients by means of the notion of a connection defined independently of any locally-Euclidean considerations. In this way, the associated de Rham complex can be satisfied by various possible algebra sheaves of coefficients modulo some well-defined cohomological conditions. We emphasize that the prominent role in the context of ADG is played by the algebra sheaf of coefficients, interpreted as a "functional coordinate arithmetic" [14,17] (see also [18–22]), meaning that all geometric objects involved in the formalism are locally expressed in terms of its sections. In this way, an algebra sheaf of coefficients is not constrained ab initio to be a smooth one, restricting the geometric solution space within the spectrum of a smooth manifold. More generally, a suitable algebra sheaf of coefficients turns out to be an algebra sheaf of generalized functions, including distributions, defined by Rosinger in the context of solutions to non-linear partial differential equations [23,24].

Concerning general relativity, which is formulated using the CDG of smooth manifolds, the possibilities offered by ADG bear a remarkable physical significance. In particular, there arises the possibility of re-assessing the global problems of general relativity related to the existence of singularities, where the metric breaks down, from the perspective of appropriate generalized algebra sheaves of coefficients. In this manner, the validity of Einstein's equations may be extended beyond differential manifolds, under the condition that the covariance properties of all tensorial physical quantities are maintained under these extensions, expressed in terms of the new sheaves of coefficients. From a physical viewpoint, this approach would allow one to obtain solutions in terms of distribution-like sheaves corresponding to non-punctual localization properties, which would nevertheless still satisfy the field equations. This clearly vindicates the following critical remark of Weyl [25]: "While topology has succeeded fairly well in mastering continuity, we do not yet understand the inner meaning of the restriction to differential manifolds. Perhaps one day physics will be able to discard it".

The possibility of obtaining extended admissible solution spaces in terms of generalized algebra sheaves of coefficients is based on the fact that the validity of the de Rham complex, in its sheaf-theoretic guise, is not restricted exclusively to the coordinatization of the tensorial physical quantities by smooth coefficients $\mathbb{C}^\infty$, as is actually the case when the $\mathbb{R}$-spectrum of the coefficients is a smooth manifold. Thus, we may consider distribution-like sheaves of coefficients satisfying the validity of the de Rham complex and, therefore, formulate and solve the field equations in terms of these distribution coefficients instead of the smooth ones. More precisely, this is the case if the following sequence of $\mathbb{R}$-linear sheaf morphisms:

$$\mathbb{A} \to \Omega^1(\mathbb{A}) \to \ldots \to \Omega^n(\mathbb{A}) \to \ldots \tag{3}$$

is a complex of $\mathbb{R}$-vector space sheaves, identified as the sheaf-theoretic de Rham complex of $\mathbb{A}$.

In this case, if the cohomological condition expressing the Poincaré Lemma, $Ker(d^0) = \mathbb{R}$ is satisfied with respect to the algebra sheaf $\mathbb{A}$ and requiring that $\mathbb{A}$ is a soft algebra sheaf, viz. any section over any closed subset of X can be extended to a global section, we obtain that the sequence:

$$\mathbf{0} \to \mathbb{R} \to \mathbb{A} \to \Omega^1(\mathbb{A}) \to \ldots \to \Omega^n(\mathbb{A}) \to \ldots \tag{4}$$

is an exact sequence of $\mathbb{R}$-vector space sheaves. Thus, the sheaf-theoretic de Rham complex of the algebra sheaf $\mathbb{A}$ constitutes an acyclic resolution of the constant sheaf $\mathbb{R}$.

The physical interpretation of this fact is the following: First of all, the essential feature of the localization method, utilizing coefficients from algebra sheaves instead of global algebras, is that the sheaf-theoretic de Rham complex is actually an acyclic resolution of the constant sheaf of the reals coordinatizing the events. For instance, referring to the CDG of smooth manifolds, the de Rham complex, expressed in terms of local smooth coefficients and their differential forms of higher orders, provides such an acyclic resolution of the constant sheaf $\mathbb{R}$. What has been uncovered by ADG is that the smooth algebra sheaf $\mathbb{C}^\infty(X))$ is not unique in this respect. More concretely, any other soft algebra sheaf $\mathbb{A}$ constituting an acyclic resolution of the constant sheaf $\mathbb{R}$ is a viable source of coefficients for the coordinatization of the tensors, maintaining at the same time all of their covariance properties in terms of the new local coefficients. This crucial fact essentially questions the uniqueness of the role of local smooth coefficients for formulating the means of dynamical variability. In other words, it questions the unique role of smooth manifold geometric spectrums as domains of validity of the field equations.

The idea to address the problem of singularities from the perspective of ADG has been proposed already, for instance in [11,12]. More concretely, in particular relation to the issue of spacetime singularities, Mallios and Rosinger [23,24] have applied ADG using as an algebra sheaf of coefficients, a variety of the so-called "spacetime foam algebras", and by Raptis [26], building up on prior work by Mallios and Raptis [27], using as a sheaf of coefficients "differential incidence algebras" defined over a locally-finite poset substitute of a continuous manifold.

Our present proposal constitutes a twist of perspective in comparison to these works, which is actually implemented by physical criteria of suitability going beyond the satisfaction of the cohomological conditions. Our quest is related to the possibility of using a particular type of a "spacetime foam algebra" as a kind of a distribution-like sheaf of coefficients, distinguished on physical grounds, for extending the domain of validity of Einstein's field equations. For this purpose, from the whole variety of "spacetime foam algebras", we distinguish only the "nowhere dense generalized function algebra" as bearing physical significance in relation to the field equations of general relativity. This is based on a physical criterion determining which properties should be characterized as intrinsic to the gravitational field and eventually deciding what should be generic with respect to its function or not. This physical criterion refers to the viable possibility of expressing the gravitational field sources via the instantiation of these generalized algebra sheaves of coefficients. Our rationale is based on the idea that in an intrinsically dynamically-variable theory, like general relativity, it should be the pertinent physical conditions or the sources of the field themselves that determine the type of these extensions as solutions to the field equations.

4. Coping with Spacetime Singularities: Conceptual and Technical Aspects

In the classical differential geometric formulation of general relativity, spacetime is represented as a connected, paracompact and Hausdorff four-dimensional $\mathbb{C}^\infty$ manifold X, endowed with a pseudo-Riemannian metric of the Lorentzian signature, which is obtained as a solution of Einstein's field equations. A spacetime manifold is considered to be without singularities if the coefficients of the metric tensor field are at least of class $\mathbb{C}^2$ and X is geodesically complete with respect to the metric, meaning that all timelike geodesic curves can be extended to arbitrary length [28,29]. Consequently, a spacetime manifold is considered to be singular if there exist incomplete geodesic curves, or equivalently finite affine length geodesics that cannot be extended. A spacetime singularity delimits a locus where the behavior of the metric tensor coefficients become ill-defined with respect to the smooth characterization of the manifold. Usually, the singular locus is identified as a locus where the spacetime curvature blows up. We note that the localization at a spacetime point-event is meaningful if the metric coefficients are smooth, or at least of class $\mathbb{C}^2$ in a neighborhood of this point.

The usual way to cope with a spacetime singularity is to consider it as a singular spacetime boundary rather than a locus within spacetime. For instance, a spacetime boundary may be defined in terms of a set of incomplete curves S. This takes place by the imposition of an appropriate equivalence relation $\sim$ on the set S, such that the quotient set $S/\sim := \partial X$ is interpreted as the singular boundary of X. The criterion of equivalence is determined by the choice of those equivalence classes, which are forced to play the role of ideal points in the extension of X by ∂X. There have been proposed various possible choices, for example Geroch's "g-boundary" or Schmidt's "b-boundary", but it is always assumed that X is topologically dense in $X \bigsqcup \partial X$ [30,31]. Following this approach, Heller and Sasin have shown that Einstein's field equations can be formulated in the extension of X by ∂X, that is on $X \bigsqcup \partial X$ defined as an "Einstein structured space" [32]. Actually, this is the Gelfand spectrum of a sheaf of Einstein algebras, which constitutes the sheaf-theoretic localization of an Einstein algebra, a notion proposed initially by Geroch in his attempt to re-formulate general relativity in algebraic terms without invoking directly a spacetime manifold background [1]. In particular, it has been proven that the closed Friedmann world model and the Schwarzschild solution, combined with Schmidt's "b-boundary" construction, fit nicely in the sheaf-theoretic context of an "Einstein structured space". In turn, this has been a first indication that the validity of Einstein's equations may be extended to bigger domains incorporating singular loci, which are not smooth manifolds anymore. It has been also pointed out that some sorts of singularities can also appear when there exists a transition to the quantum gravity regime. More concretely, the smooth manifold structure of spacetime can break down, and the possible validity of Einstein's equations should be sought for in further extended and generalized non-smooth spectrums of appropriate sheaves of algebras, where the singularities are not necessarily forced to some type of spacetime boundary.

From a broader conceptual perspective, the issue of singularities in general relativity as impossibilities of extending smooth metric solutions of Einstein's equations necessitates the coordinatization of all of the tensorial quantities by distributional coefficients effecting a type of topological coarse-graining over singular loci and, thus, localizing the point-event stratum in their terms. Under the proviso that these distributional coefficients form algebra sheaves fulfilling all of the required cohomological conditions, the means of extending local distributional solutions generalizes the standard method of extending timelike geodesic curves in a smooth manifold. The physical significance of this generalization is that the domain of validity of the field equations can be extended beyond the notion of a smooth manifold. Not only this, but additionally, these distinguishable extensions may be associated intrinsically with the gravitational field, under the constraint that sources of the field itself giving rise to singularities can be expressed topologically in the terms of distribution-like algebra sheaves.

In this state of affairs, the smoothness assumption can be retained, at best, only locally and certainly far from singular loci. Mathematically, there should exist an embedding of the algebra sheaf of smooth functions into a distribution-like algebra sheaf of coefficients qualified as a solution of the extended field equations. An illuminating way to think of the proposed approach in non-technical terms is that coping efficiently with singularities requires a process of folding out of the smooth point-event manifold background. This viewpoint has been emphasized by von Müller [33], according to whom the process of folding out into a "statu-nascendi" level should be considered in the context of a whole new categorical apparatus qualifying its intrinsic characteristics in contradistinction to the event stratum. In this manner, we suggest that the existence of a distribution-like sheaf of coefficients as a solution of the field equations within an appropriately-extended domain characterized by some generic gravitational criterion paves the way for understanding the precise nature of this folding out of the smooth point-event stratum.

The possibility of extending the formulation of Einstein's equations in the case of non-smooth spectrums using the sheaf-theoretic technique of localization in the context of ADG is of major significance. We note that non-smooth spectrums of algebra sheaves do not require the consideration of singularities as ideal points on the boundary of a smooth manifold. In other words, singular loci are allowed to be located, according to specific topological criteria, within a manifold. Of course, a natural requirement should be that the exclusion of singular loci would recast Einstein's equation in the familiar form in terms of smooth coefficients. However, clearly in the case that Einstein's equations become meaningfully extended over singular loci, then the coefficients of the metric and curvature tensors cannot be smooth any more. Therefore, from a smooth perspective, a singularity functions as an obstruction to the extension of a local solution to the field equations expressed in terms of smooth coefficients. Thus, more precisely, a singular locus plays the role of a cohomological obstruction to the extendibility of a local smooth solution. This criterion incorporates and generalizes sheaf-cohomologically the initial definition of singular behavior in terms of non-extendibility of geodesics. Essentially, the reason is that the notion of extendibility of local solutions is of a sheaf-theoretic nature, recalling for instance the well-known procedure of analytic continuation.

There are two important physical consequences emanating from the possibility of formulating Einstein's equations in terms of generalized non-smooth sheaves of coefficients. The first is related to the natural question concerning the criterion of depicting a particular sheaf of algebras for this purpose. The second is related to a possible re-evaluation of the status of the energy-momentum source term in Einstein's equations, which currently is not implemented by any process of geometrization.

Regarding the first, the required physical condition is the following: Since the formulation of Einstein's equations can be extended over singular loci, it should precisely be the nature and specification of these singular loci that would determine the appropriate sheaf of coefficients, such that a solution can be expressed eventually in terms of these coefficients. In the non-singular case, we know already that a solution can be expressed in terms of smooth coefficients. In other words, we already know that if no singularity is present, the spacetime metric, obtained as a particular solution of the vacuum Einstein equations, for example, is always expressible in terms of smooth coefficients, i.e., in terms of the sheaf of algebras $\mathbb{C}^{\infty}(X)$. Hence, we expect that in the presence of a particular type of a singular locus over which Einstein's equations hold in terms of a distribution-like sheaf of coefficients, there exists a metric solution expressed in terms of these coefficients. Not only this, but additionally, since the knowledge of the metric solution is completely expressible in terms of these coefficients, considered as unknowns when plugged into the equations, the specification of a singular locus should force a corresponding algebra sheaf as the solution. In other words, the nature of a singular locus should determine the differentiability properties of a metric solution in the case that Einstein's equations can be extended over this locus. As we stressed previously, the physical association of singular loci with sources of the gravitational field itself, giving rise to distinguishable extensions of the standard smooth manifold spacetime model of general relativity, implies that sources can be expressed topologically after all, if solutions of the field equations are expressed in terms of appropriate distribution-like algebra sheaves.

In this context, the physical significance of ADG is that it determines rigorously the criteria that these algebra sheaves of coefficients have to satisfy, such that Einstein's equations can be satisfied over various sorts of singular loci, expressed in terms of these coefficients. Not surprisingly, these criteria are of a cohomological nature. Essentially, they determine viable algebra sheaves of coefficients by the requirement that they are soft, and thus acyclic, such that the validity of the de Rham complex remains intact. In turn, the basic idea is that the Poincaré lemma should remain in force, viz. closed differential forms expressed in these generalized coefficients should be locally exact as in the smooth case, so that the differential geometric mechanism can be extended over singularities without breaking down. We will present a general form of these algebra sheaves consisting of distribution-like coefficients in the sequel. According to Clarke, the answer to many of the problems related to singularities "involve detailed considerations of distributional solutions to Einstein's equations, leading into an area that is only starting to be explored ..." [28]. We propose that the extension of the validity of Einstein's equations over singular loci in terms of appropriate sheaves of algebras, which are generally non-smooth, sheds new light on the problem of singular behavior in general relativity.

Regarding the second physical consequence, it is instructive to remind that the energy-momentum source term in the smooth formulation of Einstein's equations is not of any geometric nature. The energy-momentum tensor attributes the source of curvature entirely to matter (including the cosmological dark energy), as it does not incorporate the stress-energy associated with the gravitational field itself. There is an underlying assumption that spacetime is somehow empty unless it is filled in by matter, expressed in terms of the smooth coefficients of the energy-momentum tensor. This is the reason that when the energy-momentum part is zero, then the equations are called vacuum equations. Now, the validity of Einstein's equations over singular domains in terms of generalized non-smooth algebra sheaves casts serious doubts on this assumption. Namely, the form of Einstein's equations with the vanishing non-geometric second part may turn out to be the fundamental form of these equations. The reason is that sources of the gravitational field itself might be implemented in terms of non-smooth algebra sheaves, and thus, what is called a vacuum is not empty at all, precisely because it engulfs these sources. This idea is not actually as controversial as it sounds, if we take seriously into account that all classical experimental tests of general relativity involve a vanishing energy-momentum tensor, and thus, what they really verify is the equation $\mathbf{R}_{\mu\nu} = 0$. This issue has been also pointed out and argued for extensively, from a non sheaf-theoretic point of view, by Vishwakarma [34], who conducted a careful analysis based on the observational tests of the theory. In the sequel, we will discuss this issue in more detail from a geometrodynamical perspective in light of the particular form of distribution-like algebra sheaves.

5. Spacetime Extensions in Terms of Singularity-Free Distributional Algebra Sheaves

It is physically reasonable to expect that an admissible commutative algebra sheaf of coefficients in terms of which Einstein's equations may be extended over a singular locus should be distribution-like. For example, we may think of a matter distribution confined to a submanifold of spacetime whose density is integrable over this submanifold. In the context of a linear field theory, this should be naturally modeled in terms of a linear distribution. Unfortunately, this is not possible in the context of general relativity, which is a non-linear theory. In other words, Schwarz's linear distributions are not suitable candidates for expressing the information of singular loci.

The unsuitability of linear distributions rests on the fact that the space $\mathbb{D}'$ they form is only a linear space, but it is not an algebra. This is characterized as the "Schwarz impossibility" and may be formulated as follows: There is no symmetric bilinear morphism:

$$\circ : \mathbb{D}'(V) \times \mathbb{D}'(V) \ni (S, T) \to S \circ T \in \mathbb{D}'(V)$$

so that $S \circ T$ is the usual point-wise product of continuous functions, when $S, T \in \mathbb{C}^0(V)$. Equivalently, $\mathbb{D}'(V)$ is not closed under any multiplication that extends the usual multiplication of continuous functions, where V is an open subset X. Since all of the involved arguments are of a local character, without loss of generality, we may simply consider V as an open subset of $\mathbb{R}^4$.

A physically natural way to bypass "Schwarz impossibility" is to assume the existence of an embedding morphism $\mathbb{D}'(V) \hookrightarrow \mathbb{A}(V)$, which embeds the vector space of distributions $\mathbb{D}'(V)$ as a vector subspace in $\mathbb{A}(V)$, where $\mathbb{A}(V)$ is the quotient algebra:

$$\mathbb{A}(V) = \mathbb{K}(V)/\mathbb{I}, \tag{5}$$

and $\mathbb{K}(V)$ is a subalgebra in $\mathbb{C}^{\infty}(V)^{\Lambda}$, for some index set Λ, whereas $\mathbb{I}$ is an ideal in $\mathbb{K}(V)$. This approach was initiated by Rosinger [35,36] and developed further in [37–40].

We will restrict ourselves to a certain subclass of this type of algebras, namely the unital, associative and commutative algebras of generalized functions, whose suitably-defined ideals can engulf algebraically the information of singular loci. These algebras, introduced by Rosinger [36], have been formed in such a way as to express generalized solutions of non-linear partial differential equations. We may describe the generation of these algebras locally as follows:

Let $V \subseteq \mathbb{R}^4$ be an open set and $L = (\Lambda, \leq)$ be a right-directed partial order on some specified index set Λ. That is, for all λ, $\mu \in \Lambda$, there exists $\nu \in \Lambda$, such that $\lambda, \mu \leq \nu$. With respect to the usual componentwise operations, $\mathbb{C}^\infty(V)^\Lambda$ is a unital and commutative algebra over the reals. We define the following ideal $\mathbb{I}_L$ in $\mathbb{C}^\infty(V)^\Lambda$, whose physical meaning will be described in the sequel:

$$\mathbb{I}_L(V) = \left\{ \phi = (\phi_\lambda)_{\lambda \in \Lambda} \;\middle|\; \begin{array}{ll} \exists & \Gamma \subset V \text{ closed nowhere dense } : \\ \forall & x \in [V \setminus \Gamma] \text{ being dense } : \\ \exists & \lambda \in \Lambda : \\ \forall & \mu \in \Lambda,\ \mu \geq \lambda : \\ & \phi_\mu(x) = 0,\ \partial^p \phi_\mu(x) = 0 \end{array} \right\} \tag{6}$$

In the above definition, we think of Γ as a singular locus in $\mathbb{R}^4$, characterized as a closed and nowhere dense subset relative to the open set $V \subseteq \mathbb{R}^4$, such that its complement $V \setminus \Gamma$ in V is dense. The unital and commutative algebra $\mathbb{C}^\infty(V)^\Lambda$ contains smooth functions ϕ_λ indexed by the set Λ and defined over V, to be thought of as diagrams or sequences of Λ-indexed smooth functions. The requirement of the right-directed partial order on the specified index set Λ, which is denoted by $L = (\Lambda, \leq)$, is technically necessary in order that the above set forms actually an ideal in $\mathbb{C}^\infty(V)^\Lambda$. Now, the ideal $\mathbb{I}_L(V)$ in $\mathbb{C}^\infty(V)^\Lambda$ includes all of these sequences of smooth functions ϕ_λ that vanish asymptotically outside the singular locus Γ together with all of their partial derivatives. Therefore, intuitively speaking, the ideal of the form $\mathbb{I}_L(V)$ incorporates all of these sequences of smooth functions indexed by Λ whose support covers the singular locus Γ, whereas they vanish outside it. In this manner, the information of the singular locus Γ is encoded in the ideal $\mathbb{I}_L(V)$ in $\mathbb{C}^\infty(V)^\Lambda$. Hence, the quotient commutative algebra $\mathbb{A}_L(V) = \mathbb{C}^\infty(V)^\Lambda / \mathbb{I}_L(V)$ is an algebra of residues of sequences of smooth functions modulo the singular information ideal $\mathbb{I}_L(V)$.

A natural question in the above context refers to the requirement that the complement $V \setminus \Gamma$ of the singular locus Γ in V should be dense. The necessity of this requirement can be understood by the fact that we wish to obtain an embedding ι of the algebra of smooth functions $\mathbb{C}^\infty(V)$ into the algebra of generalized functions $\mathbb{A}_L(V)$:

$$\iota : \mathbb{C}^\infty(V) \hookrightarrow \mathbb{A}_L(V) = \frac{\mathbb{C}^\infty(V)^\Lambda}{\mathbb{I}_L(V)} \tag{7}$$

such that:

$$\varphi \hookrightarrow \iota(\varphi) = \Delta(\varphi) + [\mathbb{I}_L(V)] \tag{8}$$

where $\Delta_\Lambda |_V : \mathbb{C}^\infty(V) \to \mathbb{C}^\infty(V)^\Lambda$ is the diagonal morphism with respect to Λ, defined for an open set V as follows:

$$\Delta_\Lambda(\varphi) |_V = \left\{ \Delta(\varphi) = (\varphi_\lambda)_{\lambda \in \Lambda} \mid \varphi_\lambda = \varphi,\ \forall \lambda \in \Lambda, \varphi \in \mathbb{C}^\infty(V) \right\}.$$

Hence, for every smooth function φ in $\mathbb{C}^\infty(V)$, the diagonal image $\Delta(\varphi)$ of φ in $\mathbb{C}^\infty(V)^\Lambda$ is a sequence of smooth functions all identical to φ, indexed by Λ. The embedding ι is feasible according to the above, if and only if the ideal $\mathbb{I}_L(V)$ satisfies the off diagonality condition:

$$\mathbb{I}_L(V) \cap \Delta_\Lambda |_V = \{0\}. \tag{9}$$

Therefore, it remains to show that if the complement $V \setminus \Gamma$ of the singular locus Γ in V is dense, according to the specification in (6), then the ideal $\mathbb{I}_L(V)$ actually satisfies the above off diagonality condition. Therefore, we suppose that $V \setminus \Gamma$ is dense in V and consider a smooth function χ in $\mathbb{C}^\infty(V)$. If $\Delta_\Lambda(\chi)\mid_V := \Delta(\chi)$ belongs to the ideal $\mathbb{I}_L(V)$, then the asymptotic vanishing condition in (6) implies that $\chi = 0$ in $V \setminus \Gamma$, and therefore, we must have $\chi = 0$ in V because $V \setminus \Gamma$ is dense in V by hypothesis. Thus, it follows that the ideal $\mathbb{I}_L(V)$ satisfies the off diagonality condition (9), as required.

Conclusively, there exists a canonical injective homomorphism of commutative algebras, or equivalently, an embedding ι of the algebra of smooth functions $\mathbb{C}^\infty(V)$ into the algebra of generalized functions $\mathbb{A}_L(V)$:

$$\iota : \mathbb{C}^\infty(V) \hookrightarrow \mathbb{A}_L(V) = \frac{\mathbb{C}^\infty(V)^\Lambda}{\mathbb{I}_L(V)} \tag{10}$$

Furthermore, in view of (6), it follows immediately that the partial differential operators:

$$\partial^p : \mathbb{C}^\infty(V)^\Lambda \ni \phi = (\phi_\lambda) \mapsto \partial^p\phi = (\partial^p\phi_\lambda) \in \mathbb{C}^\infty(V)^\Lambda$$

satisfy the inclusion:

$$\partial^p(\mathbb{I}_L(V)) \subseteq \mathbb{I}_L(V). \tag{11}$$

Thus, the standard partial derivative operators on $\mathbb{C}^\infty(V)$ extend to $\mathbb{A}_L(V)$:

$$\partial^p : \mathbb{A}_L(V) \ni [\phi + \mathbb{I}_L(V)] \mapsto [\partial^p\phi + \mathbb{I}_L(V)] \in \mathbb{A}_L(V), \tag{12}$$

We conclude that the embedding of commutative algebras (10) extends to an embedding of differential algebras. Therefore, the following diagram commutes:

$$\begin{array}{ccc} \mathbb{C}^\infty(V) & \xrightarrow{\partial^p} & \mathbb{C}^\infty(V) \\ \downarrow & & \downarrow \\ \mathbb{A}_L(V) & \xrightarrow{\partial^p} & \mathbb{A}_L(V) \end{array}$$

We emphasize that the embedding (10) preserves not only the algebraic structure of $\mathbb{C}^\infty(V)$, but also its differential structure. The off diagonality condition (9) implies also the existence of an injective, linear morphism:

$$\mathbb{D}'(V) \hookrightarrow \mathbb{A}_L(V). \tag{13}$$

Therefore, the differential algebra $\mathbb{A}_L(V)$ contains the space of distributions as a linear subspace; see [38] (pp. 234–244), where those algebras that admit linear embeddings of distributions are characterized in terms of such off diagonality conditions. However, in contradistinction with (10), the embedding (13) does not commute with partial derivatives, and thus, the partial derivatives on $\mathbb{A}_L(V)$ do not, in general, coincide with distributional derivatives, when restricted to $\mathbb{D}'(V)$.

Finally, it is crucial to observe that a subset of a topological space is closed and nowhere dense if and only if it satisfies this condition locally. This is the key idea used to prove that the algebras of generalized functions $\mathbb{A}_L(V)$ form actually sheaves of commutative algebras, which additionally, are soft and flasque or flabby [23,24]. Thus, they are characterized as cohomologically-appropriate sheaves of coefficients according to ADG. More precisely, the distribution-like soft algebra sheaves of the form $\mathbb{A}_L$ constitute an acyclic resolution of the constant sheaf of the reals coordinatizing the events. Thus, we conclude that the de Rham complex can be rigorously expressed in terms of these coefficients instead of the smooth ones, and consequently, Einstein's equations can be formulated with respect to coefficients from the algebra sheaf $\mathbb{A}_L$ instead of the smooth ones from $\mathbb{C}^{\infty}$. Consequently, the validity of Einstein's equations can be extended over singular loci in a covariant manner by utilizing coefficients from the sheaf $\mathbb{A}_L$ for expressing all involved differential geometric tensorial quantities. Reciprocally, according to the intended physical interpretation of these algebra sheaves, pertaining to expressing sources of the gravitational field in terms of closed and nowhere dense subsets, the presence of a singular locus forces an algebra sheaf of the form $\mathbb{A}_L$ as coefficients with respect to which Einstein's equations retain their validity over this locus and do not break down, like in the case of insisting to use indiscriminately-smooth coefficients.

For the sake of completeness, it is instructive to remind that the softness property of the sheaves of the form $\mathbb{A}_L$ means that any section over any closed subset can be extended to a global section. Thus, these types of sheaves characterize cohomologically the topological property of paracompactness by means of acyclicity. Equivalently, soft sheaves are acyclic over a paracompact topological space. Moreover, sheaves of the form $\mathbb{A}_L$ are not only soft, but they are flasque or flabby, as well, which is a local property. This means that the restriction morphism of sections in the sheaf definition is an epimorphism. Hence, in this case, we can always extend any local section by zero to obtain a global section of $\mathbb{A}_L$.

We may recapitulate by pointing out that the first basic idea involved in the construction of distribution-like algebra sheaves of coefficients, in their role to coordinatize solutions of non-linear partial differential equations, is to model a singular locus Γ in $\mathbb{R}^4$ as a closed and nowhere dense subset relative to an open set $V \subseteq \mathbb{R}^4$, such that its complement $V \setminus \Gamma$ in V is dense. The second basic idea is to express such a closed and nowhere dense singular locus as an ideal in an algebra sheaf constructed as an extension of the smooth one over a partially-ordered set. In this manner, the ideal expressing algebraically a singular locus contains diagrams of locally-defined smooth functions indexed by Λ whose support covers the singular locus Γ, whereas they vanish outside it. Then, it can be shown that the quotient commutative algebra sheaf $\mathbb{A}_L(V) = \mathbb{C}^{\infty}(V)^{\Lambda} / \mathbb{I}_L(V)$ is an algebra sheaf of residues of diagrams of smooth functions modulo the closed nowhere dense singular ideal $\mathbb{I}_L(V)$.

It is instructive to emphasize that the algebra of global sections of the sheaf $\mathbb{A}_L(V)$ contains the space of Schwarz distributions $\mathbb{D}'(V)$ only as a linear subspace and not as a commutative subalgebra. For example, Dirac's delta, considered as a distribution, is represented in terms of a generalized function whose pertinent closed and nowhere dense set is an one-point set. It is well known that the square of the delta distribution is not a distribution itself, since the operation of point-wise multiplication of distributions is not well-defined in $\mathbb{D}'(V)$. Notwithstanding this fact, the representative generalized function may be unproblematically squared providing a legitimate generalized function without being a linear distribution itself. Clearly, by the rules of the construction of these commutative algebras of generalized functions, arbitrary nonlinear continuous operations may be applied to a generalized function giving another generalized function in the same algebra. In passing, it is also worth pointing out that the linear space of Schwarz distributions does not give rise to a flasque vector sheaf in contradistinction to the case of the embedding sheaf $\mathbb{A}_L(V)$, a property that is crucial for the global extendibility of all standard local differential geometric constructions.

In the sequel, we are going to propose a concrete class of closed and nowhere dense sets modeling the boundaries of singular loci and forming a topological link in 3D space. Conceptually, this essentially means that the semantics of folding out a local smooth event stratum into a singular domain may be associated with the formation of some topological link configuration and its concomitant algebraic expression in terms of an algebra sheaf of the type $\mathbb{A}_L$. At the final stage, we have to examine if this algebra sheaf satisfies the cohomological conditions necessary for expressing the differential geometric mechanism of general relativity in these terms instead of the globally-smooth ones. This turns out to be actually the case, and therefore, algebra sheaves of the type $\mathbb{A}_L$ can be used legitimately to express the metric solution of Einstein's field equations extended now over singularities.

The important consequence is that we can retain not only the validity, but not the form and covariance property of Einstein's equations even over singular loci. The reason is that all physical quantities can be still transformed according to a tensor law for any arbitrary admissible coordinate transformation. The difference in comparison to the smooth case is that the coordinates are allowed to be non-standard or non-smooth, while at the same time, all of the machinery of differential geometry can be applied with respect to them. In particular, while the coefficients of the tensorial physical quantities are non-smooth, all of the usual differential-geometric constructions can be carried out as in the smooth case. The only price to be paid for this generalization is the rejection of the fixed absolute smooth manifold background of the theory. We consider this fact as physically non-disturbing, since the essence of general relativity is in the covariant formulation and validity of Einstein's equations and not on the existence of a smooth background manifold. In particular, what we gain from such a generalization is not only that Einstein's equations can be extended covariantly over singular loci, but also that the solution of these equations in terms of coefficients from a sheaf of the form $\mathbb{A}_L$ are free of singularities.

6. Topological Links in Geometrodynamics

According to the paradigm of geometrodynamics [41], we may foliate a spacetime manifold X into three-dimensional spacelike leaves Σ_t by utilizing an one-parameter family of embeddings $\varepsilon_t : \Sigma \hookrightarrow X$, such that $\varepsilon_t(\Sigma) = \Sigma_t$. In the geometrodynamical formulation, the three-dimensional Riemannian manifold (Σ, h) is thought of as dynamically evolving, where the corresponding metric at time t, $h_t = \varepsilon_t{}^* g$, is derived by pulling back the spacetime metric g via ε_t. It is implicitly assumed that all three-dimensional spacelike leaves Σ_t are mutually disjoint, such that the Lorentzian manifold $(\mathbb{R} \times \Sigma, \varepsilon^* g)$ represents X, where the leaves of the considered foliation correspond to the constant time hypersurfaces.

The geometrodynamical picture is instructive for our purposes because it shows that active gravitational mass may emerge from purely topological considerations taking into account the constraints imposed by Einstein's field equations in the vacuum [42]. From a physical perspective, this may be interpreted in a novel way according to Wheeler's insight referring to "mass without mass" [43,44] as follows: Localized configurations of topologically-singular loci in open sets of a spacetime manifold restricted to closed nowhere dense subsets amount to active gravitational mass/energy in their complementary open dense subsets. In particular, if we consider that the Lorentzian manifold $(\mathbb{R} \times \Sigma, \varepsilon^* g)$ represents X, the singular loci may be localized within the three-dimensional manifold Σ. In this context, if Σ has a non-trivial topology, Gannon's theorem [45] implies that spacetime is geodesically incomplete and, thus, singular. The simplest way to implement a non-trivial topology on Σ is via the hypothesis of non-simple connectivity. More precisely, the existence of singular loci in Σ, localized in closed nowhere dense subsets makes Σ a multiple-connected topological space and, thus, topologically different from $\mathbb{R}^3$. We may recapitulate our conclusion up to now by asserting that the existence of singular loci in closed nowhere dense subsets of Σ, making it a multiply-connected topological space, implies active gravitational mass/energy in the complementary open dense subsets. Moreover, according to the "positive mass theorem" considered in the vacuum case, this gravitational mass/energy is non-zero and strictly positive. In passing, we would like to stress that Gannon's theorem should be conceived of as a significant generalization of the Penrose–Hawking singularity theorems [29], in the sense of replacing the usual geometric hypothesis of closed trapped surfaces in Σ by the more general applicable topological hypothesis of the non-simple connectivity of Σ.

In the same vein of ideas, we may also consider the system of Einstein–Maxwell equations without sources for the Maxwell field and, in this way, address from our perspective the alternative Wheeler's insight referring to "charge without charge" [43,44]. This has been originally tied to the assumption that Σ is orientable and bears the standard wormhole topology, that is homotopically equivalent to $S^1 \times S^2 - \{point\}$, such that the magnetic flux lines thread through the wormhole. In this case, the homology class of all two-spheres containing both of the wormhole mouths has zero charge, whereas the two individual wormhole mouths may be considered as having equal and opposite charges. In this context, a wormhole may be thought of in terms of a one-dimensional homology class in spacetime. From the general results of low-dimensional geometric topology [46], we know that every homology class of a four-dimensional spacetime can be represented by an embedded submanifold. Using the geometrodynamic foliation, we may restrict this representation to Σ. In this manner, we can instantiate a higher-order wormhole solution, for example by considering an appropriate two-dimensional homology class.

We are going to outline a general method of generating these types of solutions guided by the form of the algebra sheaves $\mathbb{A}_L$ incorporating gravitational properties defined on dense open sets of X and by restriction to dense open sets of Σ. For this purpose, we may consider a singular locus with boundary in $\mathbb{R}^3$ or in its compactification S^3, which is excised from $\mathbb{R}^3$ or S^3. We consider a singular locus as a singular disk cut off from S^3, which may be visualized in terms of a cone whose apex is at infinity and whose base lies at the boundary of the singular locus. A singular disk of this form excised from S^3 gives rise to a two-dimensional relative homology class of S^3, which may be interpreted according to the above as a two-dimensional embedded compact submanifold. The circular boundary of this singular disk is a closed and nowhere dense subset with respect to an open set of S^3. Analogously, we may consider the excision of more than one singular disks from S^3, such that their circular boundaries collectively define a closed and nowhere dense subset of an open set of S^3. We propose to think of these circular singular boundaries as giving rise to topological links.

The notion of a topological link is based on the underlying idea of connectivity among a collection of topological circles, called simply loops [47]. We consider that a loop is a tame closed curve. The property of tameness means that a closed curve can be deformed continuously and without self-intersections into a polygonal one, that is a closed curve formed by a finite collection of straight-line segments. Given this qualification, a loop is characterized by the following properties: First, it is a one-dimensional object. Second, it is bounded, meaning that it is contained in some sphere of sufficiently large radius. Third, a single cut at a point cannot separate a loop into two pieces, whereas any set of two cuts at two different points does separate a loop into two pieces. Moreover, a loop is called knotted if it cannot be continuously deformed into a circle without self-intersection. We only consider unknotted tame closed curves. A topological N-link is a collection of N loops, where N is a natural number. Regarding the connectivity of a collection of N loops, the crucial property is the property of the splittability of the corresponding N-link. We say that a topological N-link is splittable if it can be deformed continuously, such that part of the link lies within B and the rest of the link lies within C, where B, C denote mutually-exclusive solid spheres (balls). Intuitively, the property of splittability of an N-link means that the link can come at least partly apart without cutting. Complete splittability means that the link can come completely apart without cutting. On the other side, non-splittability means that not even one of the involved loops, or any pair of them, or any combination of them, can be separated from the rest without cutting.

According to our hypothesis, a collection of circular singular boundaries defining a closed and nowhere dense subset of an open set of S^3 gives rise to a topological link in S^3. We may now replace the loop components of such a topological link by open non-intersecting tubular neighborhoods such that the complement of the link in S^3 can be given by the structure of a three-dimensional compact and oriented manifold with a boundary. Clearly, this space is homologically equivalent to the original one since it is just its deformation retracted. Next, we may consider an ordering of the loops $l_1, l_2, \ldots l_N$ constituting the link, or equivalently, an ordering of their tubular neighborhoods $\lambda_1, \lambda_2, \ldots \lambda_N$. Then, if we take λ_i, λ_j, together with their ordering, we define the relative homology class σ_{ij} that is represented by the compact oriented embedded submanifold whose two boundary components lie on the total boundary, that is the first one in $\partial\lambda_i$ and the second in $\partial\lambda_j$. The orientation is defined as being negative on the first boundary component and positive on the second, so that we have a path from λ_i to λ_j in this case.

7. The Borromean Rings as a Universal Nowhere Dense Singular Link

According to the formalism of geometrodynamics, we consider the Lorentzian manifold $(\mathbb{R} \times \Sigma, \varepsilon^* g)$ as a representative of X, where the singularities are localized within the three-dimensional manifold Σ. We remind that, according to Gannon's theorem, if Σ is multiple-connected as a topological space, then spacetime is geodesically incomplete. According to our previous analysis, a collection of circular singular boundaries defining a closed and nowhere dense subset of an open set of S^3 gives rise to a topological link in S^3. Moreover, this implies the existence of active gravitational mass/energy in the complementary open dense subsets, which is non-zero and strictly positive.

In this context, it is important to examine if there exists a universal way via which we can obtain the three-dimensional manifold Σ by the information incorporated in a topological link in S^3 representing the singular boundaries, forming collectively a closed and nowhere dense subset. This sheds more light on the role of the algebra sheaves $\mathbb{A}_L$ utilized to express gravitational properties defined on dense open sets of X and by restriction to dense open sets of Σ and is guiding in our quest of exploring generalized wormhole-types of solutions based on topological links and their associated homology classes.

It turns out that a universal way to obtain Σ by using a topological link in S^3 representing the singular boundaries, according to the above, actually exists and is based on the notion of a universal topological link. In view of the type of solutions we are interested in, such a universal link is defined by the Borromean rings. In particular, using methods of geometric topology, it can be shown that any compact oriented three-dimensional manifold Σ without boundary can be obtained as the branched covering space of the three-sphere S^3 with the branch set the Borromean rings [48]. In this manner, the Borromean rings constitute a universal topological link.

The notion of a branched covering space is a generalization of the standard notion of a covering space, characterized as a local homeomorphism bearing the unique path lifting and homotopy lifting property [49]. More precisely, a branched covering space of the three-sphere S^3 is considered as a map from Σ to S^3 such that this map is a covering space after we delete or exclude a locus of points, called the branched locus. The universality property says that Σ can be obtained in this way if the branched locus is formed by the Borromean rings, considered as a closed and nowhere dense set with respect to an open set in S^3 in our setting. In a well-defined sense, this branched covering space provides the geometric representation of an algebra sheaf of the form $\mathbb{A}_L$ restricted to the three spatial dimensions, where the closed and nowhere dense subset formed by the Borromean rings is localized. We may extend this closed and nowhere dense subset to four dimensions by considering a timelike axis perpendicular to the Borromean rings, which plays the role of a three-fold symmetry axis of rotation.

The Borromean rings consist of three rings localized in S^3, which are linked together in such a way that each of the rings lies completely over one of the other two, and completely under the other, as is shown in the pictures below:

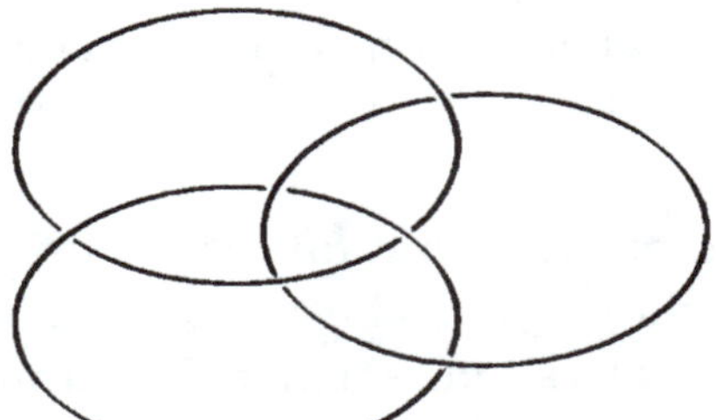

This particular type of topological linking displayed by the Borromean rings is called the Borromean link and is characterized by the following distinguishing property: if any one of the rings is removed from the Borromean link, the remaining two come completely apart. It is important to emphasize that the rings should be modeled in terms of unknotted tame closed curves and not as perfectly circular geometric circles. The adjective topological means that they can be deformed continuously under the constraint that the particular type of linkage forming the Borromean configuration is preserved.

From the viewpoint of the theory of topological links, the Borromean link constitutes an interlocking family of three loops, such that if any one of them is cut at a point and removed, then the remaining two loops become completely unlinked [47,49–52]. In more precise terms, the Borromean link is characterized topologically by the property of splittability as follows: The Borromean link is a non-splittable three-link (because it consists of three loops), such that every two-sub-link of this three-link is completely splittable. It is clear that it is a non-splittable three-link because not even one of the three loops, or any pair of them, can be separated from the rest without cutting. A two-sub-link is simply any sub-collection of two loops obtained by erasing the loop that does not belong to this sub-collection. Since the Borromean link is characterized by the property that if we erase any one of the three interlocking loops, then the remaining two loops become unlinked, it is clear that every two-sub-link of the non-splittable three-link is completely splittable.

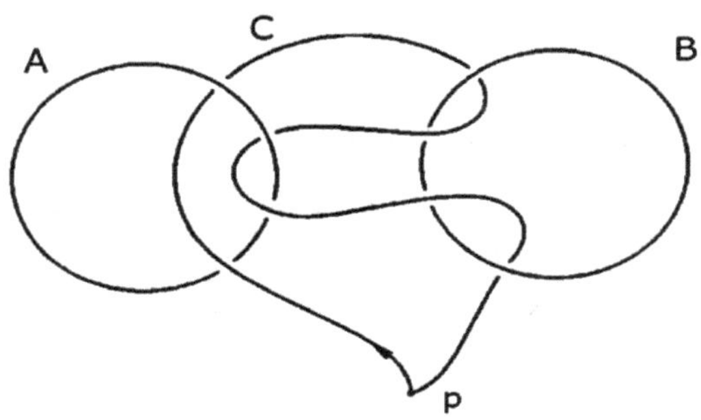

In our context, we conclude that if a triad of circular singular boundaries defining a closed and nowhere dense subset of an open set of S^3 are connected in the form of the Borromean topological link, then Σ as a compact oriented three-dimensional manifold can be obtained as the branched covering space of the three-sphere S^3 with the branch set these Borromean-linked boundaries. Based on these findings, we would like to explore their semantics in relation to the instantiation of a higher-order wormhole solution. For this purpose, we remind that the standard wormhole solution is thought of in terms of a one-dimensional homology class in a space homotopically equivalent to $S^1 \times S^2 - \{point\}$. In our framework, we do not need to impose a particular topology on Σ ab initio, since it can now be derived universally as the branched covering space of S^3 over the branch nowhere dense subset of singular boundaries forming a Borromean link. The fact that the Borromean link is a non-splittable three-link, such that every two-sub-link of this three-link is completely splittable, is characterized in homology theory by a non-vanishing triple Massey product, where all pairwise intersection products of one-dimensional homology classes vanish, reflecting the fact that the components of the Borromean link are not pairwise linked. If we denote the components of the Borromean link $\mathcal{B}$ by $\lambda_1, \lambda_2, \lambda_3$, the triple Massey product [49] is expressed as a two-dimensional cohomology class in the dense complement of $\mathcal{B}$ in S^3, that is it defines a non-trivial class in $H^2(S^3 \backslash (\lambda_1 \sqcup \lambda_2 \sqcup \lambda_3))$.

8. Conclusions

The main purpose of this communication, one hundred years after Einstein's formulation of the general theory of relativity, has been an invitation to re-think the validity domain of the field equations. The primary motivations emanate from three distinct sources: The first comes from Clarke's assertion concerning the problem of singularities, according to which, the answers "involve detailed considerations of distributional solutions to Einstein's equations, leading into an area that is only starting to be explored ...". The second comes from Weyl's critical remark regarding the role of a background differential manifold, according to which, "while topology has succeeded fairly well in mastering continuity, we do not yet understand the inner meaning of the restriction to differential manifolds. Perhaps one day physics will be able to discard it". The third comes from Wheeler's ideas regarding the notions of "mass without mass" and "charge without charge" in the vacuum, which can be given a more precise mathematical formulation in topological terms.

The sheaf-theoretic re-formulation of the usual differential geometric framework of smooth manifolds points to the conclusion that there exist distinguishable extensions of the standard smooth manifold spacetime model of general relativity, which are utilized by appropriate extensions of the sheaf of coefficients parameterizing all tensorial physical quantities of the theory. The criteria of the suitability of these extensions are determined by sheaf-cohomological means and maintain the standard covariance properties of the theory in domains, including singular loci. We have presented and discussed in detail a concrete distribution-like sheaf of coefficients incorporating singularities in closed and nowhere dense subsets of an open set of a four-dimensional spacetime. An instructive way to think of these generalized algebra sheaves of coefficients refers to the role of a singularity as an obstruction to the existence of a solution to the field equations, expressed in terms of smooth coefficients. Thus, more generally, a singular locus may be thought of as a cohomological obstruction to the extendibility of a local smooth solution. This criterion incorporates and generalizes sheaf-cohomologically the initial definition of singular behavior in terms of the non-extendibility of geodesics. Essentially, the reason is that the notion of the extendibility of some local solution is of a sheaf-theoretic nature.

At a further stage involving the formalism of geometrodynamics, the existence of singular loci in closed and nowhere dense subsets of a spatial hypersurface, making it a multiply-connected topological space, implies active gravitational mass/energy in the complementary open dense subsets. Moreover, according to the "positive mass theorem" considered in the vacuum case, this gravitational mass/energy is non-zero and strictly positive. We show that it is enough for this purpose to consider singular boundaries forming closed and nowhere dense subsets and forcing a multiple-connected topology, which in turn implies that spacetime is geodesically incomplete. In view of expressing generalized wormhole solutions in this context, we propose that closed singular boundaries may form topological links. In this manner, using the results of geometric topology, we point out that the Borromean topological link is characterized as a universal link. Since this link is characterized cohomologically by a higher order invariant, it may be associated with a generalized wormhole model, which reinforces Wheeler's ideas in geometrodynamics.

Finally, we express the hope that the proposed approach paves the way for a further technical and semantical refinement of the following two of Einstein's fundamental insights in building up general relativity, which have not been addressed in satisfactory completeness up to the present:

> "Under the influence of the ideas of Faraday and Maxwell the notion developed that the whole of physical reality could perhaps be represented as a field whose components depend on four space-time parameters. If the laws of this field are in general covariant, that is, are not dependent on a particular choice of coordinate system, then the introduction of an independent (absolute) space is no longer necessary. That which constitutes the spatial character of reality is then simply the four-dimensionality of the field. There is then no "empty" space, that is, there is no space without a field." [53]

> "A field theory is not yet completely determined by the system of field equations. Should one admit the appearance of singularities? . . . It is my opinion that singularities must be excluded. It does not seem reasonable to me to introduce into a continuum theory points (or lines etc.) for which the field equations do not hold . . ." [54]

In a nutshell, regarding the first, the utilization of distribution-like sheaves of coefficients extending the smooth one over singularities, and thus, extending the domain of validity of the field equations beyond globally-smooth manifolds, shows in agreement with geometrodynamics that active gravitational mass/energy may emerge from purely topological considerations taking into account the constraints imposed by the field equations in the vacuum. These topological considerations pertain to the modeling of singularities in terms of closed and nowhere dense sets, such that their complements who bear the induced active gravitational mass/energy are open and dense. In this manner, the vacuum can be legitimately considered as a structural quality of the field itself. Regarding the second, it is indeed unreasonable to consider singular loci in a continuum theory, where the field equations do not hold. The existence of distribution-like sheaves of coefficients provides precisely the means to bypass this problem by coordinatizing all of the tensorial quantities in their terms, extending the smooth ones and, therefore, extending the domain of validity of the field equations.

References

1. Geroch, R. Einstein algebras. *Commun. Math. Phys.* **1972**, *26*, 271–275.
2. Mallios, A. On geometric topological algebras. *J. Math. Anal. Appl.* **1993**, *172*, 301–322.
3. Mallios, A. The de Rham-Kähler complex of the Gelfand sheaf of a topological algebra. *J. Math. Anal. Appl.* **1993**, *175*, 143–168.
4. Grothendieck, A. Sur quelques points d' algèbre homologique. *Tôhoku Math. J.* **1957**, *9*, 119–221.
5. Grothendieck, A. *A General Theory of Fiber Spaces with Structure Sheaf*; Univ. Kansas: Lawrence, KS, USA, 1958.
6. Mallios, A. *Geometry of Vector Sheaves: An Axiomatic Approach to Differential Geometry, Vol I: Vector Sheaves, General Theory*; Kluwer Academic Publishers: Dordrecht, The Netherlands, 1998.
7. Mallios, A. *Geometry of Vector Sheaves: An Axiomatic Approach to Differential Geometry, Vol II: Geometry Examples and Applications*; Kluwer Academic Publishers: Dordrecht, The Netherlands, 1998.
8. Vassiliou, E. *Geometry of Principal Sheaves*; Kluwer Academic Publishers: Dordrecht, The Netherlands, 2004.
9. Selesnick, S.A. Line bundles and harmonic analysis on compact groups. *Math. Z.* **1976**, *146*, 53–67.
10. Mallios, A. *Modern Differential Geometry in Gauge Theories: Vol. 1. Maxwell Fields*; Birkhäuser: Boston, MA, USA, 2006.
11. Mallios, A. *Modern Differential Geometry in Gauge Theories: Vol. 2. Yang-Mills Fields*; Birkhäuser: Boston, MA, USA, 2009.
12. Mallios, A. Quantum gravity and "singularities". *Note Mat.* **2006**, *25*, 57.
13. Mallios, A. Geometry and physics today. *Int. J. Theor. Phys.* **2006**, *45*, 1552–1588.
14. Mallios, A. A-invariance: An axiomatic approach to quantum relativity. *Int. J. Theor. Phys.* **2008**, *47*, 1929–1948.
15. Raptis, I. A dodecalogue of basic didactics from applications of abstract differential geometry to quantum gravity. *Int. J. Theor. Phys.* **2007**, *46*, 3009–3021.
16. Mallios, A. On localizing topological algebras. *Contemp. Math.* **2004**, *341*, 79.
17. Mallios, A. On algebra spaces. *Contemp. Math.* **2007**, *427*, 263.
18. Epperson, M.; Zafiris, E. *Foundations of Relational Realism: A Topological Approach to Quantum Mechanics and the Philosophy of Nature*; Lexington Books: Lanham, MD, USA, 2013.
19. Mallios, A.; Zafiris, E. *Differential Sheaves and Connections: A Natural Approach to Physical Geometry*; World Scientific: Singapore, 2016.
20. Zafiris, E. Boolean coverings of quantum observable structure: A setting for an abstract differential geometric mechanism. *J. Geom. Phys.* **2004**, *50*, 99–114.
21. Zafiris, E. Interpreting observables in a quantum world from the categorical standpoint. *Int. J. Theor. Phys.* **2004**, *43*, 265–298.
22. Zafiris, E. Quantum observables algebras and abstract differential geometry: The topos-theoretic dynamics of diagrams of commutative algebraic localizations. *Int. J. Theor. Phys.* **2007**, *46*, 319–382.
23. Mallios, A.; Rosinger, E.E. Abstract differential geometry, differential algebras of generalized functions, and de Rham cohomology. *Acta Appl. Math.* **1999**, *55*, 231.
24. Mallios, A.; Rosinger, E.E. Space-time foam dense singularities and de Rham cohomology. *Acta Appl. Math.* **2001**, *67*, 59–89.
25. Weyl, H. *Philosophy of Mathematics and Natural Science*; Princeton Univ. Press: Princeton, NJ, USA, 2009.
26. Raptis, I. Finitary-algebraic "resolution" of the inner Schwartzschild singularity. *Int. J. Theor. Phys.* **2006**, *45*, 79–128.

27. Mallios, A.; Raptis, I. Finitary, causal and quantal vacuum einstein gravity. *Int. J. Theor. Phys.* **2003**, *42*, 1479–1619.
28. Clarke, C.J.S. *The Analysis of Space-Time Singularities*; Cambridge University Press: Cambridge, UK, 1993.
29. Hawking, S.W.; Ellis, G.F.R. *The Large Scale Structure of Space-Time*; Cambridge University Press: Cambridge, UK, 1973.
30. Bosshard, B. On the b-boundary of the closed friedmann models. *Commun. Math. Phys.* **1976**, *46*, 263–268.
31. Schmidt, B.G. A new definition of singular points in general relativity. *Gen. Relat. Grav.* **1971**, *1*, 269–280.
32. Heller, M.; Sasin, W. Structured spaces and their application to relativistic physics. *J. Math. Phys.* **1995**, *36*, 3644–3662.
33. Von Müller, A. *The Forgotten Present. In: Re-Thinking Time at the Interface of Physics and Philosophy*; von Müller, A., Filk, T., Eds.; Springer: Heidelberg, Germany, 2015; pp. 1–46.
34. Vishwakarma, R.G. Mysteries of $R_{ik} = 0$: A novel paradigm in Einstein's theory of gravitation. *Front. Phys.* **2014**, *9*, 98–112.
35. Rosinger, E.E. *Distributions and Nonlinear Partial Differential Equations*; Springer: Berlin, Germany, 1978.
36. Rosinger, E.E. Nonlinear Partial Differential Equations, Sequential and Weak Solutions. In *North Holland Mathematics Studies*; Elsevier: Amsterdam, The Netherlands, 1980.
37. Rosinger, E.E. Generalized Solutions of Nonlinear Partial Differential Equations. In *North Holland Mathematics Studies*; Elsevier: Amsterdam, The Netherlands, 1987.
38. Rosinger, E.E. Nonlinear Partial Differential Equations, an Algebraic View of Generalized Solutions. In *North Holland Mathematics Studies*; Elsevier: Amsterdam, Netherlands, 1990.
39. Rosinger, E. How to solve smooth nonlinear PDEs in algebras of generalized functions with dense singularities. *Appl. Anal.* **2001**, *78*, 355–378.
40. Rosinger, E.E. Differential algebras with dense singularities on manifolds. *Acta Appl. Math.* **2007**, *95*, 233–256.
41. Misner, C.W.; Thorne, K.S.; Wheeler, J.A. *Gravitation*; W. H. Freeman and Company: New York, NY, USA, 1970.
42. Arnowitt, R.; Deser, S.; Misner, C.W. The dynamics of general relativity. In *Gravitation: An Introduction to Current Research*; Witten, L., Ed.; John Wiley and Sons: New York, NY, USA; London, UK, 1962; pp. 227–265.
43. Misner, C.W.; Wheeler, J.A. Classical physics as geometry: Gravitation, electromagnetism, unquantized charge, and mass as properties of empty space. *Ann. Phys.* **1957**, *2*, 525–603.
44. Wheeler, J.A. On the nature of quantum geometrodynamics. *Ann. Phys.* **1957**, *2*, 604–614.
45. Gannon, D. Singularities in nonsimply connected space-times. *J. Math. Phys.* **1975**, *16*, 2364–2367.
46. Scorpan, A. *The Wild World of 4-Manifolds*; Americal Mathematical Society: Providence, RI, USA, 2005.
47. Kawauchi, A. *A Survey of Knot Theory*; Springer: Berlin, Germany, 1996.
48. Hilden, H.M.; Losano, M.T.; Montesinos, J.M.; Whitten, W. On universal groups and 3-manifolds. *Invent. Math.* **1987**, *87*, 441–456.
49. Hatcher, A. *Algebraic Topology*; Cambridge University Press: Cambridge, UK, 2002.
50. Cromwell, P.; Beltrami, E.; Rampichini, M. The borromean rings. *Math. Intell.* **1998**, *20*, 53–62.
51. Debrunner, H. Links of brunnian type. *Duke Math. J.* **1961**, *28*, 17–23.
52. Lindström, B.; Zetterström, H.-O. Borromean circles are impossible. *Am. Math. Mon.* **1991**, *98*, 340–341.
53. Jammer, M. *Concepts of Space: The History of Theories of Space in Physics; Foreword by Albert Einstein*, 3rd ed.; Dover: Mineola, NY, USA, 1993.
54. Einstein, A. *The Meaning of Relativity*, 5th ed.; Princeton University Press: Princeton, NJ, USA, 1956.

9

The Status of Cosmic Topology after Planck Data

Jean-Pierre Luminet [1,2]

[1] Laboratoire d'Astrophysique de Marseille (LAM), CNRS UMR 7326, F-13388 Marseille, France
[2] LUTH (Observatoire de Paris), CNRS UMR 8102, F-92195 Meudon, France; jean-pierre.luminet@lam.fr

Academic Editors: Stephon Alexander, Jean-Michel Alimi, Elias C. Vagenas and Lorenzo Iorio

Abstract: In the last decade, the study of the overall shape of the universe, called Cosmic Topology, has become testable by astronomical observations, especially the data from the Cosmic Microwave Background (hereafter CMB) obtained by WMAP and Planck telescopes. Cosmic Topology involves both global topological features and more local geometrical properties such as curvature. It deals with questions such as whether space is finite or infinite, simply-connected or multi-connected, and smaller or greater than its observable counterpart. A striking feature of some relativistic, multi-connected small universe models is to create multiples images of faraway cosmic sources. While the last CMB (Planck) data fit well the simplest model of a zero-curvature, infinite space model, they remain consistent with more complex shapes such as the spherical Poincaré Dodecahedral Space, the flat hypertorus or the hyperbolic Picard horn. We review the theoretical and observational status of the field.

Keywords: cosmology; general relativity; topology; cosmic microwave background; Planck telescope

---

1. Introduction

The idea that the universe might have a non-trivial topology and, if sufficiently small in extent, display multiple images of faraway sources, was first discussed in 1900 by Karl Schwarzschild (see [1] for reference and English translation). With the advent of Einstein's general relativity theory (see, e.g., the recent historical overview in [2]) and the discoveries of non-static universe models by Friedmann and Lemaître in the decade 1922–1931, the face of cosmology definitively changed. While Einstein's cosmological model of 1917 described space as the simply-connected, positively curved hypersphere S^3, de Sitter in 1917 and Lemaître in 1927 used the multi-connected projective sphere P^3 (obtained by identifying opposite points of S^3) for describing the spatial part of their universe models. In 1924, Friedmann [3] pointed out that Einstein's equations are not sufficient for deciding if space is finite or infinite: Euclidean and hyperbolic spaces, which in their trivial (*i.e.*, simply-connected) topology are infinite in extent, can become finite (although without an edge) if one identifies different points—an operation which renders the space multi-connected. This opens the way to the existence of phantom sources, in the sense that at a single point of space an object coexists with its multiple images. The whole problem of cosmic topology was thus posed, but as the cosmologists of the first half of 20th century had no experimental means at their disposal to measure the topology of the universe, the vast majority of them lost all interest in the question. A revival of interest in multi-connected cosmologies arose in the 1970s, under the lead of theorists who investigated several kinds of topologies (see [4] for an exhaustive review and references, in which the term "Cosmic Topology" was coined). However, most cosmologists either remained completely ignorant of the possibility, or regarded it as unfounded, until the 1990s when data on the CMB provided by space telescopes gave access to the largest possible volume of the observable universe. Since then, hundreds of articles have considerably enriched the field of theoretical and observational cosmology.

2. A Theoretical Reminder

At very large scale our Universe seems to be correctly described by one of the Friedmann-Lemaître (hereafter FL) models, namely homogeneous and isotropic solutions of Einstein's equations, of which the spatial sections have constant curvature. They fall into three general classes, according to the sign of curvature. In most cosmological studies, the spatial topology is assumed to be that of the corresponding simply-connected space: the hypersphere, Euclidean space or 3D-hyperboloid, the first being finite and the other two infinite. However, there is no particular reason for space to have a trivial topology: the Einstein field equations are local partial differential equations which relate the metric and its derivatives at a point to the matter-energy contents of space at that point. Therefore, to a metric element solution of Einstein field equations, there are several, if not an infinite number, of compatible topologies, which are also possible models for the physical universe. For example, the hypertorus T^3 and the usual Euclidean space E^3 are locally identical, and relativistic cosmological models describe them with the same FL equations, even though the former is finite in extent while the latter is infinite. Only the boundary conditions on the spatial coordinates are changed. The multi-connected FL cosmological models share exactly the same kinematics and dynamics as the corresponding simply-connected ones; in particular, the time evolutions of the scale factor are identical.

In FL models, the curvature of physical space (averaged on a sufficiently large scale) depends on the way the total energy density of the universe may counterbalance the kinetic energy of the expanding space. The normalized density parameter Ω_0, defined as the ratio of the actual energy density to the critical value that strictly Euclidean space would require, characterizes the present-day contents (matter, radiation and all forms of energy) of the universe. If Ω_0 is greater than 1, then the space curvature is positive and the geometry is spherical; if Ω_0 is smaller than 1, the curvature is negative and geometry is hyperbolic; eventually Ω_0 is strictly equal to 1 and space is locally Euclidean (currently said flat, although the term can be misleading).

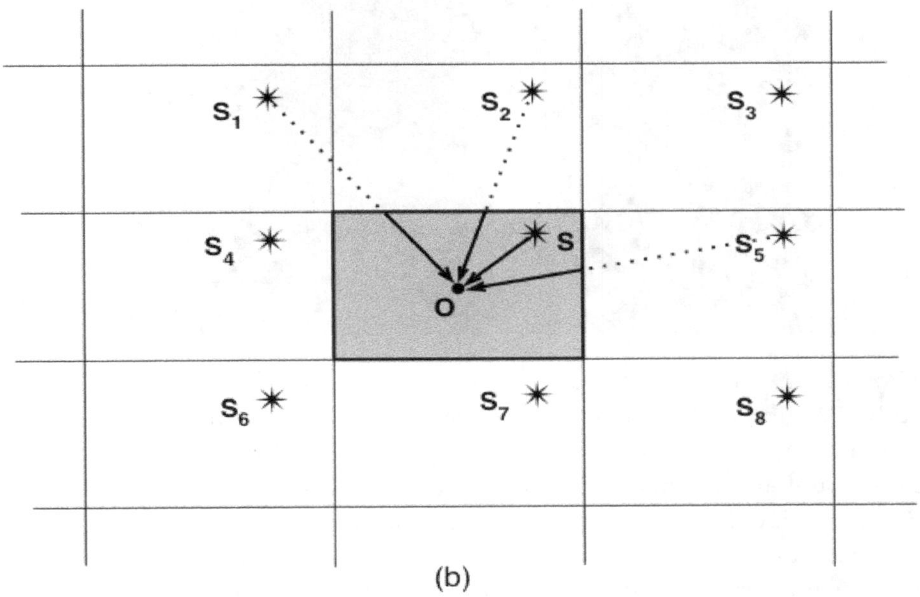

Figure 1. The Illusion of the Universal Covering Space. In the case of a 2D torus space, the fundamental domain, which represents real space, is the interior of a rectangle, whose opposite edges are identified. The observer O sees rays of light from the source S coming from several directions. He has the illusion of seeing distinct sources S_1, S_2, S_3, *etc.*, distributed along a regular canvas which covers the UC space—an infinite plane.

Independently of curvature, a much discussed question in the history of cosmology (and also philosophy) is to know whether space is finite or infinite in extent. Of course no physical measure can ever prove that space is infinite, but a sufficiently small, finite space model could be testable. Although the search for space topology does not necessarily solve the question of finiteness, it provides many multi-connected universe models of finite volume.

The effect of a non-trivial topology on a cosmological model is equivalent to considering the observed space as a simply-connected 3D-slice of space-time, known as the "universal covering space" (hereafter UC) being filled with repetitions of a given shape, the "fundamental domain", which is finite in some or all directions, for instance a convex polyhedron; by analogy with the two-dimensional case, we say that the fundamental domain tiles the UC space.

There is a subgroup of isometries acting on the UC which produces its tiling by these copies (for such group action, see the basic works [5,6]). Physical fields repeat their configuration in every copy and thus can be viewed as defined on the UC space, but subject to periodic boundary conditions, which are the matching rules between neighbouring tiles. The copies around a fixed one carry the multiple images of objects from the cosmos. By analogy with crystallography, the UC plays the role of the macroscopic crystal, the cosmos plays the role of the fundamental unit (see Figure 1). But in contrast to crystallography, the UC in topology can be Euclidean, spherical or hyperbolic. For the flat and hyperbolic geometries, there are infinitely many copies of the fundamental domain; for the spherical geometry with a finite volume, there are a finite number of tiles.

Figure 2. The Poincaré Dodecahedral Space can be described as the interior of a spherical ball whose surface is tiled by 12 curved regular pentagons. When one leaves through a pentagonal face, one returns to the ball through the opposite face after having turned by 36°. As a consequence, the space is finite but without boundaries, therefore one can travel through it indefinitely. One has the impression of living in a UC space 120 times larger, paved with dodecahedra that multiply the images like a hall of mirrors. The return of light rays that cross the walls produces optical mirages: a single object has several images. This numerical simulation calculates the closest phantom images of the Earth, which would be seen in the UC space (Image courtesy of J. Weeks).

There are seventeen multi-connected Euclidean spaces (for an exhaustive study, see [7]), the simplest of which being the hypertorus T^3, whose fundamental domain is a parallelepiped of which opposite faces are identified by translations. Seven of these spaces have an infinite volume, ten are of finite volume, six of them being orientable hypertori. All of them could correctly describe the spatial part of the flat universe models, as they are consistent with recent observational data which constrain the space curvature to be very close to zero. Note that in current inflationary scenarios for the big bang, one can always have a nearly flat universe at present without fine-tuning the initial value of the spatial curvature, while considering exactly flat models corresponds to fine-tuning the initial curvature to be strictly zero.

In spaces with non-zero curvature, the presence of a length scale (the curvature radius) precludes topological compactification at an arbitrary scale. The size of the space must now reflect its curvature, linking topological properties to the total energy density Ω_0. All spaces of constant positive curvature are finite whatever be their topology. The reason is that the universal covering space—the simply-connected hypersphere S^3—is itself finite. There is a countable infinity of spherical spaceforms (for a complete classification, see [8]), but there is only a finite set of "well-proportioned" topologies, *i.e.*, those with roughly comparable sizes in all directions, which are of a particular interest for cosmology. As a now celebrated example, let us mention the Poincaré Dodecahedral Space (hereafter PDS), obtained by identifying the opposite pentagonal faces of a regular spherical dodecahedron after rotating by 36° in the clockwise or counterclockwise direction around the axis orthogonal to the face, depending on which handedness the physical nature favors [9], see Figure 2. Its volume is 120 times smaller than that of the hypersphere with the same curvature radius.

Eventually there is also an infinite (but non countable) number of hyperbolic manifolds, with finite or infinite volumes. Their classification is not well understood, but the volumes of the compact hyperbolic space forms are bounded below by $V = 0.94271$ (in units of the curvature radius), which correspond to the so-called Weeks manifold.

The computer program *CurvedSpaces* [10] is especially useful to depict the rich structure of multi-connected manifolds with any curvature.

3. Probing Cosmic Topology

The observable universe is the interior of a sphere centered on the observer and whose radius is that of the cosmological horizon—roughly the radius of the last scattering surface (hereafter LSS), presently estimated at 14.4 Gpc. Cosmic Topology aims to describe the shape of the *whole* universe. One could think that the whole universe is necessarily greater than the observable one, as it would obviously be the case if space was infinite, for instance the simply-connected flat or hyperbolic space. Then the observable universe would be an infinitesimal patch of the whole universe and, although it has long been the standard "mantra" of many theoretical cosmologists, this is not and will never be a testable hypothesis.

The whole universe can also be finite (without an edge), e.g., a hypersphere or a closed multi-connected space, but greater than the observable universe. In that case, one easily figures out that if whole space widely encompasses the observable one, no signature of its finiteness will show in the experimental data. But if space is not too large, or if space is not globally homogeneous (as is permitted in many space models with multi-connected topology), and if the observer occupies a special position, some imprints of the space finiteness could be observable.

Surprisingly enough, the whole space could be smaller than the observable universe, due to the fact that space can be both multi-connected, have a small volume and produce topological lensing. This is the only case where there are a lot of testable possibilities, whatever the curvature of space.

The present observational constraints on the Ω_0 parameter favor a spatial geometry that is nearly flat with a 0.4% margin of error [11]. Note that the constraints on the curvature parameter can be looser if we consider a general form of dark energy (not the cosmological constant), which leaves rooms to consider positively or negatively curved cosmological models that are usually regarded as being excluded. However, even with the curvature so severely constrained by cosmological data, there are still possible multi-connected topologies that support positively curved, negatively curved, or flat metrics. Sufficiently "small" universe models would generate multiple images of some light sources, in such a way that the hypothesis of multi-connected topology can be tested by astronomical observations. The smaller the space, the easier it is to observe the multiple images of luminous sources in the sky (generally not seen at the same age, except for the CMB spots). Note, however, the coincidence problem that occurs in order to get an observable non-trivial topology: for flat space, we need to have the topology scale length near the horizon scale, while for curved spaces, the curvature radius needs to be near the horizon scale. However, there are so many other, non-explained coincidence problems in standard cosmology that it should not deviate our attention from the possibility of a detectable topology.

How do the present observational data constrain the possible multi-connectedness of the universe and, more generally, what kinds of tests are conceivable (see [12] for a non-technical book about all aspects of topology and its applications to cosmology)?

Different approaches have been proposed for extracting information about the topology of the universe from experimental data. One approach is to use the 3D distribution of astronomical objects such as galaxies, quasars and galaxy clusters: if the whole universe is finite and small enough, we should be able to see "all around" it because the photons might have crossed it once or more times. In such a case, any observer might recognize multiple images of the same light source, although distributed in different directions of the sky and at various redshifts, or to detect specific statistical properties in the distribution of faraway sources. Various methods of "cosmic crystallography", initially proposed in [13], have been widely developed by other groups ([14] and references therein). However, for plausible small universe models, the first signs of topological lensing would appear only at pretty high redshift, say $z \approx 2$. The main limitation of cosmic crystallography is that the presently available catalogs of observed sources at high redshift are not complete enough to perform convincing tests for topology. For instance, looking for nontoroidal topological lensing, [15] applied the crystallographic method to the SDSS quasar sample; though they found no robust signature, cosmological interpretation of the result was prohibited by the data incompleteness and by the uncertainty in quasar physics. On the other hand, [16] proposed to use deep surveys of distant (at redshifts $z \sim 6$) starburst galaxies for an independent test of the cubic hypertorus model. Their calculation showed that even photometric redshifts would suffice in this purpose, which makes their strategy a realistic and interesting one.

The other approach uses the 2D CMB maps (for a review, [17]). The last scattering surface (LSS) from which the CMB is released represents the most distant source of photons in the universe, hence the largest scales with which the topology of the universe can be probed.

The idea that a small universe model could lead to a suppression of power on large angular scales in the fluctuation spectrum of the CMB had been proposed in the 1980s [18]: in some way, space would be not big enough to sustain long wavelengths. After the release of COBE data in 1992 and the higher resolution and sensitivity of WMAP (2003), there were indeed indications of low power on large scales which could have had a topological origin, and many authors used it to constrain the models. The best fits between theoretical power spectra computed for various topologies and the observed one were obtained with the positively curved Poincaré Dodecahedral Space [9,19,20] and the flat hypertorus [21]. In addition, it was shown [22] that the low-order multipoles tended to be relatively weak in "well-proportioned" spaces, *i.e.*, whose dimensions are approximately equal in all directions. Some globally inhomogeneous topologies can also reproduce the large-angle CMB power suppression if the location of the observer is so adjusted that his fundamental domain becomes well-proportioned [23]. However, this possibility was not borne out by detailed real- and harmonic-space analyses in two dimensions, so that the arguments based on the angular power spectrum and favoring small universe models failed [24]. In any case, to gain all the possible information from the correlations of CMB anisotropies, one has to consider the full covariance matrix rather than just the power spectrum.

Indeed the main imprint of a non trivial topology on the CMB is well-known in the case when the characteristic topological length scale (called the injectivity radius) is smaller than the radius of the LSS: the crossings of the LSS with its topological images give rise to pairs of matched circles of equal radii, centered at different points on the CMB sky, and exhibiting correlated patterns of temperature variations [25], see Figure 3. For instance, the PDS model predicts six pairs of antipodal circles with an angular radius comprised between 5° and 55° (sensitively depending on the cosmological parameters) and a relative phase of 36°.

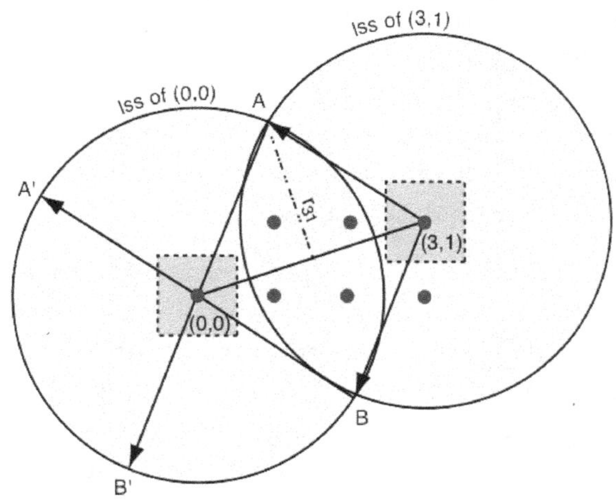

Figure 3. The circles-in-the-sky method is illustrated here in a 2D torus space. The fundamental domain is a square (with a dotted outline), all of the red points are copies of the same observer. The two large circles (which are normally spheres in a three-dimensional space) represent the last scattering surfaces (LSS) centered on two copies of the same observer. One is in position (0,0), its copy is in position (3,1) in the universal covering space. The intersection of the circles is made up of the two points A and B (in three dimensions, this intersection is a circle). The observers (0,0) and (3,1), who see the two points (A,B) from two opposite directions, are equivalent to a unique observer at (0,0) who sees two identical pairs (A,B) and (A′,B′) in different directions. In three dimensions, the pairs of points (A,B) and (A′,B′) become a pair of identical circles, whose radius r_{31} depends on the size of the fundamental domain and the topology.

4. Results and Discussion

Such "circles-in-the-sky" searches have been looked for in WMAP maps by several groups, using various statistical indicators and massive computer calculations, and interpreting their results differently. Some authors [26] claimed that most of non-trivial topologies, including PDS and T^3, were ruled out: they searched for antipodal or nearly antipodal pairs of circles in the WMAP data and found no such circles. However, their analysis could not be applied to more complex topologies, for which the matched circles deviate strongly from being antipodal. On the other hand, other groups claimed to have found hints of multi-connected topology, using different statistical indicators [27–29].

Most studies have emphasized searches for fundamental domains with antipodal correlations. The search for matched circle pairs that are not back-to-back has nevertheless been carried out recently, with no obvious topological signal appearing in the seven-year WMAP data [30]. The statistical significance of such results still has to be clarified. In any case, a lack of nearly matched circles does not exclude a multi-connected topology on scale less than the horizon radius: detectable topologies may produce circles of small radii which are statistically hard to detect and current analysis of CMB sky maps could have missed even antipodal matching circles, because various effects may damage or even destroy the temperature matching.

Other methods for experimental detection of non-trivial topologies have thus been proposed and used to analyze the experimental data, such as the multipole vectors and the likelihood (Bayesian) method. The latter ameliorates some of the spoiling effects of the temperature correlations such as the integrated Sachs-Wolfe and Doppler contributions [31].

The most up-to-date study based on CMB temperature correlations used the Planck 2013 intensity data [32]. In that work, they applied two techniques: first, a direct likelihood calculation of (a very few) specific topological models; second, a search for the expected repeated "circles in the sky", calibrated by simply-connected simulations. Both of these showed that the scale of any possible topology must exceed roughly the comoving distance to the LSS, χ_{rec}. For the cubic torus, they found that the radius of the largest sphere inscribed in the topological fundamental domain must be $R_i > 0.92\ \chi_{rec}$. The matched-circle limit on topologies predicting back-to-back circles larger than 15° in angular radius and assuming that the relative orientation of the fundamental domain and mask allows its detection was $R_i > 0.94\ \chi_{rec}$ at the 99% confidence level.

Finally, it is now widely understood that the polarization of the CMB can provide a lot of additional informations for reconstructing the cosmological model. Riazuelo *et al.* [33] were the first to show how the polarization could also be used to put additional constraints on space topology and a little bit tighter than those coming from temperature intensity. Maps of CMB polarization from the 2015 release of Planck data [34] provided the highest-quality full-sky view of the LSS available to date. However their study specialized only to flat spaces with cubic toroidal (T^3) and slab (T^1) topologies. These searches yield no detection of a compact topology with a scale below the diameter of the LSS. More precisely, $R_i > 0.97\ \chi_{rec}$ for the T^3 cubic torus and $R_i > 0.56\ \chi_{rec}$ for the T^1 slab.

5. Conclusions

The overall topology of the universe has become an important concern in astronomy and cosmology. Even if particularly simple and elegant models such as the PDS and the hypertorus are now claimed to be ruled out at a subhorizon scale, many more complex models of multi-connected space cannot be eliminated as such. In addition, even if the size of a multi-connected space is larger (but not too much) than that of the observable universe, we could still discover an imprint in the CMB, even while no pair of circles, much less ghost galaxy images, would remain. The topology of the universe could therefore provide information on what happens outside of the cosmological horizon [35].

Whatever the observational constraints, a lot of unsolved theoretical questions remain. The most fundamental one is the expected link between the present-day topology of space and its quantum origin, since classical general relativity does not allow for topological changes during the course of cosmic evolution. Theories of quantum gravity should allow us to address the problem of a quantum origin of space topology. For instance, in quantum cosmology, the question of the topology of the universe is completely natural. Quantum cosmologists seek to understand the quantum mechanism whereby our universe (as well as other ones in the framework of multiverse theories) came into being, endowed with a given geometrical and topological structure. We do not yet have a correct quantum theory of gravity, but there is no sign that such a theory would *a priori* demand that the universe have a trivial topology. Wheeler first suggested that the topology of space-time might fluctuate at a quantum level, leading to the notion of a space-time foam [36]. Additionally, some simplified solutions of the Wheeler-de Witt equations for quantum cosmology show that the sum over all topologies involved in the calculation of the wavefunction of the universe is dominated by spaces with small volumes and multi-connected topologies [37]. In the approach of brane worlds in string/M-theories, the extra-dimensions are often assumed to form a compact Calabi-Yau manifold; in such a case, it would be strange that only the ordinary, large dimensions of our 3-brane would not be compact like the extra ones. However, still at an early stage of development, string quantum cosmology can only provide heuristic indications on the way multi-connected spaces would be favored.

Acknowledgments: The author thanks the anonymous referees for suggesting additional references and improvements of the original manuscript.

References

1. Starkman, G.D. Topology and Cosmology. *Class. Quantum Gravity* **2008**, *15*, 2529–2538. [CrossRef]
2. Iorio, L. Editorial for the Special Issue 100 Years of Chronogeometrodynamics: The Status of the Einstein's Theory of Gravitation in Its Centennial Year. *Universe* **2015**, *1*, 38–81. [CrossRef]

3. Friedmann, A. Über die Möglichkeit einer Welt mit konstanter negativer Krümmung des Raumes. *Zeitsfricht Phys.* **1924**, *21*, 326–332. (English Translation in Bernstein, J.; Feinberg, G. *Cosmological Constants. Papers in Modern Cosmology*; Columbia University Press: New York, NY, USA, 1986; pp. 59–65.)
4. Lachièze-Rey, M.; Luminet, J.-P. Cosmic topology. *Phys. Rep.* **1995**, *254*, 135–214. [CrossRef]
5. Wolf, J.A. *Spaces of Constant Curvature*; Publish or Perish Inc.: Wilmington, DC, USA, 1984.
6. Kramer, P. Topology of Platonic Spherical Manifolds: From Homotopy to Harmonic Analysis. *Symmetry* **2015**, *7*, 305–326. [CrossRef]
7. Riazuelo, A.; Weeks, J.; Uzan, J.-P.; Lehoucq, R.; Luminet, J.-P. Cosmic microwave background anisotropies in multiconnected flat spaces. *Phys. Rev. D* **2004**, *69*, 103518. [CrossRef]
8. Gausmann, E.; Lehoucq, R.; Luminet, J.-P.; Uzan, J.-P.; Weeks, J. Topological lensing in spherical spaces. *Class. Quantum Gravity* **2001**, *18*, 5155–5186. [CrossRef]
9. Luminet, J.-P.; Weeks, J.; Riazuelo, A.; Lehoucq, R.; Uzan, J.-P. Dodecahedral space topology as an explanation for weak wide-angle temperature correlations in the cosmic microwave background. *Nature* **2003**, *425*, 593–595. [CrossRef] [PubMed]
10. Geometrygames. Available online: http://www.geometrygames.org (accessed on 8 January 2016).
11. Hinshaw, G.; Larson, D.; Komatsu, E.; Spergel, D.N.; Bennett, C.L.; Dunkley, J.; Nolta, M.R.; Halpern, M.; Hill, R.S.; Odegard, N.; *et al*. Nine-year WMAP Observations: Cosmological Parameter Results. *Astrophys. J. Suppl.* **2013**, *208*, 19.
12. Luminet, J.-P. *The Wraparound Universe*; AK Peters: New York, NY, USA, 2008.
13. Lehoucq, R.; Lachièze-Rey, M.; Luminet, J.-P. Cosmic crystallography. *Astron. Astrophys.* **1996**, *313*, 339–346.
14. Fujii, H.; Yoshii, Y. An improved cosmic crystallography method to detect holonomies in flat spaces. *Astron. Astrophys.* **2011**, *529*, A121. [CrossRef]
15. Fujii, H.; Yoshii, Y. A Search for Nontoroidal Topological Lensing in the Sloan Digital Sky Survey Quasar Catalog. *Astrophys. J.* **2013**, *773*, 152–160. [CrossRef]
16. Roukema, B.; France, M.; Kazimierczak, T.; Buchert, T. Deep redshift topological lensing: Strategies for the T^3 candidate. *Mon. Not. R. Astron. Soc.* **2014**, *437*, 1096–1108. [CrossRef]
17. Levin, J. Topology and the cosmic microwave background. *Phys. Rep.* **2002**, *365*, 251–333. [CrossRef]
18. Fagundes, H.V. The Quadrupole Component of the Relic Radiation in a Quasi-Hyperbolic Cosmological Model. *Astrophys. Lett.* **1983**, *23*, 161.
19. Caillerie, S.; Lachièze-Rey, M.; Luminet, J.-P. A new analysis of the Poincaré dodecahedral space model. *Astron. Astrophys.* **2007**, *476*, 691–696. [CrossRef]
20. Roukema, B.F.; Bulinski, Z.; Gaudin, N.E. Poincaré dodecahedral space parameter estimates. *Astron. Astrophys.* **2008**, *492*, 657–673. [CrossRef]
21. Aurich, R. A spatial correlation analysis for a toroidal universe. *Class. Quantum Gravity* **2008**, *25*, 225017. [CrossRef]
22. Weeks, J.; Luminet, J.-P.; Riazuelo, A.; Lehoucq, R. Well-proportioned universes suppress the cosmic microwave background quadrupole. *Mon. Not. R. Astron. Soc.* **2004**, *352*, 258–262. [CrossRef]
23. Aurich, R.; Lustig, S. How well proportioned are lens and prism spaces? *Class. Quantum Gravity* **2012**, *29*, 175003. [CrossRef]
24. Bielewicz, P.; Riazuelo, A. The study of topology of the Universe using multipole vectors. *Mon. Not. R. Astron. Soc.* **2009**, *396*, 609–623. [CrossRef]
25. Cornish, N.J.; Spergel, D.N.; Starkman, G.D. Circles in the sky: Finding topology with the microwave background radiation. *Class. Quantum Gravity* **1998**, *15*, 2657–2670. [CrossRef]
26. Cornish, N.J.; Spergel, D.N.; Starkman, G.D.; Komatsu, E. Constraining the Topology of the Universe. *Phys. Rev. Lett.* **2004**, *92*, 201302. [CrossRef] [PubMed]
27. Roukema, B.; Lew, B.; Cechowska, M.; Marecki, A.; Bajtlik, S. A Hint of Poincaré Dodecahedral Topology in the WMAP First Year Sky Map. *Astron. Astrophys.* **2004**, *423*, 821–831. [CrossRef]
28. Lew, B.; Roukema, B. A test of the Poincaré dodecahedral space topology hypothesis with the WMAP CMB data. *Astron. Astrophys.* **2008**, *482*, 747–753. [CrossRef]
29. Aurich, R.; Lustig, S.; Steiner, F. The circles-in-the-sky signature for three spherical universes. *Mon. Not. R. Astron. Soc.* **2006**, *369*, 240–248. [CrossRef]
30. Vaudrevange, P.; Starkman, G.; Cornish, N.; Spergel, D. Constraints on the topology of the Universe: Extension to general geometries. *Phys. Rev. D* **2012**, *86*, 083526. [CrossRef]

31. Kunz, M.; Aghanim, N.; Cayon, L.; Forni, O.; Riazuelo, A.; Uzan, J.-P. Constraining topology in harmonic space. *Phys. Rev. D* **2006**, *73*, 023511. [CrossRef]
32. Ade, P.A.R.; Aghanim, N.; Armitage-Caplan, C.; Arnaud, M.; Ashdown, M.; Atrio-Barandela, F.; Aumont, J.; Baccigalupi, C.; Banday, A.J.; Barreiro, R.B.; *et al.* Planck 2013 results XXVI. Background geometry and topology of the Universe. *Astron. Astrophys.* **2014**, *571*, A26.
33. Riazuelo, A.; Caillerie, S.; Lachièze-Rey, M.; Lehoucq, R.; Luminet, J.-P. Constraining Cosmic Topology with CMB Polarization. **2006**, arXiv:astro-ph/0601433.
34. Ade, P.A.R.; Aghanim, N.; Arnaud, M.; Ashdown, M.; Aumont, J.; Baccigalupi, C.; Banday, A.J.; Barreiro, R.B.; Bartolo, N.; Battaner, E.; *et al.* Planck 2015 results XVIII. Background geometry and topology. **2015**, arXiv:1502.01593.
35. Fabre, O.; Prunet, S.; Uzan, J.-P. Topology beyond the horizon: How far can it be probed? *Phys. Rev. D* **2015**, *92*, 04003. [CrossRef]
36. Wheeler, J. On the nature of quantum geometrodynamics. *Ann. Phys.* **1957**, *2*, 604–614. [CrossRef]
37. e Costa, S.S.; Fagundes, H.V. On the Birth of a Closed Hyperbolic Universe. *Gen. Relativ. Gravit.* **2001**, *33*, 1489–1494. [CrossRef]

10

Convexity and the Euclidean Metric of Space-Time

Nikolaos Kalogeropoulos

Carnegie Mellon University in Qatar, Education City, P.O. Box 24866 Doha, Qatar; nkaloger@cmu.edu or nikos.physikos@gmail.com

Academic Editors: Stephon Alexander, Jean-Michel Alimi, Elias C. Vagenas and Lorenzo Iorio

Abstract: We address the reasons why the "Wick-rotated", positive-definite, space-time metric obeys the Pythagorean theorem. An answer is proposed based on the convexity and smoothness properties of the functional spaces purporting to provide the kinematic framework of approaches to quantum gravity. We employ moduli of convexity and smoothness which are eventually extremized by Hilbert spaces. We point out the potential physical significance that functional analytical dualities play in this framework. Following the spirit of the variational principles employed in classical and quantum Physics, such Hilbert spaces dominate in a generalized functional integral approach. The metric of space-time is induced by the inner product of such Hilbert spaces.

Keywords: space-time metric; convexity; smoothness; Hilbert spaces; Banach spaces; dualities

1. Introduction

When one looks at the equations describing the four fundamental interactions of nature, then s/he immediately notices that the kinematic equations in the Lagrangian formalism involve second order derivatives with respect to space-time variables. It may be worthwhile to try to understand the reasons behind this phenomenon, which has been taken for granted as an empirical fact since the earliest days of Newtonian mechanics. Modelling of the fundamental interactions except gravity relies, at the classical level, on Classical Mechanics, on Electromagnetic Theory and its Yang-Mills/non-abelian gauge "generalizations". General Relativity can also be seen as a gauge theory whose gauge group is the diffeomorphism group (re-parametrization invariance) of the underlying topological space endowed with its metric structure.

In all of the above, and in the Lagrangian approach which we employ throughout this work, the Euler-Lagrange equations that describe the underlying dynamics can be seen to emerge from variational principles; such equations could use derivatives of arbitrarily high order and a formalism for accommodating this fact has already been developed. However, in practice, when dealing with fundamental interactions and not performing perturbative or approximate calculations that rely on series expansions, one rarely needs derivatives that are of higher than the second-order with respect to space-time variables.

The statements on the number of derivatives in the equations of dynamics can be seen to be essentially equivalent, upon partial integration, to the fact that the kinetic terms as well as relevant potential energy terms are at most quadratic with respect to first order derivatives of the fundamental variables/fields/order parameters. This in turn, allows one to use Euclidean/Riemannian concepts to model the evolution of such systems; for particle systems one uses the more familiar aspects of finite-dimensional Riemannian spaces [1], and for field theories one may have to resort to using aspects of infinite dimensional manifolds [2] which involve further subtleties.

One could use, equivalently in the simplest context, a Hamiltonian approach where the equations involve first order derivatives. Each of these two lines of approach Lagrangian versus Hamiltonian has its own advantages and drawbacks, as is well-known. See, for instance, the very recent review [3] for an approach to gravity based on first order actions including boundary terms. Without denying the advantages of the Hamiltonian or first order formalisms, we will adopt in this work the Lagrangian/second-order formalism, as already mentioned above. Our viewpoint is somewhat influenced by and may have common points with that of [4], even though the methods which we use and the results we reach are substantially different from those of that work.

We will adopt the convenient technical device of Wick-rotating to a "Euclidean" (i.e., positive-definite) signature metric. We would like to state at this point, that we do not consider positive-definite metrics "superior" in any sense, to these of Lorentzian (indefinite) signature. On the contrary, all experiments, so far, point to the latter as being the physical ones, so even if unstated, our underlying view is closer to [5] rather than [6], for instance. On the other hand, on purely formalistic grounds, the Euclidean (positive-definite) metrics are easier to work with, as they obey the positivity property and the triangle inequality that their indefinite metric (Lorentzian) counterparts are lacking.

A substantial amount of effort has been spent, in recent decades, into understanding the non-trivial features of space-time, as described by the General Theory of Relativity or other theories incorporating space-time diffeomorphism invariance. Even though considerable progress has been made toward such an understanding, it is probably fair to state that many important issues still remain unresolved [7]. It is not clear, for instance, why space-time is 4-dimensional, to what extent it is smooth and how such a smoothness arises, why its Wick-rotated metric obeys the Pythagorean theorem etc. Henceforth we will work only with positive definite (Wick-rotated) metrics of space-time and by "Euclidean" we will mean only the ones obeying the Pythagorean theorem. In linear algebraic and functional analytic language the metrics we call Euclidean would be called the l^2 metrics. Moreover, we will use the term space-time in order to keep in mind that our arguments actually purport to describe (indefinite/Lorentzian signature) space-time even if we use positive-definite (Wick-rotated) metrics throughout this work.

A potentially fruitful way toward answering why the space-time metric is Euclidean, which is the subject of this work, is to look at it through the eyes of convexity. Convexity plays a central role in many branches of Mathematics, but seems to be under-appreciated and under-utilized in gravitational Physics [8], and not only. This can also be considered as a partial motivation for looking for answers to the questions of our interest through convexity. In the context of linear spaces, convexity turns out to be dual to smoothness, a fact that we also use to support the case for the Euclidean form of the space-time metric.

In Section 2, we provide a general background and a physical interpretation, wherever feasible, from the theory of normed spaces with emphasis on properties pertinent to our arguments. In Section 3, we discuss the aspects of convexity, smoothness and their duality via a Legendre-Fenchel transform, employing in particular Clarkson's modulus of convexity and to the Day-Lindenstrauss modulus of smoothness. In Section 4, we put our, less than rigorous, argument together on how the previous results result in space-time metric that has the Euclidean form. Section 5, presents some conclusions and caveats.

2. Background and Physical Interpretation of Some Concepts

Space-times are assumed to be locally flat, to first order approximation of their metric. This is an outcome of the application of the Equivalence Principle, and a well-known fact in Riemannian/Lorentzian geometry. Therefore, we can analyze the ultralocal aspects of the features of a space-time by confining our attention to vector spaces. Most features of such tangent spaces can be captured by the various closely related functional spaces that can be defined on them. There are numerous classes of functional spaces that have been investigated during the last century, in the context of Functional Analysis, such as Lebesgue, Hardy, Bergman, Sobolev, Orlicz, Besov, Triebel-Lizorkin, etc. spaces. Most of the previously named spaces are or rely on, in one form or another, constructions and results of (usually infinite dimensional) Banach spaces. Such infinite dimensional Banach spaces will be the main vector spaces of interest in this work.

In this Section we provide some preliminary information on these and related mathematical constructions which are pertinent to the subsequent sections, where most of our arguments are developed. We attempt to stress their physical motivation and interpretation, wherever feasible, from the viewpoint we adopt in this work.

2.1. Norms on Linear Spaces

To make the exposition more readable and self-contained, we recall [9,10] that a norm on a vector space $\mathcal{V}$ defined over a field $\mathbb{F}$, usually $\mathbb{F} = \mathbb{R}$ or $\mathbb{F} = \mathbb{C}$ in most applications in Physics, is a function $\| \cdot \| : \mathcal{V} \to \mathbb{R}_+$ where $\mathbb{R}_+ = \{\lambda \in \mathbb{R} : \lambda \geq 0\}$, such that for all $x, y \in \mathcal{V}$

- Positive definiteness: $\|x\| = 0$, if and only if $x = 0$.
- Homogeneity: $\|kx\| = |k|\|x\|$, for all $x \in \mathcal{V}$, and all $k \in \mathbb{F}$.
- Triangle inequality: $\|x + y\| \leq \|x\| + \|y\|$

A vector space $\mathcal{V}$ endowed with a norm, which moreover is complete in this norm, namely such that all Cauchy sequences have limits belonging to $\mathcal{V}$, is called a Banach space. Examples of Banach spaces that are explicitly used in this work are:

- c_0 is the space of all sequences $a = (a_n), n \in \mathbb{N}$ converging to zero, with $a_n \in \mathbb{F}$, and the sup-norm

$$\|a\| = \sup_n |a_n| \tag{1}$$

- $l^p, 1 \leq p < \infty$, which is the space of all sequences $a = (a_n), n \in \mathbb{N}$ with $a_n \in \mathbb{F}$ endowed with the norm

$$\|a\|_p = \left(\sum_{i=1}^{n} |a_n|^p \right)^{\frac{1}{p}} < \infty \tag{2}$$

- l^∞, the space of all bounded sequences $a = (a_n), n \in \mathbb{N}$ endowed with the supremum norm

$$\|a\| = \sup_n |a_n| \tag{3}$$

- $L^p(\mathbb{R}^n), 1 \leq p < \infty$, the space of Lebesgue integrable functions $f : \mathbb{R}^n \to \mathbb{F}$ endowed with the norm

$$\|f\|_p = \left(\int_{\mathbb{R}^n} |f|^p \right)^{\frac{1}{p}} \tag{4}$$

- $L^\infty(\mathbb{R}^n)$, the space of all essentially bounded $f : \mathbb{R}^n \to \mathbb{F}$, endowed with the norm

$$\|f\|_\infty = \inf\{C : |f| < C almost everywhere\} \tag{5}$$

The above functional spaces $L^p(\mathbb{R}^n), 1 \leq p \leq \infty$ are, strictly speaking, spaces over equivalence classes of functions, where two functions are considered equivalent ("equal") if they differ from each other, at most, in a set of measure zero. One uses the Lebesgue measure ("volume") of $\mathbb{R}^n$ in the definition of such $L^p(\mathbb{R}^n)$. In most applications in Physics, the distinction between functions and classes of equivalent functions is tacitly assumed and not explicitly stated. For completeness, we mention that a Hilbert space is a linear space endowed with an inner product $(\cdot, \cdot)$ which is, moreover, complete [9,10]. The norm which we will assume that Hilbert spaces are endowed with, is induced by their inner product by $\|x\|^2 = (x, x)$. Among the above examples of Banach spaces, l^2 and $L^2(\mathbb{R}^n)$ are Hilbert spaces. Since the spaces that we will be referring to are normed, hence metrizable, they are endowed with the topology induced by the metric. An important topological property of such topological (vector) spaces is separability: this means that such spaces contain a countable dense subset. For such metrizable spaces, being separable is equivalent to being second countable. Topological spaces that are finite or countably infinite are, obviously, separable. For Hilbert spaces separability implies the existence of a countable orthonormal basis, hence any separable infinite dimensional Hilbert space is isometric to l^2. In most cases in quantum theory, separability of the space of the Hllbert space of wave-functions is tacitly assumed. Banach spaces can also be either separable or not separable [11]. The issue of separabilty of the spaces we use is not pertinent to the arguments of this work, so it will not be encountered anywhere in the sequel.

2.2. Norm Equivalence

Naturally, one can endow a vector space $\mathcal{V}$ with many different norms. Usually the "appropriate" choice of such a norm has substantial implications for the specific predictions of the physical model built on $(\mathcal{V}, \|\cdot\|)$. In other words, the choice of two "different", in the naive sense of the word, norms will usually result in substantially different predictions of the physical quantities resulting form such calculations. Hence the choice of a norm is usually considered to be a piece of data which is initially provided by hand, in any model. In most cases such a choice of norm is not explicitly discussed, because it is assumed that a norm arising from an inner product is the physically relevant one. In this work, we would like to know why this may be the case.

Given that a vector space may be endowed with different norms, one can ask what the word "different" may actually mean. If two norms are point-wise different but still reasonably close to each other, in some particular sense, let's say with respect to a particular metric, can they still be declared as "equivalent"? Consider, for instance, the case of Hamiltonian systems of many degrees of freedom. This paragraph operates in the context of Euclidean metrics but can still be used to highlight our point of view. Suppose that one changes the metric of the phase space. If all thermodynamic quantities remain invariant under such a change of norm, is it reasonable to consider the two metrics as "equivalent" or should someone insist as treating them as "different"? Under some additional conditions, two such metrics may turn out to be equal; this is reminiscent of the (Hamburger, Stieltjes etc.) moment problem pertaining to the equality of underlying probability distributions, if the sets of moments of these distributions are equal [12]. Something similar occurs in Geometry or Analysis: quite often two metrics or norms are declared as "equivalent" even though they may be pointwise distinct.

This re-interpretation of the concept of "equivalence" can have substantial consequences for Statistical or Quantum Physics. The issue at hand can be seen as a form of "stability". Consider, for instance, a Hamiltonian system. This determines a symplectic structure on the phase space $\mathcal{M}$ of the system or, alternatively, a Poisson structure on the space of smooth functions $C^\infty(\mathcal{M})$ [13]. However, it does not generically determine a metric structure on $\mathcal{M}$. Such a metric is usually assumed to stem from the quadratic "kinetic" term of the Hamiltonian, at least for systems having a kinetic term of such form. Then we can proceed to perform an analysis of the evolution of this Hamiltonian system, quantize the system etc. based on this symplectic and metric structures. However, one should not forget that the metric structure was not "natural", or may not be unique. For this reason if it changes, especially a little bit, one would expect the physically relevant results to remain practically invariant. This can be interpreted as a form of structural stability of the underlying Hamiltonian dynamical system, even though it applies to an auxiliary piece of data, such as the metric. It is a subject of much discussion on whether such a metric structure should be assumed, and if so, to what extent it determines the statistically significant features of the system in an appropriate many-body "thermodynamic" limit [14].

Given the above considerations, one may be willing to allow variations of the assumed metric of the underlying Hamiltonian system. Then two metrics can be declared as equivalent, as also previously mentioned, if they give the same predictions of physically relevant quantities. The question is then how to find such variations of metrics, or norms. From an analytical viewpoint, one can use a central concept of Analysis, that of the limit, to determine how to answer such questions. The most crude/rough approach is to demand that two metrics/norms $\|\cdot\|_1, \|\cdot\|_2$ to be considered as equivalent if all sequences converging with respect to one of them also converge with respect to the other. This is realized when there are constants $0 < C_1 \leq C_2$ such that

$$C_1 \| \cdot \|_1 \leq \| \cdot \|_2 \leq C_2 \| \cdot \|_1 \tag{6}$$

This always holds for any two norms on finite dimensional vector spaces. However, it is not true in general for infinite dimensional functional spaces, which are at the center of our attention.

A way to interpret (6) from the viewpoint of statistical theories is as follows: physically relevant results of the microscopic or quantum dynamics should be somehow reflected or emerge in the large scale/multi-particle or thermodynamic limit. Such features should remain largely unaffected by most "small-scale" details of the system or their perturbations. This tacitly assumes that the underlying quantities characterizing geometric characteristics of the system are proportional to the effective measure(s) used in the calculations of the pertinent statistical quantities. This is clearly true for ergodic systems, but it can also be true for non-ergodic systems such as the ones whose thermodynamic behavior is conjecturally described by any of the many recently proposed entropic functionals, such as the "Tsallis entropy" [15], or the "κ-entropy" [16], for instance. So, from a geometric viewpoint, equations like (6) express this insensitivity to small-scale details. This viewpoint motivates and pervades "coarse geometry" [17] and is frequently encountered in constructions related to hyperbolic spaces [18] or groups [19]. Metrically, it is expressed by demanding invariance under quasi-isometries. In dynamical systems one can see a similar viewpoint in several occasions, an example of which is that the topological entropy of a map or flow on a metric space $(\mathfrak{X}, d)$ does not actually depend on the specific metric/distance function d, but only on the class on metrics on $\mathfrak{X}$ that induce the same topology on $\mathfrak{X}$. Then the key/desired invariance akin to (6), is the topological conjugacy [20].

2.3. *The Operator Norm and the Banach-Mazur Distance*

Before continuing, for completeness of the exposition, we state two definitions that will be extensively used in the sequel [9,10]. Let $(\mathcal{X}, \| \cdot \|_{\mathcal{X}})$ and $(\mathcal{Y}, \| \cdot \|_{\mathcal{Y}})$ be two normed spaces, over $\mathbb{R}$, $\mathbb{C}$ (or any other field, although the general case does not appear to be of any particular interest in Physics, so far) and let T be a continuous linear map $T : \mathcal{X} \to \mathcal{Y}$. If such a map exists between $\mathcal{X}, \mathcal{Y}$ which is bijective, and its inverse T^{-1} is also bijective, then $\mathcal{X}, \mathcal{Y}$ are called isomorphic. Actually less is needed: the Open Mapping Theorem guarantees that if T is bijective and bounded, so is T^{-1}. If a mapping T is an isometry, namely if

$$\|Tx\|_{\mathcal{Y}} = \|x\|_{\mathcal{X}} \tag{7}$$

then $\mathcal{X}, \mathcal{Y}$ are called isometric. Since boundedness of T is equivalent to its continuity, one can see that isometric spaces are isomorphic. Let the space of bounded linear maps from $\mathcal{X}$ to $\mathcal{Y}$ be denoted by $\mathcal{B}(\mathcal{X}, \mathcal{Y})$. This space $\mathcal{B}(\mathcal{X}, \mathcal{Y})$ can be endowed with the operator (sup-) norm which for $T \in \mathcal{B}(\mathcal{X}, \mathcal{Y})$ is given by the following equivalent definitions

$$\|T\| = \sup_{x \in \mathcal{X}} \{\|Tx\| : \|x\| \leq 1\} = \sup_{x \in \mathcal{X}} \{\|Tx\| : \|x\| = 1\} = \sup_{x \in \mathcal{X}} \left\{ \frac{\|Tx\|}{\|x\|} : \|x\| \neq 0 \right\} \tag{8}$$

It turns out that if $\mathcal{X}$ is a normed space and if $\mathcal{Y}$ is a Banach space, then $\mathcal{B}(\mathcal{X}, \mathcal{Y})$ endowed with the operator norm (8) is a Banach space too. We know that every normed space can be isometrically embedded in a Banach space. So, most of the pertinent features of general normed spaces are contained in Banach spaces, therefore we can use the latter believing that we are not losing important aspects of the flexibility or generality of the former, for applications in Physics. Given this, our question is then reduced to asking, why among all Banach spaces, the inner product (Hilbert) spaces are the ones describing most fundamental aspects of nature most accurately, so far as we know today.

In the spirit of norm-equivalence, discussed in Subsection 2.2, one can ask how close, or how far, from each other are two linear spaces. From a metric viewpoint they are identical, if they are isometric.

We want to have a "reasonable" distance function that measures how far from each other they may be, if they are not isometric. Defining such a distance function is clearly a matter of choice which ideally, for our purposes, should also reflect some desirable physical properties. It appears that the classical Banach-Mazur distance has properties fitting such requirements. It is defined as follows. Let $\mathcal{X}, \mathcal{Y}$ be two isomorphic Banach spaces. The Banach-Mazur distance between them is defined as

$$d_{BM}(\mathcal{X}, \mathcal{Y}) = \inf_T \{\|T\| \cdot \|T^{-1}\|\} \tag{9}$$

where $T : \mathcal{X} \to \mathcal{Y}$ is an isomorphism. If $\mathcal{X}, \mathcal{Y}$ are not isomorphic, then their Banach-Mazur distance is infinite, by definition. We can immediately see that

$$d_{BM}(\mathcal{X}, \mathcal{Y}) \geq 1 \tag{10}$$

and that for isometric spaces such a distance is exactly equal to one. The converse is also true, but only for finite-dimensional Banach spaces. Therefore, the Banach-Mazur distance is actually a distance function on the set of equivalence classes of normed spaces (where "equivalence" is defined as "isometry"), but in a multiplicative sense, namely for three isomorphic linear spaces $\mathcal{X}, \mathcal{Y}, \mathcal{Z}$, it satisfies

$$d_{BM}(\mathcal{X}, \mathcal{Z}) \leq d_{BM}(\mathcal{X}, \mathcal{Y}) \cdot d_{BM}(\mathcal{Y}, \mathcal{Z}) \tag{11}$$

To get to the usual triangle inequality instead of (11), we have to consider the logarithm of d_{BM}. The Banach-Mazur distance is invariant under invertible linear maps T, namely

$$d_{BM}(T\mathcal{X}, T\mathcal{Y}) = d_{BM}(\mathcal{X}, \mathcal{Y}) \tag{12}$$

In some sense the Banach-Mazur distance expresses the minimum distortion that any isomorphism between two linear spaces can possibly entail.

Before closing this Subsection, one cannot help but distinguish between the functional spaces arising in a theory developed on space-time and the underlying form of the metric of space-time itself. These are clearly two quite distinct classes of spaces that have to be treated independently. However, one can hope that if and when a reasonably testable model of quantum gravity is found, then its Hilbert (or more generally, functional) space of "wave-functions" will induce the observable metric of space-time. So from now on, the working assumption will be that such quantum mechanical Hilbert spaces induce the space-time metric. Therefore, we should address the question about what is so special about inner product (Hilbert) spaces among the class of all Banach spaces. Since we will be working with spaces of functions, we will focus on infinite dimensional Banach spaces in the sequel. To be more concrete, we will have in mind spaces seen frequently in applications such as the spaces of p-summable sequences (l^p), or of Lebesgue p-integrable functions on $\mathbb{R}^n$ (L^p) endowed with their cardinality or their induced Euclidean measure "volume" respectively, appearing in (1)–(5).

2.4. Reflexive and Super-Reflexive Spaces

To proceed in determining desirable properties of the Banach spaces of functions on $\mathbb{R}^n$, we consider the following. It is widely believed among many, or even most, quantum gravity practitioners that spacetime properties such as its topology, smoothness, metric etc. should be "derived" from a quantum theory of gravity rather than be put in the models by hand. This seems to be a widespread belief, regardless of the exact approach to quantum gravity that someone follows. It is based in the fact that macroscopic properties of systems can be derived from their quantum counterparts and rely, to some extent, in the great separation of scales and numbers of degrees of freedom between the microscopic and the macroscopic scales. This is the main reason, as accepted today, of why Statistical Mechanics works so well in providing accurate predictions for systems with many degrees of freedom.

Independently of such physical considerations, there has been a recent surge of activity in Geometry and Analysis purporting to better understand first order calculus properties of non-smooth spaces [21]. An influential work in this direction has been [22], where conditions were given for the existence of a differentiable structure on metric measure spaces, based on Lipschitz maps, which also have the doubling property and admit a Poincaré inequality. Among the numerous works that clarified, elaborated and generalized [22], we could point out [23–26]. In these works, the differentiable structure appears naturally as a result of the more "primitive" assumptions stated above and presented in [22]. What is pertinent to our purposes is that [22] discovered that Sobolev spaces of functions on such metric measure spaces, which seem to be the most relevant from a physical viewpoint, turn out to be reflexive. Moreover they admit a uniformly convex norm. We will elaborate on the first condition in this Subsection, and the second in the next Section.

Reflexive Spaces. Let $\mathcal{X}$ be a Banach space and let $B_{\mathcal{X}}$ indicate its closed unit ball, namely

$$B_{\mathcal{X}} = \{x \in \mathcal{X} : \|x\| \leq 1\} \tag{13}$$

The dual of $\mathcal{X}$, denoted by $\mathcal{X}'$, is the space of (real or complex valued) continuous linear functionals of $\mathcal{X}$, namely an element of $\mathcal{B}(\mathcal{X}, \mathbb{R})$ or $\mathcal{B}(\mathcal{X}, \mathbb{C})$. Examples of such dual spaces are $(c_0)' = l^1, (l^1)' = l^\infty, (L^p(\mathbb{R}^n))' = L^q(\mathbb{R}^n), 1 < p < \infty, p^{-1} + q^{-1} = 1$ etc. In case $\mathcal{X}$ is finite-dimensional, the closed unit ball of its dual $B_{\mathcal{X}'}$ is the polar body of the unit ball of $\mathcal{X}$, namely

$$B_{\mathcal{X}'} = B^{\circ}_{\mathcal{X}} \tag{14}$$

Generalizing, one can define the bi-dual of $\mathcal{X}$ as the dual of $\mathcal{X}'$. These dualities induce a natural linear mapping $F : \mathcal{X} \to (\mathcal{X}')'$ given by

$$F(f) = f(x) \tag{15}$$

for $x \in \mathcal{X}$ and $f \in \mathcal{X}'$ being its dual. There is no a priori reason why the double dual of $\mathcal{X}$ should be equal to $\mathcal{X}$. Usually $\mathcal{X} \subset (\mathcal{X}')'$. A simple example of this inclusion is that the dual of c_o is l^1 and the dual of l^1 is l^∞. Therefore the inclusion $F : c_o \to l^\infty$ is not surjective. If, however, it happens that under the canonical map F

$$\mathcal{X} = (\mathcal{X}')' \tag{16}$$

then the Banach space $\mathcal{X}$ is called reflexive. It is important that $\mathcal{X}$ is isometric to $(\mathcal{X}')'$ under the canonical embedding (15). It is possible for a non-reflexive space to be isometric to its bi-dual; an example is provided by James' space. Obviously every finite-dimensional Banach space is reflexive, due to the rank-nullity theorem. It should be immediately noticed that reflexivity is a topological property and not a property of the norm of a particular space. This is quite important given the fact that sometimes we will consider renormings of Banach spaces for the reasons mentioned in Section 2.2.

An example of reflexive Banach spaces are the Lebesgue spaces $L^p(\mathbb{R}^n), 1 < p < \infty$. Reflexive Banach spaces have numerous desirable properties: one can mention, for instance

1. the dual of reflexive space is reflexive.
2. the closed subspaces of reflexive spaces are reflexive.
3. the quotient spaces of reflexive spaces are reflexive, etc.

The question that comes to mind is whether there are physical reasons why a linear space should be reflexive. This is unclear in our opinion, beyond the desirable mathematical properties previously mentioned. It is not clear, for instance, what would be the physical consequences if the bi-dual of a Banach space were dense, rather than surjective, under the canonical mapping (15).

Discrete spacetime and"internal" symmetries, such as parity, time-inversion and charge conjugation whose "double dual" is the identity, namely idempotent operations, have played and continue to play an important role in several branches of Physics. Despite this, it is not clear to us if the linear functional duality (16) has or reflects some deeper physical origin. One would certainly not want to preclude spaces such as $L^1(\mathbb{R}^n)$ or $L^\infty(\mathbb{R}^n)$ from being used as functional spaces in applications due to their non-reflexivity. The actual question that is relevant to our purposes is whether such functional spaces have anything to do with the determination of the Euclidean metric of space-time. The work of [22] and subsequent developments seem to point out that reflexive (Sobolev) spaces may have some special geometric significance under the assumptions of his work. For this reason when combined with the implications of reflexivity for convexity and smoothness, we will restrict our attention to reflexive Banach spaces only, in the sequel.

Finite representability. One may be able to demand a stronger property along the lines of reflexivity, from physically relevant Banach spaces, for the purposes of determining the space-time metric, First a definition: a Banach space $\mathcal{X}$ is finitely representable in a Banach space $\mathcal{Z}$ if for every $\epsilon > 0$ and for every finite-dimensional subspace $\mathcal{X}_0 \subset \mathcal{X}$ there is a subspace $\mathcal{Z}_0 \subset \mathcal{Z}$ such that $d_{BM}(\mathcal{X}_0, \mathcal{Z}_0) < 1 + \epsilon$. This essentially means that any finite-dimensional subspace of $\mathcal{X}$ can be represented, almost isometrically, in $\mathcal{Z}$. Equivalently, one controls the distortion of the embedding of every finite-dimensional subspace of $\mathcal{X}$ into $\mathcal{Z}$. From a physical viewpoint the above definition may be of interest, since it is at the confluence of two ideas: one has to do with the fact that based on quantum physics, or on the statistical interpretation of theories of many degrees of freedom, one may have to reconsider or even dispense with the concept of strict, "point-wise", equality. Instead one should think much more along the lines of probabilistic equivalence, something that of course needs further qualifications. From such a perspective though, an approximate rather than strict, demand for isometry such as required in the definition of finite representability of a finite dimensional linear space is not unreasonable. The second idea relies on the fact that in any physical measurement we have a finite number of pieces of data on which to rely. As a result, the infinite dimensional spaces are excellent mathematical models, but from a very pragmatic perspective we see only their finite subspaces and then we mentally and technically extrapolate to the infinite dimensional counterparts. From this viewpoint, properties of finite dimensional vector spaces is the most of what someone can realistically expect to have to deal with in physical applications.

Super-reflexive spaces. A Banach space $\mathcal{X}$ is called super-reflexive if every Banach space which is finitely representable in $\mathcal{X}$ is reflexive. Equivalently, a Banach space $\mathcal{X}$ is super-reflexive if no non-reflexive Banach space $\mathcal{Y}$ is finitely representable in $\mathcal{X}$. Examples of super-reflexive spaces, pertinent to our discussion, are the Lebesgue spaces $L^p(\mathbb{R}^n), 1 < p < n$. Super-reflexive spaces are reflexive, but the converse is not true. Super-reflexive spaces have numerous desirable properties, from a physical viewpoint, some of which will be encountered in the next Section as they are pertinent to convexity and smoothness properties. One property is that if a Banach space is isomorphic to a super-reflexive space then it is itself super-reflexive. Another useful property is is that a Banach space $\mathcal{X}$ is super-reflexive if and only if its dual $\mathcal{X}'$ is super-reflexive. The super-reflexivity of Banach spaces is a property which allows the structure of infinite dimensional Banach spaces to be determined by the embedding properties of its finite-dimensional subspaces. Since c_0 and l^1 are not reflexive Banach spaces, they are not super-reflexive either. For completeness, we mention there are reflexive spaces are not necessarily super-reflexive: indeed consider a Banach space such that $L^\infty(\mathbb{R}^n)$ is finitely representable in it; then it cannot be not super-reflexive. In closing, one would like to notice that super-reflexivity, very much like reflexivity, is a topological property: as it does not really depend on the specific norm with which the underlying linear space is endowed.

Super-properties. One can be more general at this point and talk about "super-properties", a term that we will occasionally use in the sequel. These were defined by R.C. James in [27]. Here, we follow the excellent "pedestrian" exposition of [28]. Consider a property P that is valid on a Banach space $\mathcal{X}$. Consider two finite-dimensional subspaces $\mathcal{Y}, \mathcal{Z} \subset \mathcal{X}$ and numbers $n_P(\mathcal{Y}), n_P(\mathcal{Z})$ respectively such that

$$n_P(\mathcal{Y}) \to n_P(\mathcal{Z}) \quad \text{as} \quad d_{BM}(\mathcal{Y}, \mathcal{Z}) \to 1 \tag{17}$$

Probably the most straightforward case of this occurrence is when a relation like

$$n_P(\mathcal{Z}) \leq d_{BM}(\mathcal{Y}, \mathcal{Z}) n_P(\mathcal{Y}) \tag{18}$$

holds. Then the Banach space $\mathcal{X}$ has the property P when

$$n_P(\mathcal{X}) = \sup_{\mathcal{Y}} n_P(\mathcal{Y}) < \infty \tag{19}$$

where the supremum is taken over all finite dimensional subspaces of $\mathcal{X}$. Whether a property P holds for the Banach space $\mathcal{X}$ evidently depends on the family $\mathfrak{F}(\mathcal{Y})$ of all finite-dimensional spaces $\mathcal{V}$ such that

$$\forall \epsilon > 0, \exists \mathcal{Y} \subset \mathcal{X} : d_{BM}(\mathcal{V}, \mathcal{Y}) < 1 + \epsilon \tag{20}$$

In this terminology, finite representability of $\mathcal{Y}$ in $\mathcal{X}$ amounts to $\mathfrak{F}(\mathcal{Y}) \subset \mathfrak{F}(\mathcal{X})$. The property P is called a super-property if whenever a Banach space $\mathcal{X}$ has P, then every Banach space $\mathcal{Y}$ finitely representable in $\mathcal{X}$ also has P.

The Radon-Nikodým property. An additional property which is quite desirable at the technical level, and which is extensively used in Statistical Mechanics, where it is usually taken for granted, is the Radon-Nikodým property. It basically provides a way to make a transition between two different measures in a measure space. From a certain viewpoint, it can be seen as a generalization of the change of variables formula employed in multivariable calculus integration. For "practical purposes", it states that one can use a function alongside the volume of a manifold as equivalent to any absolutely continuous measure. Such a density function is the micro-canonical distribution employed extensively in equilibrium Statistical Mechanics. This mathematical result extends to vector-valued measures as follows: consider a probability space (Ω, μ) with a σ- algebra of (Borel) sets Σ, $U \in \Sigma$ and a Banach space $\mathcal{X}$. Let the vector-valued measure $\nu : \Sigma \rightarrow \mathcal{X}$ be countably additive and of bounded variation. Then there is a (Bochner) integrable function $f : \Omega \rightarrow \mathcal{X}$ such that

$$\nu(U) = \int_U f d\mu \tag{21}$$

For details and the generalization of this concept, from a geometric viewpoint, see [24]. As stated previously, this is a convenient technical theorem which allows one to use continuous functions f as a form of a "derivative" $d\nu/d\mu$. Such functions play an important role in Statistical Mechanics and quantum/thermal field theories as they do allow the coarse-grained distributions of interest to be treated as continuous rather than as discrete variables, something that many times simplifies the calculations. Moreover, it is certainly true that thermodynamic potentials such as the entropy have the drawback of being coordinate-dependent for continuous distributions; the only known way around such a difficulty is the use of reference measures which invariably lead to Radon-Nikodým derivatives. Hence it may be a relief to know that reflexive Banach spaces $\mathcal{X}$ do obey the Radon-Nikodým property and that the same is true for super-reflexive spaces: the obey the super Radon-Nikodým property.

3. Convexity and Smoothness

There are numerous characterizations, singling out inner product spaces among all Banach spaces. One can consult, for instance, the book [29] for a classical, extensive exposition and numerous results. We have also found a host of pertinent information in [30,31].

In this Section, we provide some information about one modulus of convexity and one modulus of smoothness and use them to single out Hilbert spaces among all the Banach spaces of interest.

3.1. *Why Convexity and Smoothness?*

There is little doubt that convexity is a fundamental concept, whose origin and initial developments can be traced as far as the Greek antiquity schools of Geometry such as the Pythagoreans, having far-reaching consequences in many branches of Mathematics. We are interested in aspects of convexity, in this work, mainly in the context of vector spaces [32] with our approach oriented toward the infinite dimensional cases.

Convexity enters dynamics very early, both historically and at a stage of its development. It is present as early as Newton's equations, at least. A particle trajectory in 3-dimensional space "bends" locally in the general direction of the total force acting on the particle. In other words, the total force acts toward the "convex interior", vaguely speaking, of the curve. The resulting acceleration is associated to the curvature of the trajectory in space. Such a curvature is an extrinsic concept though, namely it depends on the way the curve has been embedded in 3-dimensional space. Incidentally, the intrinsic geometry of a line is trivial. However this hints at a strong connection between curvature and convexity of the embedded curve. One can generalize this observation for higher dimensional sub-manifold embeddings. Such embeddings become crucial in the configuration or in the phase space of a system. Consider, for instance, an isolated system; its energy os constant. Hence, as is well-known, its evolution takes place in a co-dimension one sub-manifold of its configuration space. This can be seen as a simple example of an embedding. The quantity characterizing such embeddings locally, in Riemannian geometry, is the second fundamental form or its closely associated shape operator. This topic is a classical one. For an overview and many references, see the recent thesis [33]. The statement of interest, for our purposes, is that if one considers a compact hyper-surface $\mathcal{M}$ of $\mathbb{R}^n$ endowed with the induced Euclidean metric, and the second fundamental form of this embedding is positive-definite everywhere on $\mathcal{M}$, then $\mathcal{M}$ bounds a convex subset of $\mathbb{R}^n$ [34]. This provides a local characterization of convexity for embeddings and can be seen as a generalization of the kinematical framework that originated with Newton's equations.

One also expects convexity to be related to smoothness at least in the context of finite-dimensional vector spaces $\mathcal{V}$. A hand-waving argument that may illustrate this relationship is as follows: Consider a curve $\gamma : [0,1] \to \mathcal{V}$ which is locally rectifiable and which is arc-length parametrized by $s \in [0,1]$. For simplicity, and in order to make this argument more clear, assume that γ rests on a 2-plane, at least in some neighborhood $\mathcal{U}_x$ around a point x corresponding to $\gamma(s_0)$. Then consider the osculating circle of γ at x: this is a circle with a common tangent to the tangent $\frac{d\gamma(s_0)}{ds}$ whose center is along the normal line to the tangent in the 2-plane around $\mathcal{U}_x$. The radius of the osculating circle is equal to the radius of curvature $R(s_0)$ at x. The smaller the radius of curvature $R(s_0)$ the more "steeply" the curve turns. Now assume that $R(s_0) \to 0$. This will result in the formation of a corner at $\gamma(s_0)$ so γ will no longer be differentiable at s_0. At the same time, intuitively speaking, the "amount of convexity", a concept that really needs to be made precise, of $\gamma(s_0)$ will increase as $R(s_0) \to 0$. A word of caution at this point: using Clarkson's modulus of convexity which is a primary object of interest in this work, such a shrinking of the radius of curvature would leave that modulus unaffected. Still the statements regarding convexity in this paragraph were meant to be heuristic and suggestive, rather than precise, in order to visually illustrate our motivation for the use of convexity and smoothness and their inter-relation, in the rest of this work.

Convexity can also be seen at the level of Einstein's field equations of General Relativity. These are, in 4-dimensions,

$$R_{\mu\nu} - \frac{1}{2}Rg_{\mu\nu} = \frac{8\pi G}{c^4}T_{\mu\nu} \tag{22}$$

assuming that the cosmological constant $\Lambda = 0$ and $\mu, \nu = 0,1,2,3$. Consider the null energy condition [35], which is (arguably) the most fundamental of the energy conditions [36],

$$T_{\mu\nu}l^{\mu}l^{\nu} \geq 0 \tag{23}$$

where l is a null vector. (23) amounts to essentially demanding that

$$R_{\mu\nu} l^\mu l^\nu \geq 0 \tag{24}$$

(24) can be interpreted as a (mean) convexity condition in a null direction: one way is to see that the Ricci tensor involves two derivatives of the components of the metric tensor hence the non-negativity of (24) signifies convexity in analogy with the case of functions. Such convexity can also be seen in a Riemannian context. There, the non-negativity of the Ricci curvature (24) can be shown [37] to imply a generalized Brunn-Minkowski inequality, which is essentially a statement about the concavity, the convexity of the opposite function, of the volume in Euclidean and Riemannian spaces.

3.2. A Modulus of Convexity

Convex sets. To be more precise and attempt to make the exposition somewhat self-contained, we present the following well-known definitions and statements. Consider $\mathcal{V}$ to be a vector space over $\mathbb{R}$ or $\mathbb{C}$. A subset $\mathcal{A} \subset \mathcal{V}$ is called (affinely) convex if

$$\{ta_1 + (1-t)a_2, t \in [0,1]\} \subset \mathcal{A}, \forall a_1, a_2 \in \mathcal{A} \tag{25}$$

Equivalently, for every $n \in \mathbb{N}$ and for every $t_0, t_1, \ldots, t_n \in [0,1]$ such that $t_0 + t_1 + \ldots + t_n = 1$ and for every $a_0, a_1, \ldots, a_n \in \mathcal{A}$, we have $t_0 a_0 + t_1 a_1 + \ldots + t_n a_n \in \mathcal{A}$. Obviously $\mathbb{R}^n$ as well as its linear and affine subspaces are convex. The same conclusion holds about open and closed unit balls in normed vector spaces.

Convex functions. Convex functions can be considered as generalizations of convex sets. Let $\mathcal{A} \subset \mathcal{V}$ be a convex subset of the linear space $\mathcal{V}$ and let $f : \mathcal{A} \to \mathbb{R}$ be a function. The epigraph of f is defined to be the set

$$Ep(f) = \{(a,t) \in \mathcal{A} \times \mathbb{R} : f(a) \leq t\} \tag{26}$$

Then, such a function f is called convex, if $Ep(f)$ is a convex subset of $\mathcal{A} \times \mathbb{R}$. Equivalently, if for all $a, b \in \mathcal{A}$ and $t \in [0,1]$, such f satisfies the inequality

$$f(ta + (1-t)b) \leq (1-t)f(a) + tf(b) \tag{27}$$

The combination of homegeneity and the triangle inequality shows immediately that a norm $\|\cdot\|$ on $\mathcal{V}$ is a convex function. Other convex functions are the distance functions between two lines in Euclidean space ($\mathbb{R}^n$ endowed with the Euclidean metric) and, more generally, distance functions on metric spaces of negative curvature. Moreover, if in a normed (more generally: a metric) space $\mathcal{V}$ with $\mathcal{A} \subset \mathcal{V}$ and for all $a \in \mathcal{V}$, one defines

$$d_{\mathcal{A}}(a) = \inf_{b \in \mathcal{A}} \|a - b\| \tag{28}$$

when $\mathcal{A}$ is a nonempty closed convex subset of $\mathcal{V}$, then the distance function $d_{\mathcal{A}} : \mathcal{V} \to \mathbb{R}_+$ is convex.

Convex functions have many nice properties: they are semicontinuous and almost everywhere differentiable, they possess left and right derivatives (which however need not coincide), limits of sequences of convex functions defined on convex sets are convex functions, they have a unique minimum and they obey the local-to-global property, namely a locally convex function is actually (globally) convex. It is properties like these that make convex functions so useful and widespread in Physics and, in particular, in Analytical Mechanics and Thermodynamics.

A modulus of convexity. A next logical step is to find a way to quantify the extent of convexity of a set or of a function. Naturally, determining such a "modulus of convexity" is not a unique process and it involves making certain choices. Simplicity and computability in, at least, simple cases are usually good guidelines, as far as physical applications are concerned. A relatively recent list of such moduli which is quite extensive, even if not necessarily comprehensive, can be found in [38].

The oldest and most studied among the moduli of convexity is due to J.A. Clarkson [39,40]. To define it, consider the normed linear space $(\mathcal{X}, \|\cdot\|)$, let $B_{\mathcal{X}}$ indicate its closed unit ball, as before, and let $\varepsilon \in [0,2]$. The modulus of local convexity is defined by

$$\delta(x,\varepsilon) = \inf\left\{1 - \left\|\frac{x+y}{2}\right\| : y \in B_{\mathcal{X}}, \|x-y\| \geq \varepsilon\right\} \tag{29}$$

Then Clarkson's modulus of convexity of $\mathcal{X}$ denoted by $\delta_{\mathcal{X}} : [0,2] \to [0,1]$ is defined as

$$\delta_{\mathcal{X}}(\varepsilon) = \inf\{\delta(x,\varepsilon) : x \in B_{\mathcal{X}}\} \tag{30}$$

This can be equivalently expressed as

$$\delta_{\mathcal{X}}(\varepsilon) = \inf\left\{1 - \left\|\frac{x+y}{2}\right\| : x, y \in \mathcal{X}, \|x\| = \|y\| = 1, \|x-y\| = \varepsilon\right\} \tag{31}$$

It should be immediately noticed that this modulus of convexity is really a property of the 2-plane spanned by $x, y \in \mathcal{X}$ which is then inherited by $\mathcal{X}$. This should look familiar: the curvature of an n-dimensional Riemannian manifold $\mathcal{M}$ is a genuinely 2-dimensional concept which is actually formulated on its Grassmann manifold $G_{2,n}(\mathcal{M})$. The (sectional) curvature expresses the deviation of the metric of $\mathcal{M}$ from its flat counterpart [41]. The characteristic of convexity of $(\mathcal{X}, \|\cdot\|)$ is defined as [42]

$$\varepsilon_0(\mathcal{X}) = \sup\{\varepsilon \in [0,2] : \delta_{\mathcal{X}}(\varepsilon) = 0\} \tag{32}$$

Uniformly convex spaces. A Banach space $(\mathcal{X}, \|\cdot\|)$ is uniformly convex when it has non-zero modulus of convexity, namely $\delta_{\mathcal{X}}(\varepsilon) > 0, \varepsilon \in (0,2]$ or equivalently when $\varepsilon_0(\mathcal{X}) = 0$. Geometrically, the idea of the definition is simple: uniformly convex spaces have a unit ball $B_{\mathcal{X}}$ whose boundary unit sphere $S_{\mathcal{X}}$ does not contain any (affine) line segments. Roughly speaking: the further away from containing an affine segment $S_{\mathcal{X}}$ is, the higher the modulus of convexity of $\mathcal{X}$ is. It should be noticed that according the D.P. Milman [43] - B.J. Pettis [44] theorem, uniformly convex spaces are reflexive. Actually one can see that uniformly convex spaces are actually super-reflexive. Hence, if one deems reflexivity or super-reflexivity to be a desirable, or pertinent, property in an argument about the Euclidean nature of the space-time metric, as was stated above, then confining their attention to uniformly convex Banach spaces will not miss this property.

From a different viewpoint, it is known from the work of P. Enflo [45] and R.C. James [46] that $\mathcal{X}$ is super-reflexive if and only if it has an equivalent uniformly convex norm. "Equivalence" in this theorem is meant to be understood in the sense described in Subsection 2.2. Therefore, insisting on having super-reflexivity of the underlying functional spaces that determine/induce the space-time metric usually has to allow for a change of their norm to a uniformly convex one, if this is feasible. This change may have substantial physical implications for the underlying model. If however, as mentioned above, one is only interested in large-scale "coarse" phenomena that ignore spatially "small" details and arise as a result of statistical averaging of many degrees of freedom, then one believes that such a renorming will leave the macroscopic quantities of interest unaffected.

Modulus of convexity of Lebesgue spaces. Explicitly calculating the modulus of convexity for specific Banach spaces has proved to be more difficult than one might have naively anticipated. This is one reason why so many different moduli of convexity have been defined over the decades, after Clarkson's work [38]. For completeness, we mention that for $p = 1$ and for $p = \infty$, $\delta_{L^p} = 0$ as these two Banach spaces are not uniformly convex. The fact that the other Lebesgue spaces $L^p(\mathbb{R}^n), 1 < p < \infty$ are uniformly convex was already known to [39]. For a simpler and more recent proof, see [47]. However, the explicit asymptotic form of their modulus of convexity was determined by [48] who relied on the inequalities bearing his name [48,49] to reach his result. Hanner proved that

$$\delta_{L^p}(\varepsilon) = \begin{cases} (p-1)\frac{\varepsilon^2}{8} + o(\varepsilon^2), & \text{if } 1 < p \leq 2 \\ \frac{1}{p}\left(\frac{\varepsilon}{2}\right)^p + o(\varepsilon^p), & \text{if } 2 \leq p < \infty \end{cases} \tag{33}$$

One cannot fail to notice the "phase transition" in the asymptotic behavior of this modulus of convexity occurring around $p = 2$, namely around the Hilbert space case. An inevitable question is how generic such a behavior might be, among all Banach spaces. If it is, then it would be an initial indication that Hilbert spaces may be "special", from a convexity viewpoint. It turns out that this is essentially true, and this is the statement on which the main argument of the present work relies. This is the realization that among all Banach spaces, the Hilbert space is the "most" convex. To be more precise, it was proved in [50] that among all Banach spaces $\mathcal{X}$, the Hilbert spaces $\mathcal{H}$ have the largest modulus of convexity, namely

$$\delta_{\mathcal{X}}(\varepsilon) \leq \delta_{\mathcal{H}}(\varepsilon) \tag{34}$$

where

$$\delta_{\mathcal{H}}(\varepsilon) = \frac{2 - \sqrt{4 - \varepsilon^2}}{2} \tag{35}$$

It is also interesting to notice that the results of [50] when combined with those of [51,52] prove that if a linear space $\mathcal{X}$ is such that the equality in (34) holds, then $\mathcal{X}$ is an inner product space.

Modulus of convexity and equivalences. It should be noted that the modulus of convexity $\delta_{\mathcal{X}}(\varepsilon)$ is not necessarily itself a convex function of ε. What we know, for instance, is that for an infinite dimensional uniformly convex Banach space $\mathcal{X}$, we have that

$$\delta_{\mathcal{X}}(\varepsilon) \leq C\varepsilon^2 \tag{36}$$

For a Hilbert space we have (35) which is, obviously, compatible with (36). So, in a quantitative sense, the closer $\delta_{\mathcal{X}}$ is to ε^2 for a normed space $\mathcal{X}$, the closer $\mathcal{X}$ is being maximally convex, hence the closer it is to being a Hilbert space $\mathcal{H}$. T. Figiel [53] showed that one can consider as a modulus of convexity instead of $\delta_{\mathcal{X}}(\varepsilon)$ the greatest convex function $\tilde{\delta}_{\mathcal{X}}(\varepsilon)$, namely

$$\tilde{\delta}_{\mathcal{X}}(\varepsilon) \leq \delta_{\mathcal{X}}(\varepsilon) \tag{37}$$

Then [53]

$$c_1 \delta_{\mathcal{X}}(c_2 \varepsilon) \leq \tilde{\delta}(\varepsilon) \tag{38}$$

for constants $c_1 > 0, c_2 > 0$.

The re-definition employed in [53] is reminiscent of the equivalence of growth functions in geometric group theory when one has to decide whether such a growth function is exponential, polynomial or has an intermediate growth rate [54]. The reason for such similarity is quite clear: in geometric group theory, as in our case, one really cares about equivalences that may distort distances, but leave large-scale details of the structure invariant. The metric equivalence employed there is that of quasi-isometric maps, which distorts distances between two metric spaces $(\mathfrak{X}_1, d_1), (\mathfrak{X}_2, d_2)$ according to

$$\frac{1}{c_1} d_1(x,y) - c_2 \leq d_2(f(x), f(y)) \leq c_1 d_1(x,y) + c_2, \qquad \forall x, y \in \mathfrak{X}_1 \tag{39}$$

where $c_1 > 1$ and $c_2 > 0$. The difference between (6) and (39) is the presence of c_2 in (39) which completely ignores structures whose length is smaller than c_2. We have previously employed such equivalences in our work in [55,56] in our attempt to determine the dynamical basis of a power-law entropy, something that may have implications for the derivation of the metric of space-time from microscopic models of quantum gravity. The use of such maps acquires even greater significance when one considers that it is intimately related to properties of (Gromov) hyperbolic spaces. After a $3+1$ decomposition, 3-dimensional manifolds represent space-like hyper-surfaces in spacetimes. Following the results of the Thurston geometrization program (see [57] for an overview), it seems that "most" of the 3-dimensional manifolds are hyperbolic. Hence quasi-isometries and similar ideas may be relegated to a central role in determining classical/long-distance structures of the (4-dimensional) spacetimes from their microscopic/quantum foundations.

3.3. A Modulus of Smoothness and a Duality

A quantity intimately related to convexity, in the context of normed linear spaces, is smoothness. In line with convexity, one also needs a way to quantify the amount of smoothness of a linear space. Once more, there is no unique way of how to go about constructing such a modulus of smoothness and actually several such moduli of smoothness have been constructed over the years [38].

A modulus of smoothness. The oldest and most studied modulus of smoothness is due to M.M. Day [40] and J. Lindenstrauss [58]. The modulus of smoothness of a normed space $(\mathcal{X}, \|\cdot\|)$ is a function $\rho_{\mathcal{X}} : [0, \infty) \to \mathbb{R}$ which is defined by

$$\rho_{\mathcal{X}}(t) = \sup\left\{\frac{\|x+ty\| + \|x-ty\|}{2} - 1, x, y \in B_{\mathcal{X}}\right\} \tag{40}$$

This can be alternatively defined as

$$\rho_{\mathcal{X}}(t) = \sup\left\{\frac{\|x+y\| + \|x-y\|}{2} - 1 : x \in B_{\mathcal{X}}, \|y\| \leq t\right\} \tag{41}$$

or as

$$\rho_{\mathcal{X}}(t) = \sup\left\{\frac{\|x+y\| + \|x-y\|}{2} - 1, x, y \in S_{\mathcal{X}}\right\} \tag{42}$$

One defines the coefficient of smoothness of $\mathcal{X}$ by

$$\rho_0(\mathcal{X}) = \lim_{t\to 0^+} \frac{\rho_{\mathcal{X}}(t)}{t} \tag{43}$$

Uniformly smooth spaces. The Banach space $(\mathcal{X}, \|\cdot\|)$ is called uniformly smooth if $\rho_0(\mathcal{X}) = 0$. One sees immediately that uniform smoothness is a point-wise property and is essentially 2-dimensional, as is the case for uniform convexity. In a pictorial sense, this modulus of smoothness captures the fact that the unit sphere $S_{\mathcal{X}}$ of the Banach space $\mathcal{X}$ is smooth, i.e., that it has not corners. More issue on this issue is discussed below.

Before continuing it may be worth developing a pictorial idea about uniform convexity and uniform smoothness. Consider, for instance, a square with rounded corners. This is not uniformly convex, since it includes line segments in its boundary, but it lacks corners, so it is uniformly smooth. Consider now the, assumed non-empty and not a point, intersection of two equal radius disks: it is uniformly convex as its boundary contains no line segments, but it is not uniformly smooth since at the two intersection points of the two circles which are the boundaries of the two disks, the figure has corners.

Modulus of smoothness of Lebesgue spaces. Unlike the modulus of convexity, the modulus of smoothness is a convex function, essentially by definition. The asymptotic behavior of the modulus of smoothness of the Lebesgue spaces $L^p(\mathbb{R}^n)$ were also calculated in [48]. O. Hanner found that

$$\rho_{L^p}(t) = \begin{cases} \frac{t^p}{p} + o(t^p), & \text{if } 1 < p \leq 2 \\ \\ \frac{p-1}{2}t^2 + o(t^2), & \text{if } 2 \leq p < \infty \end{cases} \tag{44}$$

The same comments and thoughts regarding the phase transition around $p = 2$ apply here as in the case of the modulus of convexity as stated above. The counterpart of (34) was also established, and it states that that for any Banach space $\mathcal{X}$ one has

$$\rho_{\mathcal{X}}(t) \geq \rho_{\mathcal{H}}(t) \tag{45}$$

Therefore Hilbert spaces are the least smooth among all Banach spaces. Moreover, the exact same statement as the one following (35) applies for the modulus of smoothness: if a Banach space $\mathcal{X}$ obeys

$$\rho_{\mathcal{X}}(t) = \rho_{\mathcal{H}}(t) \tag{46}$$

then $\mathcal{X}$ is an inner product space. The modulus of smoothness, very much like for the modulus of convexity, of a Hilbert space is a result of the validity of the parallelogram equality and it is given by

$$\rho_{\mathcal{H}}(t) = \sqrt{1+t^2} - 1 \tag{47}$$

The analogue of (36) is that for an infinite dimensional Banach space $\mathcal{X}$ one has

$$\rho_{\mathcal{X}}(t) \geq ct^2 \tag{48}$$

The Milman-Pettis theorem established, in addition to uniform convexity, that uniformly smooth spaces are reflexive. Even though the dual of a super-property may not be a super-property, using the "convexity-smoothness" duality of the next paragraph, one sees that uniformly smooth spaces are indeed super-reflexive. Analogous things can be stated about the uniform smoothability of a Banach space as were stated about their uniform convexifiability, with equivalent norms.

A duality and the Legendre-Fenchel transforms. One observes form the above that Clarkson's modulus of convexity and the Day-Lindenstrauss modulus of smoothness appear to behave very much like dual concepts. The fact is that this suspected duality is true. More precisely, [58] proved that a Banach space $\mathcal{X}$ is uniformly convex if and only if its dual $\mathcal{X}'$ is uniformly smooth. The exact relation between the corresponding moduli is given by the Legendre-Fenchel transform

$$\rho_{\mathcal{X}'}(t) = \sup\left\{\frac{\varepsilon t}{2} - \delta_{\mathcal{X}}(\varepsilon), 0 \leq \varepsilon \leq 2\right\} \tag{49}$$

or, equivalently,

$$\rho_{\mathcal{X}}(t) = \sup\left\{\frac{\varepsilon t}{2} - \delta_{\mathcal{X}'}(\varepsilon), 0 \leq \varepsilon \leq 2\right\} \tag{50}$$

Therefore, every theorem valid for convexity has a dual analogue valid for smoothness. We see, for instance, that

$$\rho_0(\mathcal{X}') = \frac{1}{2}\varepsilon_0(\mathcal{X}), \qquad \rho_0(\mathcal{X}) = \frac{1}{2}\varepsilon_0(\mathcal{X}') \tag{51}$$

which shows that a Banach space $\mathcal{X}$ is uniformly convex if and only if its dual $\mathcal{X}'$ is uniformly smooth. Continuing along the lines of the quantification of convexity and smoothness, one can state the following: consider the Banach space $\mathcal{X}$. It turns out that

$$\delta_{\mathcal{X}}(\varepsilon) \geq C\varepsilon^q, \qquad q \geq 2 \tag{52}$$

in which case $\mathcal{X}$ is called a $q-$convex space. According to a theorem of Figiel and Assouad, this is equivalent to the existence of some constant α such that

$$\frac{1}{2}\left(\|x+y\|^q + \|x-y\|^q\right) \geq \|x\|^q + \alpha\|y\|^q, \qquad \forall x, y \in \mathcal{X} \tag{53}$$

For smoothness, the corresponding statement is that for the Banach space $\mathcal{Y}$ one has

$$\rho_{\mathcal{Y}}(t) \leq ct^p, \qquad 1 < p \leq 2 \tag{54}$$

in which case Y is called $p-$smooth. Then, according to a theorem of Pisier and Assouad, this is equivalent to the existence of some constant β such that

$$\frac{1}{2}\left(\|x+y\|^p + \|x-y\|^p\right) \leq \|x\|^p + \beta\|y\|^p, \qquad \forall x, y \in \mathcal{Y} \tag{55}$$

Based on the Legendre-Fenchel duality noted above, one can also state that a Banach space $\mathcal{X}$ is q-convex if and only if its dual $\mathcal{X}'$ is p-smooth, where q qnd p are harmonic conjugates, namely

$$\frac{1}{p} + \frac{1}{q} = 1 \tag{56}$$

3.4. Smoothness, Derivatives and Equivalences

The word "smoothness" is associated, in a typical Physicist's mind, with the concept of the derivative. It may be of interest to know that the same can be stated for the cases of the Banach spaces of our interest. Using infinite dimensional spaces though introduces additional complexities and also possible counter-intuitive features that should be carefully accounted for. For one, there are several possible definitions for the (directional) derivative in a normed space. We only need the definition of the Frechét derivative in this work [31]. Let $\mathcal{X}, \mathcal{Y}$ be Banach spaces and let $f : \mathcal{X} \to \mathcal{Y}$. Then f is called Frechét differentiable at $x \in \mathcal{X}$, if there is a bounded linear operator $A_x : \mathcal{X} \to \mathcal{Y}$ such that the following limit exists uniformly for $y \in S_{\mathcal{X}}$

$$A_x y = \lim_{t \to 0} \frac{f(x+ty) - f(x)}{t} \tag{57}$$

If this limit indeed exists, then it is called the Frechét derivative of f at x and is indicated by $D_f(x)$. The pertinent statement is that if $\mathcal{X}$ is a uniformly smooth Banach space then its norm $f(x) = \|x\|$ is Frechét differentiable for every $x \in \mathcal{X}\backslash\{0\}$. In such a case, one can see that the linear approximation to f at x through $D_f(x)$ is valid, namely we can write

$$f(x+y) = f(x) + D_f(x)y + o(\|y\|) \tag{58}$$

If a function is Frechét differentiable at a point, it turns out that it is also continuous at that point, which is the Banach space analogue of a well-known and frequently used result of elementary calculus.

In addition to the above, it turns out [31] that for a Banach space $\mathcal{X}$ the following statements are equivalent:

- $\mathcal{X}$ is super-reflexive.
- $\mathcal{X}$ admits an equivalent, uniformly convex norm, whose modulus of convexity satisfies, for some $q \geq 2$,
$$\delta_{\mathcal{X}}(\varepsilon) \geq c_1 \varepsilon^q \tag{59}$$
- $\mathcal{X}$ admits an equivalent, uniformly smooth norm, whose modulus of smoothness satisfies, for some $1 < p \leq 2$
$$\rho_{\mathcal{X}}(t) \leq c_2 t^p \tag{60}$$

Moreover, Asplund [59] showed that a space which admits two equivalent norms one of which is uniformly convex and the other uniformly smooth, admits a third one which is equivalent to the previous two and which has both properties. As a special case, one can state that a super-reflexive space admits an equivalent norm which is both uniformly convex and uniformly smooth.

3.5. *Type, Co-Type and Moduli*

Rademacher functions. The powers q and p in the lower bound of the modulus of convexity (52) and in the upper bound in the modulus of smoothness (54), respectively, have a nice geometric-probabilistic interpretation. To formulate it, we need to use the Rademacher functions which are defined as $r_i : [0,1] \to \pm 1$ by

$$r_i(t) = \text{sign}[\sin(2^i \pi t)], \qquad i \in \mathbb{N} \tag{61}$$

We see that the Rademacher functions can be interpreted as a sequence of identically distributed random variables on $[0,1]$ endowed with its Lebesgue measure, taking values ± 1, each with probability 0.5. It may be worth noting that for vectors $x_i, i = 1, \dots n$ in a Banach (or more generally in a normed) space $(\mathcal{X}, \|\cdot\|)$, one has

$$\int_0^1 \left\| \sum_{i=1}^n r_i(t) x_i \right\|^p dt = \mathbb{E}\left(\left\| \sum_{i=1}^n \epsilon_i x_i \right\|^p \right) \tag{62}$$

where the expectation value $\mathbb{E}$ is taken over $\epsilon_i = \pm 1$.

Type. Suppose now that for any finite number n and any choice of vectors $x_i, i = 1, \dots, n$ of $\mathcal{X}$ there is a constant $C_p > 0$ such that

$$\left(\int_0^1 \left\| \sum_{i=1}^n r_i(t) x_i \right\|^p dt \right)^{\frac{1}{p}} \leq C_p \left(\sum_{i=1}^n \|x_i\|^p \right)^{\frac{1}{p}} \tag{63}$$

then the Banach space $\mathcal{X}$ is said to have type p. The best constant $C_p(\mathcal{X})$ in the definition (63) is called type-p constant of $\mathcal{X}$. Consider A.I. Khintchine's inequality for $a_i, a_i \in \mathbb{R}, i = 1, \dots, n, 0 < p < \infty$, the Rademacher functions $r_i(t)$, and constants A_p, B_p, namely

$$A_p \left(\int_0^1 \left| \sum_{i=1}^n a_i r_i(t) \right|^p dt \right)^{\frac{1}{p}} \leq \left(\int_0^1 \left| \sum_{i=1}^n a_i r_i(t) \right|^2 dt \right)^{\frac{1}{2}} \leq B_p \left(\int_0^1 \left| \sum_{i=1}^n a_i r_i(t) \right|^p dt \right)^{\frac{1}{p}} \tag{64}$$

If we assume that all vectors are equal in (63), then by using (64) we see that that $p \leq 2$. By using the triangle inequality, one can also see that $p \geq 1$. Therefore the only allowed values for a type p Banach space are $1 \leq p \leq 2$. This is equivalent to stating that the elements of the family $\mathfrak{F}(\mathcal{Z})$ of finite-dimensional subspaces $\mathcal{Z} \subset \mathcal{X}$ satisfy

$$\sup_{\mathfrak{F}(\mathcal{Z})} C_p(\mathcal{Z}) < \infty \tag{65}$$

which shows that $\mathcal{X}$ having type p is a super-property. One can also see that if $p' < p$ then type p implies type p'.

Co-type. With similar notation as for type, a Banach space $\mathcal{Y}$ has co-type q if there is a constant $C'_q > 0$ such that

$$\left(\int_0^1 \left\| \sum_{i=1}^n r_i(t) x_i \right\|^q dt \right)^{\frac{1}{q}} \geq C'_q \left(\sum_{i=1}^n \|x_i\|^q \right)^{\frac{1}{q}} \tag{66}$$

for any $n \in \mathbb{N}$ and for any set of vectors $x_i \in \mathcal{Y}$. Again, the best constant $C'_q(\mathcal{Y})$ in (66) is called the co-type q constant for $\mathcal{Y}$. This again is equivalent to stating that the elements of the family $\mathfrak{F}(\mathcal{W})$, for all finite-dimensional subspaces $\mathcal{W} \subset \mathcal{Y}$ satisfy

$$\sup_{\mathfrak{F}(\mathcal{W})} C_q(\mathcal{W}) < \infty \tag{67}$$

which also shows that $\mathcal{Y}$ having co-type q is a super-property. By using Khintchine's inequality again, one can see that the co-type of any Banach space that has one has to be $q \geq 2$. One can also see that if $q' > q$, then co-type q implies co-type q'.

Type and co-type properties. As an explicit example of type and co-type we know [30] that the Lebesgue spaces $L^p(\mathbb{R}^n)$ have

- Type p and co-type 2, if $1 \leq p \leq 2$
- Type 2 and co-type p, if $2 \leq p < \infty$

One again cannot fail to notice the "phase transition" in the behavior of type and co-type of $L^p(\mathbb{R}^n)$ taking place around $p = 2$. Hilbert spaces are singled out in a stronger sense: indeed, using the parallelogram equality (38) below, one can see that a Hilbert space has type 2 and co-type 2. That the converse is also true, is a non-trivial result due to S. Kwapień [60].

The type and co-type of a Banach space $\mathcal{X}$ can be seen in a variety of ways: one way is to state that they are a way of quantifying how far $\mathcal{X}$ is from being a Hilbert space. The comparison of the definitions of type and co-type with the parallelogram equality that only Hilbert spaces satisfy

$$\|x+y\|^2 + \|x-y\|^2 = 2(\|x\|^2 + \|y\|^2) \tag{68}$$

is quite suggestive. In view of the last paragraph of this Subsection (see below) the same can be stated by comparing (68) to (53) and (55) which alternatively quantify the moduli of convexity and smoothness of $\mathcal{X}$.

Now, one can inquire about the robustness of the concepts of type and co-type. J.P. Kahane's inequality [61] is a vector-valued extension of A.I. Khintchine's inequality (64) and states, with the above notation, that for every $0 < p < q < \infty$ there is a constant $C(p,q) > 0$ such that

$$\left(\int_0^1 \left\|\sum_{i=1}^n r_i(t)x_i\right\|^p dt\right)^{\frac{1}{p}} \leq \left(\int_0^1 \left\|\sum_{i=1}^n r_i(t)x_i\right\|^q dt\right)^{\frac{1}{q}} \leq C(p,q)\left(\int_0^1 \left\|\sum_{i=1}^n r_i(t)x_i\right\|^p dt\right)^{\frac{1}{p}} \tag{69}$$

This shows that the type p and co-type q are properties that are maintained under an equivalent norm. Moreover, instead of using Rademacher functions in the definitions, something which is technically convenient, one could use centered Bernoulli random variable, Gaussian random variables etc. with just a change in the values of the constants in the definitions [62].

Type, co-type and moduli. The relation of the type and co-type of a Banach space with its moduli of convexity and smoothness is contained in the following theorem due to T. Figiel, G. Pisier [63]: Let $\mathcal{X}$ be a uniformly convex Banach space with modulus of convexity satisfying $\delta_{\mathcal{X}}(\varepsilon) \geq C\varepsilon^q$ for some $q \geq 2$. Then $\mathcal{X}$ has co-type q. Let $\mathcal{Y}$ be a uniformly smooth Banach space whose modulus of smoothness satisfies $\rho_{\mathcal{Y}}(t) \leq ct^p$ for some $1 < p \leq 2$. Then $\mathcal{Y}$ has type p. Therefore the moduli of smoothness and convexity of a Banach space are bounded by the type and co-type of that Banach space, assuming that the latter exist. It may be worth noticing at this point the behavior of type and co-type under duality: It is known that when a Banach space $\mathcal{X}$ has type p, then its dual $\mathcal{X}'$ has co-type q where p and q are harmonic conjugates of each other. However the converse is not true without one additional assumption. The accurate statement is that if a Banach space $\mathcal{X}$ has non-trivial type, and co-type q, then its dual $\mathcal{X}'$ has type p, where p and q are harmonic conjugates of each other. For excellent expositions of the type and co-type of normed spaces, including proofs of all of the above statements, one may consult [30,31,62].

4. The Space-Time Metric from Variational Principles

One cannot fail to notice from the content of the previous Sections, the unique role that inner product (Hilbert) spaces $\mathcal{H}$ play among all Banach spaces $\mathcal{X}$ and, in particular, the distinct role of $L^2(\mathbb{R}^n)$ among all $L^p(\mathbb{R}^n)$. Such Hilbert spaces are, at the same time, the most convex and least smooth among all Banach spaces. They are the only Lebesgue spaces that have the same type and co-type 2. In addition, they are super-reflexive. Moreover such Hilbert spaces are the only self-dual Lebesgue spaces under harmonic conjugation, a basic fact reflecting properties of the polarity of convex bodies and of the Legendre-Fenchel transformations [64].

Functional integrals and variational principles. Using extremal (more accurately: stationary) properties of functionals under infinitesimal variations subject to appropriate boundary conditions has been a fundamental aspect of Classical and Quantum Physics since the time of Maupertuis, D'Alembert and Lagrange at least [65], if not earlier. In particular, a large number of works in Quantum Physics have used and continue to use as starting point, especially for calculational purposes, the stationary phase or saddle point approximation which rely on the vanishing variation under small perturbations of a judiciously chosen functional (the "classical action" $\mathcal{S}$) [66], an approach that can be traced back to an original idea of P.A.M. Dirac [67]. In this path-integral approach, as is very well-known, one starts with the path-integral/canonical partition function as the primary object encoding the statistically significant properties of the system

$$\mathcal{Z} = \int e^{-\mathcal{S}}[\mathcal{D}\phi] \tag{70}$$

where ϕ corresponds to the variables ("fields") in the action $\mathcal{S}$ to be integrated over and $[\mathcal{D}\phi]$ represents an appropriate integration measure, which may not rigorously be known on whether it even exists, but whose ad hoc choice (usually being of Gaussian form) allows concrete calculations to be performed and eventually the results derived from it to be compared with experimental data. In the case of quantum gravity the most immediate choice for $\mathcal{S}$ is usually taken to be the Einstein-Hilbert or the Palatini actions, a Chern-Simons action, of their discretized counterparts etc., each of which may be augmented with boundary terms and/or topological terms etc.

Other entropies and robustness. Before proceeding, we would like to have a short digression. During the last three decades, there have been several functionals that have been proposed purporting to capture the collective/thermodynamic properties of systems of many degrees of freedom. One motivation for the formulation of such functionals, such as the "Tsallis entropy" [15] or the "κ-entropy" [16] is to determine the thermodynamic properties of systems with long-range interactions. From the Newtonian viewpoint, gravity clearly falls in this category, as well as Maxwell's theory of electrodynamics etc. Assuming that such functionals may prove to be applicable to a path-integral formulation of aspects of semi-classical or even quantum gravity, the arguments of the present work will still hold without any major modifications. The minor modification needed in case such functionals are pertinent, is to use in (70) another convex function instead of the exponential one, something akin to the aptly named "q-exponential" [15], whose form will have to be determined. One could possibly use the maximum entropy principle subject to appropriate constraints, for such a purpose. A second minor, for our purposes, modification may be to substitute some other measure in the place of the often used Gaussian measure in (70). Beyond these points, we expect the above analysis to still be valid. Another point that will most likely change is the rate of convergence to the limit of the saddle-point approximation, which is intimately related to the probability of dealing with space-time geometries that may be non-Euclidean. This in the spirit of theories having a statistical interpretation, where even when the "classical" limit is known, it is the form of the "semi-classical" contributions/corrections that is used to distinguish between several competing models purporting to describe the same physical phenomenon.

Generalized path integrals. Going back to our argument, in the spirit of the path-integral, an often discussed but still unsolved question is whether one should extend (70) by considering additional contributions by summing over more "primitive", than the metric, structures such over all topological, piecewise-linear, differentiable etc. structures. Most of the treatments to quantum gravity that we are aware of, address the issue of a possible sum in the right-hand-side of (70) over all topologies. Then the modification of (70) states that the partition function of quantum gravity should be

$$\mathcal{Z}_{top} = \sum_{\text{topologies}} \int e^{-\mathcal{S}} [\mathcal{D}\phi] \tag{71}$$

Such possible summation over all topologies has presented insurmountable difficulties, which can be credited in large part for the eventual demise of the dynamical triangulation approach to quantum gravity [68], where one is faced with hard problems of Gödelian type indecisive propositions.

On the other hand, (71) can be used as as starting point for a similar question, where the summation is not over all topologies of 4-dimensional manifolds, which are of course locally spaces endowed with a Euclidean metric. One could instead ask for summation over all metrics that can be placed on the underlying topological or uniform structure of the space whose classical limit will eventually be a space-time. This however is would be a very broad set, hence a very difficult to analyze class of metrics. To be closer to something manageable, and also be in accordance with the equivalence principle demanding the local approximation of the underlying structure with linear spaces, one may wish to consider locally, only p-integrable Banach metrics, namely metrics/norms induced from $L^p(\mathbb{R}^n)$. This effectively generalizes the underlying space-time structure from that of a Riemannian to something akin to a Finslerian space. Therefore one could write a modified path-integral/canonical partition function, instead of (71), as

$$\mathcal{Z}_B = \sum_{p \geq 1} \int e^{-\mathcal{S}} [\mathcal{D}\phi] \tag{72}$$

with the summation being over all metrics of a space(-time) which are locally induced by the $L^p(\mathbb{R}^n)$. In conventional path-integral approaches to quantum gravity such a summation does not arise, since one has already determined to only use Euclidean metrics on the space-time underlying manifold.

One could use the lack of renormalizability of the path-integral expressions like (70) around their saddle points, namely around a fixed background metric, to argue against such an approach [69]. After all, if (70) gives rise to non-renormalizable interactions around Minkowski space, this should force us to believe that it precludes (72) from being more successful to that end. Such a criticism would be misguided though. Lack of renormalizability of (70) takes into account only metrics of Riemannian (quadratic) form on the underlying space and the saddle point is calculated within the set of such metrics. What we propose is to enlarge such a set to the more general locally p-integrable Banach metrics. Since our argument is quite empirical/"phenomenological" rather than fundamental/dynamical, the issue of renormalizability of the underlying path-integral does not even enter our considerations.

Demanding a summation over such p-integrable metrics as in (72) may be reasonable, or not, but only after someone can properly write a finite classical action $\mathcal{A}$ for them. How exactly to do this is not clear to us at this stage. There are synthetic definitions of the Ricci and even the scalar curvature for metric measure spaces with very little regularity [70]. One could use them and alongside a general minimal cost transportation to possibly argue in this direction. What is sorely lacking in such cases though is the formalism that could accommodate expressions for the non-gravitational fields capturing the essence of the stress-energy tensor and recasting it in such a synthetic framework. Therefore a dynamical argument, which would be the most desirable, in favor of Hilbert spaces and their induced Euclidean metrics does not seem to be feasible, in any obvious way, at this stage.

A "kinematic" approach via smoothness. Given the above difficulties, we have to resort to ad hoc decisions in order to proceed. To the extent of our knowledge, there has never been a variational principle of "maximum convexity" or a principle of "minimal smoothness" that would single out Hilbert spaces among all Banach spaces. The path-integral/partition function approach to quantum Physics can be interpreted as suggesting that all allowed possibilities in quantum evolution should be considered in calculating quantities of interest, each possibility however being assigned with a different weight factor. Following this viewpoint one can extend/stretch the domain of this interpretation to allow not only for a set of Riemannian metrics to contribute to the evolution of a gravitational system but also consider a broader class of possible metrics. To keep things close to the familiar territory of Riemannian/Lorentzian metrics we have used induced metrics on space-time locally induced by $L^p(\mathbb{R}^n)$, as was mentioned before. The familiar picture of space-time appears then as the classical limit of a theory of quantum gravity, and so are its associated properties like smoothness, etc.

Given the assumed irregularity/granularity of space-time at a fundamental level (expressed through spin networks in loop quantum gravity, partially ordered sets with discrete measures in the causal set approach, simplicial approximations and Regge calculus in causal dynamical triangulations etc.) it may not be out of place to assume that nature chooses the least smooth class of metrics among such induced metrics from $L^p(\mathbb{R}^n), 1 \leq p < \infty$. Therefore it is the Hilbert space metric/norm $L^2(\mathbb{R}^n)$ that provides the only extremal, hence dominant, contribution in an "extended" path-integral approach (72) which, in turn, induces its properties to the classical space-time limit of the quantum gravitational theory. In short, the induced metric of space-time inherits its Euclidean character from that of $L^2(\mathbb{R}^n)$ which dominates the path-integral (72), by being the least smooth.

Convexity and predictability. A somewhat complementary argument for Hilbert spaces and the induced Euclidean metric form on space-time, can be made based on convexity and predictability. As stated in the previous sections, the Hilbert spaces $\mathcal{H} = L^2(\mathbb{R}^n)$ are the most convex among all $L^p(\mathbb{R}^n)$. Contrast the behavior of the norm/metric of $\mathcal{H}$ to those of the family of $L^p(\mathbb{R}^n)$ that are the least convex. These are $L^1(\mathbb{R}^n)$ and $L^\infty(\mathbb{R}^n)$ which are neither uniformly convex nor uniformly smooth, nor are they reflexive, so they have been largely ignored in most part of this work. Nevertheless, use of these two spaces can help make this argument more transparent.

Consider, for concreteness, the $L^1(\mathbb{R}^n)$ space, or to be more intuitive, the metric induced by the related l^1 norm on $\mathbb{R}^2$. This metric, for $\vec{x} = (x_1, x_2), \vec{y} = (y_1, y_2)$ where the coordinates are considered with respect to a Cartesian system, is given by

$$d(\vec{x}, \vec{y}) = |x_1 - x_2| + |y_1 - y_2| \tag{73}$$

Then one can see that there is an infinity of geodesics connecting $\vec{x}$ and $\vec{y}$. This by itself is not a drawback: after all, the north and south poles of a sphere with the induced metric from the Euclidean space are also connected by infinite geodesics (the meridians). The definition of geodesics is not a problem either: in metric geometry [71] they can be defined to be the isometric images of the unit interval. The problem exists because many of the geodesics between $\vec{x}$ and $\vec{y}$ in the (73) are branching: geodesics that have initially a common segment can separate after a while. If one assumes strong locality in a theory of gravity, whose metric is even "Euclidianized" (made positive-definite) after a Wick rotation, then this presents a problem with predictability. Since the theory does not possess any "memory" in its formulation, how then one can make any prediction based on the behavior of geodesics which largely encode the underlying geometry, if the theory has branching geodesics? If, for instance, the action $\mathcal{S}$ or the resulting kinematic equations possessed some form "memory" as in the case of systems being modelled by fractional derivatives [72–74], then the use of geometric structures with branching geodesics might not pose a serious problem to predictability. Knowing this, one may wish to stay as far away as possible from using metrics that may allow for the possibility of branching geodesics. It turns out that the Hilbert space $L^2(\mathbb{R}^n)$ is the furthest away from resembling $L^1(\mathbb{R}^n)$ which has branching geodesics, at least when one uses the modulus of convexity to quantify such a difference.

The counter-argument to the above is that $L^1(\mathbb{R}^n)$ has branching geodesics exactly because it is not uniformly convex. If someone chose any other $L^p(\mathbb{R}^n), p \neq 1$ then this problem would not exist. This is largely correct. However it assumes very much like many occasions in classical Physics that some objects can be well-approximated by point particles, which in the absence of (non-gravitational) forces move along causal (in the Lorentzian signature framework) geodesics. When the quantum nature of such an object comes into consideration though, even in a fixed, classical background space-time, this statement would not be accurate. The uncertaintly principle would prevent such point-like structures from existing; this introduces substantial technical complications for any operator in the assumed Fock space of a quantum gravity theory following the canonical approach, such as loop gravity for instance: the operator has to be "smeared", namely to act on test not at a particular point but on an appropriately chosen neighborhood of it before applying the canonical commutation relations [75].

The result is that a wave-function will sweep out a tube, for short times, rather than a line, in such a space-time as it evolves. For predictability purposes, since the Schrödinger, the Klein-Gordon, the Dirac etc. equations involve usual derivatives, as opposed to being integro-differential equations that may signify that memory effects are taken into account, it is quite important for such tubes not to have a branching property. Naively speaking and without getting into any details, we believe that this goal has the best chances of being realized, if the underlying space has a metric which is as far away from having branching geodesics as possible, which again brings us back to favoring the use of a Hilbert space.

The space-time metric from Hilbert spaces. Going from $L^2(\mathbb{R}^n)$ to the metric of space-time itself is quite straightforward, in principle. Consider as the linear spaces of interest to be appropriate $A \subset \mathbb{R}^n$. For the case of point particles this will be the tangent to the particles's space-time evolution trajectory. We can then confine ourselves to the analysis of a subspace of $L^2(\mathbb{R}^n)$ which is comprised of the characteristic functions χ_A of such subspaces. Then the quadratic metric on $L^2(\mathbb{R}^n)$ gets induced on such A which acquires itself a quadratic metric. Use the equivalence principle and "patch together" such A endowed with their Euclidean metrics to form the space-time of interest. This is a kinematic construction. The dynamics is provided by Einstein's equations, after a Wick rotation back to indefinite signature metrics. This transition between metrics of different signatures may involve several subtleties which may have to be addressed at that stage, but this is outside the scope of the present work.

5. Discussion and Outlook

In this work we have presented a non-rigorous, conjectural argument that aims to explain why the metric of space-time, after being suitably Wick-rotated, obeys the Pythagorean theorem. Such a metric can be seen as descending from the metrics of appropriate Banach spaces which provide a reasonable kinematical framework of a mesoscopic description/quasi-classical limit of models of quantum gravity. The advantage of this approach is also its disadvantage: it is kinematical with ad hoc aspects. It does not attempt to delve into the actual realm of the approaches to quantum gravity, however it is motivated by and uses some of the common points of such approaches. We relied on aspects of General Relativity and the Einstein equations for some motivation, used the spirit of the variational principles and path-integrals to formulate our approach in the spirit of classical and quantum Physics and then for our proposal we used standard results from functional analysis, convexity and the theory of normed spaces.

As in any, partly ad hoc and conjectural, work the virtues of the current work may be seen as out-weighting their shortcomings, or vice versa. The value of work like the present, which may appear to purport to justify the "obvious", may lie in the ideas involved in it and in the methods used to reach the conclusions. Much more importantly from a physical viewpoint, it may also point out to reasons about the inapplicability of its conclusions should pertinent experiments refute our currently held ideas about space-time properties, or should one observe such exceptions at, or beyond, the galaxy cluster or the Planck scales.

We would like to add that there should be some skepticism regarding the role of the Wick rotation of the space-time signature to a positive-definite one. There is little doubt that the Wick rotation has been very successful in obtaining results in perturbative quantum field theories, or more generally when employing saddle point approximations, which may otherwise be ill-defined or inaccessible in a covariant approach to such relativistic theories. It is not clear, to us at least, to what extent employing such Euclidean (positive-definite) metrics is equivalent to purely Lorentzian arguments and results, in particular in regimes outside the possible domain of validity of such saddle-point approximations. It is well-known now that many fundamental results of Riemannian geometry (such as the Hopf-Rinow theorem etc.) are either not valid in Lorentzian geometry or may become valid when appropriate modifications are made. Such discrepancies become even more pronounced when one considers the topological and causal formalism of space-times and other such statements of Lorentzian geometry which have no obvious analogue in the Riemannian case [35,76,77]. Since our results rely on positive-definite metrics of the underlying (linear) spaces, we can only be skeptical about their applicability in the physical, indefinite (Lorentzian) signature, case.

A technical point that may be worth addressing before closing, is the fact that all the analysis in this work uses in a very essential way the particular moduli of convexity and smoothness whose definitions and properties have been stated above. Naturally, these are mathematical constructions and cannot conceivably be unique or automatically be considered the "most useful" among their peers. And actually they are not. As mentioned in the opening paragraphs of this work, there are several, generally inequivalent, moduli encoding convexity and smoothness [38]. If history is any guide, new such moduli will keep being defined, their properties being examined and their values will be calculated in concrete cases in the future. We have chosen the above two moduli because they appear to us to be the simplest, the most intuitive and also the most developed. It is theoretically possible that other moduli may single out other spaces, as opposed to Hilbert ones, but we have not been able to find any such results in the existing literature, nor can we see any overwhelming physical reason for their implementation.

Acknowledgments: This work was, largely, performed when the author was a faculty member at the Weill Cornell Medicine-Qatar.

References

1. Pettini, M. Geometry and Topology in Hamiltonian Dynamics and Statistical Mechanics. In *Interdisciplinary Applied Mathematics*; Springer: New York, NY, USA, 2007; Volume 33.
2. Kriegl, A.; Michor, P.W. The Convenient Setting of Global Analysis. In *Mathematical Surveys and Monographs*; American Mathematical Society: Providence, RI, USA, 1997; Volume 53.
3. Corichi, A.; Rubalcava-García, I.; Vukašinac, T. Actions, topological terms and boundaries in first order gravity: A review. *Int. J. Mod. Phys. D* **2016**, *25*, 1630011.
4. Kar, A.; Rajeev, S.G. A Non-Riemannian Metric on Space-Time Emergent from Scalar Quantum Field Theory. *Phys. Rev. D* **2012**, *86*, 065022.
5. Sorkin, R.D. Is the spacetime metric Euclidean rather than Lorentzian? In *Recent Research in Quantum Gravity*; Dasgupta, A., Ed.; Nova Science Publishers Inc.: New York, NY, USA, 2013.
6. Bojowald, M. Canonical Relativity and the Dimensionality of the World. In *Relativity and the Dimensionality of the World*; Petkov, V., Ed.; Springer: New York, NY, USA, 2007; pp. 137–152.
7. Nawarajan, D.; Visser, M. Global properties of physically interesting Lorentzian spacetimes. *arXiv* **2016**, arXiv:1601.03355.
8. Gibbons, G.W.; Ishibashi, A. Convex Functions and Spacetime Geometry. *Class. Quant. Grav.* **2001**, *18*, 4607–4628.
9. Rudin, W. *Functional Analysis*, 2nd ed.; McGraw-Hill Inc.: New York, NY, USA, 1991.
10. Lieb, E.H.; Loss, M. *Analysis*, 2nd ed.; American Mathematical Society: Providence, RI, USA, 2001.
11. Hájek, P.; Santalucía, V.M.; Vanderwerff, J.; Zizler, V. *Biorthogonal Systems in Banach Spaces*; Springer: New York, NY, USA, 2008.
12. Simon, B. The classical moment problem as a self-adjoint finite difference operator. *Adv. Math.* **1998**, *137*, 82–203.
13. Weinstein, A. The local structure of Poisson manifolds. *J. Diff. Geom.* **1983**, *18*, 523–557.
14. Klauder, J. Enhanced quantum procedures that resolve difficult problems. *Rev. Math. Phys.* **2015**, *27*, 1530002.
15. Tsallis, C. *Introduction to Nonextensive Statistical Mechanics: Approaching a Complex World*; Springer: New York, NY, USA, 2009.

16. Kaniadakis, G. Theoretical Foundations and Mathematical Formalism of the Power-Law Tailed Statistical Distributions. *Entropy* **2013**, *15*, 3983–4010.
17. Roe, J. *Lectures on Coarse Geometry*; American Mathematical Society: Providence, RI, USA, 2003.
18. Bridson, M.R.; Haefliger, A. *Metric Spaces of Non-Positive Curvature*; Springer: Berlin, Germany, 1999.
19. Gromov, M. Hyperbolic Groups. In *Essays in Group Theory*; Gesten, S.M., Ed.; Springer: New York, NY, USA, 1987; pp. 75–264.
20. Katok, A.; Hasselblatt, B. *Introduction to the Modern Theory of Dynamical Systems*; Cambridge University Press: Cambridge, UK, 1995.
21. Heinonen, J. Nonsmooth calculus. *Bull. Amer. Math. Soc.* **2007**, *44*, 163–232.
22. Cheeger, J. Differentiability of Lipschitz functions on metric measure spaces. *Geom. Funct. Anal.* **1999**, *9*, 428–517.
23. Keith, S. A differentiable structure for metric measure spaces. *Adv. Math.* **2004**, *183*, 271–315.
24. Cheeger, J.; Kleiner, B. Differentiability of Lipschitz maps from metric measure spaces to Banach spaces with the Radon-Nikodým property. *Geom. Funct. Anal.* **2009**, *19*, 1017–1028.
25. Bate, D. Structure of measures in Lipschitz differentiability spaces. *J. Amer. Math. Soc.* **2015**, *28*, 412–482.
26. Cheeger, J.; Kleiner, B.; Schioppa, A. Infinitesimal structure of differentiability spaces, and metric differentiation. *arXiv* **2016**, arXiv:1503.07348.
27. James, R.C. Some self-dual properties of normed linear spaces. In *Symposium on Infinite Dimensional Topology, 1967*; Princeton University Press: Princeton, NJ, USA, 1972; pp. 159–175.
28. Maurey, B. Type, Cotype and *K*-convexity. In *Handbook of the Geometry of Banach Spaces*; Johnson, W.B., Lindenstrauss, J., Eds.; Elsevier Science B.V.: Amsterdam, The Netherlands, 2003; pp. 1299–1332.
29. Amir, D. *Characterizations of Inner Product Spaces*; Springer: Basel, Switzerland, 1986.
30. Lindenstrauss, J.; Tzafriri, L. *Classical Banach Spaces I and II*; Reprint of the 1977 and 1979 Editions; Springer: Berlin, Germany, 1996.
31. Benyamini, Y.; Lindenstrauss, J. *Geometric Nonlinear Functional Analysis*; American Mathematical Society: Providence, RI, USA, 2000.
32. Rockafellar, R.T. *Convex Analysis*; Princeton University Press: Princeton, NJ, USA, 1970.
33. Verpoort, S. The Geometry of the Second Fundamental form: Curvature Properties and Variational Aspects. Ph.D. Thesis, Katholieke Universiteit Leuven, Leuven, Belgium, May 2008.
34. Bishop, R. Infinitesimal convexity implies local convexity. *Ind. Univ. Math. J.* **1974/1975**, *24*, 168–172.
35. Chruściel, P.T.; Galloway, G.J.; Pollack, D. Mathematical general relativity: A sampler. *Bull. Amer. Math. Soc.* **2010**, *47*, 567–638.
36. Parikh, M. Two Roads to the Null Energy Condition. *Int. J. Mod. Phys. D* **2015**, *24*, 1544030.
37. Cordero-Erausquin, D.; McCann, R.J.; Schmuckenschläger, M. A Riemannian interpolation inequality à la Borell, Brascamp and Lieb. *Invent. Math.* **2001**, *146*, 219–257.
38. Fuster, E.L. Moduli and Constants: ... what a show! Available online: http://www.uv.es/llorens/Documento.pdf (accessed on 5 November 2016).
39. Clarkson, J.A. Uniformly convex spaces. *Trans. Amer. Math. Soc.* **1936**, *40*, 394–414.
40. Day, M.M. Uniform convexity in factor and conjugate spaces. *Ann. Math.* **1944**, *45*, 375–385.
41. Gromov, M. Sign and geometric meaning of curvature. *Rend. Mat. Sem. Mat. Fis. Milano Milan J. Math.* **1991**, *61*, 9–123.
42. Goebel, K. Convexity of balls and fixed point theorems for mapping with nonexpansive maps. *Comp. Math.* **1970**, *22*, 269–274.
43. Milman, D.P. On some criteria for the regularity of spaces of the type (B). *Dokl. Acad. Nauk. SSSR* **1938**, *20*, 243–246.
44. Pettis, B.J. A proof that every uniformly convex space is reflexive. *Duke Math. J.* **1939**, *5*, 249–253.
45. Enflo, P. Banach spaces which can be given an equivalent uniformly convex norm. *Israel J. Math.* **1972**, *13*, 281–288.
46. James, R.C. Super-reflexive Banach spaces. *Can. J. Math.* **1972**, *24*, 896–904.
47. Hanche-Olsen, H. On the uniform convexity of L^p. *Proc. Amer. Math. Soc.* **2006**, *134*, 2359–2362.
48. Hanner, O. On the uniform convexity of L^p and l^p. *Arkiv. Mat.* **1956**, *3*, 239–244.
49. Ball, K.; Carlen, E.A.; Lieb, E.H. Sharp uniform convexity and smoothness inequalities for trace norms. *Invent. Math.* **1994**, *115*, 463–482.

50. Nordlander, G. The modulus of convexity in normed linear spaces. *Arkiv. Mat.* **1960**, *4*, 15–17.
51. Day, M.M. Some characterizations of inner product spaces. *Trans. Amer. Math. Soc.* **1947**, *62*, 320–337.
52. Schoenberg, I.J. A remark on M.M. Day's characterization of inner product spaces and a conjecture of L.M. Blumenthal. *Proc. Amer. Math. Soc.* **1952**, *3*, 961–964.
53. Figiel, T. On the moduli of convexity and smoothness. *Stud. Math.* **1976**, *56*, 121–155.
54. Grigorchuk, R.; Pak, I. Groups of intermediate growth, an introduction. *L'Enseign. Math.* **2008**, *54*, 251–272.
55. Kalogeropoulos, N. Tsallis entropy composition and the Heisenberg group. *Int. J. Geom. Methods Mod. Phys.* **2013**, *10*, 1350032.
56. Kalogeropoulos, N. Groups, non-additive entropy and phase transitions. *Int. J. Mod. Phys. B* **2014**, *28*, 1450162.
57. Scott, P. The geometries of 3-manifolds. *Bull. London Math. Soc.* **1983**, *15*, 401–487.
58. Lindenstrauss, J. On the modulus of smoothness and divergent series in Banach spaces. *Mich. Math. J.* **1963**, *10*, 241–252.
59. Asplund, E. Averaged norms. *Israel J. Math.* **1967**, *5*, 227–233.
60. Kwapień, S. Isomorphic characterizations of inner product spaces by orthogonal series with vector valued coefficients. *Stud. Math.* **1972**, *44*, 583–595.
61. Kahane, J.P. *Some Random Series of Functions*, 2nd ed.; Cambridge University Press: Cambridge, UK, 1985.
62. Milman, V.D.; Schechtman, G. *Asymptotic Theory of Finite Dimensional Normed Spaces*; Springer: Berlin, Germany, 1986.
63. Figiel, T.; Pisier, G. Séries aléatoires dans les espaces uniformément convexes ou uniformément lisses. *C. R. Acad. Sci. Paris Ser. A* **1974**, *279*, 611–614.
64. Artstein-Avidan, S.; Milman, V. The concept of duality in convex analysis, and the characterization of the Legendre transform. *Ann. Math.* **2009**, *169*, 661–674.
65. Lanczos, C. *The Variational Principles of Mechanics*, 4th ed.; University Toronto Press: Toronto, ON, Canada, 1970.
66. Zinn-Justin, J. *Quantum Field Theory and Critical Phenomena*, 3rd ed.; Oxford University Press: Oxford, UK, 1996.
67. Dirac, P.A.M. The Lagrangian in Quantum Mechanics. In *Physikalische Zeitschrift der Sowjetunion*; Charkow, Technischer staatsverlag: Charkow, Ukraine, 1933; pp. 64–72.
68. Ambjørn, J.; Görlich, A.; Jurkiewicz, J.; Loll, R. CDT and the Search for a Theory of Quantum Gravity. *arXiv* **2013**, arXiv:1305.6680.
69. Veltman, M.J.G. Quantum Theory of Gravitation. In *Methods in Field Theory, Proceedings of the Les Houches Summer School of Theoretical Physics, Session 28, 28 July–6 September 1975*; Balian, R., Zinn-Justin, J., Eds.; North Holland Publisher: Amsterdam, The Netherlands, 1981.
70. Villani, C. Synthetic Theory of Ricci Curvature Bounds, Takagi Lectures 2015. *Jpn. J. Math.* **2016**, *11*, 219–263.
71. Burago, D.; Burago, Y.; Ivanov, S. *A Course in Metric Geometry*; American Mathematical Society: Providence, RI, USA, 2001.
72. Laskin, N. Fractional Quantum Mechanics. *Phys. Rev. E* **2000**, *62*, 3135–3145.
73. Tarasov, V.E. *Fractional Dynamics: Applications of Fractional Calculus to Dynamics of Particles, Fields and Media*; Springer: New York, NY, USA, 2011.
74. Calgagni, G. Geometry and field theory in multi-fractional spacetime. *J. High Energy Phys.* **2012**, *2012*, 65.
75. Ashtekar, A.; Lewandowski, J. Background Independent Quantum Gravity: A Status Report. *Class. Quant. Grav.* **2004**, *21*, R53.
76. Beem, J.K.; Ehrlich, P.E.; Easley, K.L. *Global Lorentzian Geometry*, 2nd ed.; Marcel Dekker Inc.: New York, NY, USA, 1996.
77. Minguzzi, E.; Sánchez, M. The causal hierarchy of spacetimes. In *Recent Developments in Pseudo-Riemannian Geometry*; Alekseevsky, D.V., Baum, H., Eds.; ESI Lectures in Mathematics and Physics; European Mathematical Society: Zürich, Switzerland, 2008; pp. 299–358.

Permissions

All chapters in this book were first published in MDPI; hereby published with permission under the Creative Commons Attribution License or equivalent. Every chapter published in this book has been scrutinized by our experts. Their significance has been extensively debated. The topics covered herein carry significant findings which will fuel the growth of the discipline. They may even be implemented as practical applications or may be referred to as a beginning point for another development.

The contributors of this book come from diverse backgrounds, making this book a truly international effort. This book will bring forth new frontiers with its revolutionizing research information and detailed analysis of the nascent developments around the world.

We would like to thank all the contributing authors for lending their expertise to make the book truly unique. They have played a crucial role in the development of this book. Without their invaluable contributions this book wouldn't have been possible. They have made vital efforts to compile up to date information on the varied aspects of this subject to make this book a valuable addition to the collection of many professionals and students.

This book was conceptualized with the vision of imparting up-to-date information and advanced data in this field. To ensure the same, a matchless editorial board was set up. Every individual on the board went through rigorous rounds of assessment to prove their worth. After which they invested a large part of their time researching and compiling the most relevant data for our readers.

The editorial board has been involved in producing this book since its inception. They have spent rigorous hours researching and exploring the diverse topics which have resulted in the successful publishing of this book. They have passed on their knowledge of decades through this book. To expedite this challenging task, the publisher supported the team at every step. A small team of assistant editors was also appointed to further simplify the editing procedure and attain best results for the readers.

Apart from the editorial board, the designing team has also invested a significant amount of their time in understanding the subject and creating the most relevant covers. They scrutinized every image to scout for the most suitable representation of the subject and create an appropriate cover for the book.

The publishing team has been an ardent support to the editorial, designing and production team. Their endless efforts to recruit the best for this project, has resulted in the accomplishment of this book. They are a veteran in the field of academics and their pool of knowledge is as vast as their experience in printing. Their expertise and guidance has proved useful at every step. Their uncompromising quality standards have made this book an exceptional effort. Their encouragement from time to time has been an inspiration for everyone.

The publisher and the editorial board hope that this book will prove to be a valuable piece of knowledge for researchers, students, practitioners and scholars across the globe.

List of Contributors

Fabrizio Tamburini
ZKM – Zentrum für Kunst und Medientechnologie, Lorentzstr. 19, D-76135 Karlsruhe, Germany

Ignazio Licata
Institute for Scientific Methodology (ISEM), Via Ugo La Malfa 153, I-90146 Palermo, Italy
School of Advanced International Studies on Theoretical and Nonlinear Methodologies of Physics, I-70124 Bari, Italy
International Institute for Applicable Mathematics and Information Sciences (IIAMIS), B.M. Birla Science Centre, Adarsh Nagar, Hyderabad 500 463, India

Malik Al Matwi
Department of Mathematical Science, Ritsumeikan University, 4-2-28, Kusatsu-Shi, Shiga 525-0034, Japan

Zack Carson and Kent Yagi
Department of Physics, University of Virginia, Charlottesville, VA 22904, USA

Øyvind Grøn
Oslo and Akershus University College of Applied Sciences, Faculty of Technology, Art and Sciences, PB 4 St. Olavs. Pl., NO-0130 Oslo, Norway

Francesco De Paolis, Mosè Giordano, Gabriele Ingrosso, Luigi Manni, Achille Nucita and Francesco Strafella
Department of Mathematics and Physics "Ennio De Giorgi", University of Salento, CP 193, I-73100 Lecce, Italy
INFN (Istituto Nazionale di Fisica Nucleare), Sezione di Lecce CP-193, Italy

José Wadih Maluf
Instituto de Física, Universidade de Brasília, C. P. 04385, 70.919-970 Brasília DF, Brazil

Ivan Debono
Paris Centre for Cosmological Physics, APC, AstroParticule et Cosmologie, Université Paris Diderot, CNRS/IN2P3, CEA/lrfu, Observatoire de Paris, Sorbonne Paris Cité, 10, rue Alice Domon et Léonie Duquet, 75205 Paris CEDEX 13, France

George F. Smoot
Paris Centre for Cosmological Physics, APC, AstroParticule et Cosmologie, Université Paris Diderot, CNRS/IN2P3, CEA/lrfu, Observatoire de Paris, Sorbonne Paris Cité, 10, rue Alice Domon et Léonie Duquet, 75205 Paris CEDEX 13, France
Physics Department and Lawrence Berkeley National Laboratory, University of California, Berkeley, 94720 CA, USA
Helmut and Anna Pao Sohmen Professor-at-Large, Hong Kong University of Science and Technology, Clear Water Bay, Kowloon, 999077 Hong Kong, China

Elias Zafiris
Parmenides Foundation, Center for the Conceptual Foundations of Science, Kirchplatz 1, Pullach, 82049 Munich, Germany

Jean-Pierre Luminet
Laboratoire d'Astrophysique de Marseille (LAM), CNRS UMR 7326, F-13388 Marseille, France
LUTH (Observatoire de Paris), CNRS UMR 8102, F-92195 Meudon, France

Nikolaos Kalogeropoulos
Carnegie Mellon University in Qatar, Education City, Doha, Qatar

Index

www.ingramcontent.com/pod-product-compliance
Lightning Source LLC
LaVergne TN
LVHW081038210926
848485LV00004B/1055